CHINA'S REGIONAL DISPARITIES: ISSUES AND POLICIES

CHINA'S REGIONAL DISPARITIES: ISSUES AND POLICIES

V.F.S. Sit
and
Lu Da-dao
(Editors)

Nova Science Publishers, Inc.
Huntington, New York

Senior Editors: Susan Boriotti and Donna Dennis
Office Manager: Annette Hellinger
Graphics: Wanda Serrano and Dorothy Marczak
Information Editor: Tatiana Shohov
Book Production: Cathy DeGregory, Kay Seymour, Jennifer Vogt and Lynette Van Helden
Circulation: Ave Maria Gonzalez, Ron Hedges and Andre Tillman

Library of Congress Cataloging-in-Publication Data
available upon request

ISBN 1-56072-915-5

227 Main Street, Suite 100
Huntington, New York 11743
Tele. 631-424-6682 Fax 631-425-5933
e-mail: Novascience@earthlink.net
Web Site: http://www.nexusworld.com/nova

Printed in the United States of America

CONTENTS

CONTRIBUTORS

Victor F.S. SIT	Professor, Department of Geography, University of Hong Kong
Da-dao LU	Professor and Former Director, Institute of Geography, Chinese Academy of Science
Jian-ming CAI	Associate Professor, Institute of Geography, Chinese Academy of Science
Jie FAN	Professor, Institute of Geography, Chinese Academy of Science
Bo-jie FU	Professor and Deputy Director, Ecological & Environmental Research Centre, Chinese Academy of Science
Shu-zhong GU	Associate Professor and Deputy Head, Department of Resource Economics, Commission for Integrated Survey of Natural Resources, Chinese Academy of Science
Fengjun JIN	Associate Professor, Institute of Geography, Chinese Academy of Science
Yok-shiu LEE	Associate Professor, Department of Geography, University of Hong Kong
Ya-qin LI	Associate Professor, Institute of Microbiology, Chinese Academy of Science
Chu-sheng LIN	Associate Professor, Department of Geography, University of Hong Kong
Hui LIU	Associate Professor, Institute of Geography, Chinese Academy of Science
Wei-dong LIU	Associate Professor, Institute of Geography, Chinese Academy of Science
Yi LIU	Professor & Deputy Director, Institute of Geography, Chinese Academy of Science
Becky, P.Y. LOO	Assistant Professor, Department of Geography, University of Hong Kong
Xiao-min PANG	Associate Professor, Institute of Geography, Chinese Academy of Science
James, Ji-xian WANG	Associate Professor, Department of Geography, University of Hong Kong

Chapter 1

REGIONAL DISPARITIES IN CHINA: BACKGROUND OF RELATED THEORIES AND PREVIOUS STUDIES

Victor F.S. Sit
Professor, Department of Geography,
University of Hong Kong

The purpose of this chapter is to lay out the background of this collection and to introduce the chapters it contained. It consists of five sections. Section one briefly presents the concerns and experience of industrialized countries as they have emerged since the 1920s. Section two provides a broad-stroke presentation of theories related to regional disparities. Section three is a short discussion on existing regional policy goals and development approaches. Section four is a brief review on existing literature which concerns with regional disparities in China, published either in the English or Chinese language. Section five is a summary of the different views in the literature on the nature of the Chinese state which has an impact on regional disparities of the country as well as the situation of regional disparities in the Peoples Republic of China (PRC) in 1949-1995. The last section introduces the chapters that follow this introduction.

CONCERNS AND EFFORTS TO REDRESS REGIONAL DISPARITIES

Being the first country in the world to have undergone the First Industrial Revolution, United Kingdom (U.K.) was also the first to experience a vigorous economic restructuring as its industrialization deepened in the 20th century. It led to the decline of the coal- and raw material-based heavy industries in the country, such as ship-building and iron and steel industries, and the rise of new fabricating industries of electrical engineering and the auto industry, etc. As a consequence, U.K.'s 'north', i.e. Scotland and Northern England suffered from industrial decline in the first decades of the 20th century, leading to a drop in average income and hectic increase in unemployment. In the reverse, the Midlands and the London Region, i.e. the 'south' driven by the new industries, further prospered. Thus, the widening regional economic disparity between the 'north' and 'south' caused public concerns as migrants moved down from the declining 'north' to resettle in the booming 'south'. Official attention towards regional economic disparities was therefore first expressed in the U.K. in the 1920s (Parsons, 1986; Lawless & Brown, 1986; McCrone, 1969; Thomas, 1978; Smith, 1989).

Starting from the 1920s, U.K. led the world as the first government that took steps to redress regional disparities through stimulating growth in the 'north' and constraining growth in the 'south' to stem the N-S migration. The steps adopted included the 'carrot' measures of special local taxation, and fiscal and infrastructural incentives applied in the 'north', and the 'stick' measures to restrict industrial and office expansion and new start-ups in the 'south'. In addition, a new town development programme and new urban and regional planning institutions and mechanisms were established, culminating in the setting up of an elaborated urban and regional planning system by the end of the 1940s (Martin, 1988; McCallum, 1976; Cameron, 1974; Lawless & Brown, 1986).

The above approach of the U.K. government lasted roughly until the end of the 1970s. Its experience has since sparked off various forms of intervention by many governments, especially those of the developed countries, with an aim to redress their regional economic disparities. For example, in 1963, the French government divided the nation into 20 regions and set up a National Land and Regional Development Agency to increase government input in regional development. In 1962, the Japanese government adopted the Comprehensive National Development Plan and in 1965, passed the National Comprehensive Development Law with the stated aim of developing the country through the progressive promotion of its 3 spatial zones. Even in the United States of America (U.S.), in 1961 and 1965, the government there enacted relevant legislation on regional development and public works and economic development programmes, e.g. the Appalachian Regional Development Act, to foster economic growth of selected regions within the country (Bradshaw, 1991; Hansen, 1990).

In more recent times, government attention and efforts towards regional disparity have increased. We can take two examples for illustration. Inspired by the awareness that market mechanisms are inadequate for correcting regional disparity, the Canadian government set up the Development of Regional and Economic Expansion. It has hence wavered between the development approach emphasizing federal transfer payments and the neo-classical approach of enforcing a free market (Savoie, 1992). In the European Union (EU), regional divergence in economic development is still serious, largely due to differences in regional sectoral specialization, productivity level, and difference in growth history. As such, since the 1980s, the EU has targeted investments and loans at subsidiary regions in order to alleviate such spatial disparities, with the intent of achieving greater equity and political and social cohesion (Paci & Piglian, 1999). A number of important measures for achieving these goals have evolved, the most notable ones include the passing of the Single European Act in 1987 and the creation of the Social Fund in 1989, which direct members of the EU to work towards more spatial equity and economic cohension by:

- achieving growth and adaptation in regional economies showing structural backwardness;
- converting declining industrial regions by developing new activities;
- combating long term unemployment;
- integrating young persons into employment patterns; and
- speeding up adjustment of the agricultural production patterns.

In 1987, the EU sanctioned a total of 7.2 billion ECU for such purposes. In 1995, such amount had reached 14.5 billion ECU. In addition, in 1989-95, the European Investment Bank had approved loans amounting to 91 billion ECU for related regional projects that costed a total of 210 billion ECU. However, the continuation and persistence of regional disparities have raised doubts on the EU efforts. Negative criticisms have been raised, including the ad hoc nature of the projects, the political and bureaucratic tussles involved in their sanction, and the lack of imagination of the projects (Schachter, Bevilacqua & Schachter, 1999).

Based on the concerns and experiences of many such countries over the past decades, many generalizations have been made over the questions of the 'what' and 'why' of regional economic disparities, as well as questions regarding the 'how' in redressing regional disparities.

REGIONAL DEVELOPMENT THEORIES

Classical economic theories have seldom paid attention to the spatial aspects of economic development. However, the existence and realities of regional economic disparities have given rise to many studies that attempt to account for it and/or to theorize it (Higgins & Savoie, 1997; Adams & Pigiliaru, 1999; Amin & Thrift, 1994; Gore, 1984; Williamson, 1965; Wang & Hu, 1999 etc.). In general, we may categorize those that have emerged since the 1950s into three major types, i.e.: -

(1) Convergence/balanced theories

These include many neoclassical theories that share a common believe that convergence, or balanced regional economic development, is a logical end result of the economic development process given that market forces, especially open trade and factor mobility, are the norm. Under perfect market conditions, poor countries/regions would grow faster as there is diminishing returns to reproductive capital and mobility of capital and technology, leading to 'unconditional convergence', i.e. dispersion of income over time (Solow 1956; Cass, 1965; Koopmen, 1965; Smith, 1975; Courchene & Melvin, 1998; Barro & Sale-I-Martin, 1991, 1992, 1995; Sachs & Warner, 1995). There are a number of variants of this basic construct, for example North (1966) emphasized the significance of base-industry export in the development of a region in his general theory of regional development, while others stressed on comparative advantages, equal exchange and price equalization on the acceleration of convergence (e.g. Romans, 1965; Ohlin, 1935). In sum, these authors believe that regional inequality is a temporary phenomenon and would disappear over time. The policy outcome of such theories is thus mainly 'laissez-faireism', i.e. the unleashing of market forces, or an urge for removing artificial barriers to trade and factor mobility.

(2) Theories of Unbalanced Regional Development

The earlier works of 'unbalanced development' include Myrdal (1957) on the effect of 'cumulative causation' which argues that market forces tend to draw economic activities to places of some initial advantages. Once growth has taken place there, it would have a direct impact on the rest of the country, either as a positive spread effect, or as a negative backwash that draws more resources to the place that has already developed. Hirchman (1958), in more or less the same vein, postulates that: in order to grow, an economy must develop one or a few regional centers first, chosen for their more advantageous development history. Thus there is a tendency of polarization in a country at the early stage of its development, but development will thence trickle down to the rest of the economy as the country enters a latter stage of development.

Both Myrdal and Hirchman's ideas have been linked to the concept of the growth pole of Perroux (1950, 1955) and the growth centre concept of Boudeville (1966) which became popular in the 1960s and 1970s as guiding principles in regional development policies (Hansen, 1972; Fu & Salih, 1978; Higgins, 1983). There are at least two common elements in these development approaches, i.e. (a) market forces may not induce convergence, hence government intervention is necessary to check them when they tend to produce polarization, and (b) the economic failure of backward regions may breed further economic failure and therefore there is a need for the government to stimulate growth in such regions in order to help them to break away from the 'self-perpetuating vortex' of underdevelopment. In these theories, growth poles/centres represent vantage points whereby cost-effective measures to stimulate growth may be concentrated.

Other theories on unbalanced growth include the 'product-life cycle' (Vernon, 1981), the spatial sequential or zonal progression concept of development (Fan, 1995) and the inverted-U model of Kuznets (1955) and Williamson (1965). Kuznets hypothesized that the relationship between income equality and level of economic development takes the form of an inverted-U. It was supported by Williamson who did a cross-sectional time-series study with a large number of countries in which he discovered 'a systematic relationship between national development level and regional inequality', i.e. there is (a) a consistent rising regional income disparity and increasing North-South dualism in the early stage of development, and (b) regional convergence and disappearance of North-South problems at the more matured stages of national growth and development. These views are supported by a number of cross-sectional and national studies, such as Mera's (1975, 1977) study on primacy in which he warned against the adoption of dispersal policies for investment and population in developing countries whose national goal is fast economic growth. Richardson's study (1980) on South Korea also shows a polarization tendency at its early stage of growth (1960-1970) and polarization-reversal in its post-1976 development.

The unbalanced regional development theories underline the effects of initial local advantages (history of development, comparative advantages) and government intervention on regional growth. These favour the North or chosen growth poles and they will create a 'backwash' or 'negative polarisation' effects on the south. They have implied that accelerating efficiency and growth should have a higher priority than reducing regional disparity in the early stage of development. This is particularly so with Williamson who believes that, even within the socialist states, governments are unlikely to sacrifice economic growth for ideals of income equality.

(3) Neo-Marxist Theories

Neo-Marxist theories on regional development rest largely on the belief that economic spatial development, especially income distribution, is the direct result of the power structure. Their contribution lies therefore in the analysis of the spatial structure at the national or even world scale in the core-periphery framework (Friedmann 1966, 1969), and the postulation of a dependency relationship between the core and the periphery (Frank, 1967; Emmanuel, 1969; Castells, 1977; Slater, 1975). In these theories, regional disparity is corollary to the power structure based on a dependency relationship and that regional inequality under capitalism is seen as necessary for capital accumulation (Soja, 1980; Harvey, 1982).

Goals and Policies in Regional Development

As gleaned from the above brief review, different theories imply different perspectives on the cause, nature and hence solutions to the issue of regional disparities. In an earlier section, we have mentioned that EU's regional development policy goals of greater economic equity and cohesion and now the EU proposes to achieve them through five measures, including attacking on unemployment, promoting industries in declining regions and enhancing adaptation to new activities in areas of structural backwardness, and how Canada has wavered between reliance on the market and more government intervention. Indeed, regional development goals are often a product of compromise between two policy goals which are often times at conflict with each other, i.e. regional parity and growth (Weng & Hsu, 1996). Okun (1975) labeled the two slightly differently as: 'efficiency' and 'fairness', and claims that the regional policy of any country represents a difficult choice between the two by the respective government. Of course there are no shortage of people who believe that 'growth' and 'efficiency' should proceed with 'parity' and 'fairness'. Neoclassical models that predict unconditional convergence under market conditions argue for the adoption of Laissez-faireism as the policy norm for regional development. Many growth theories of the unbalanced inclination have, however, implied that 'growth' and 'efficiency' have to come first, especially at the early stage of development. There are also studies that claim that regional income disparity would harm economic growth, such as Person & Tabellini's (1994) time series studies of a number of countries.

As most studies on regional development are based on market economies, they have not paid attention to the political, military, and ideological goals that exist in regional development strategies of some countries, particularly those under a socialist system or command economy. This has been criticized by Wang and Hu (1999) as a form of economic-determinism. They noted the more complicated considerations in regional development goals in countries such as China, where at times political and military goals have more weight than economics. Hansen (1988), besides, points out that regional development theories have to be established on the basis of social change and that relying on 'economic engineering' alone could not produce viable regional development policies. Therefore, there is truism in Savoie's view that no single economic theory explains regional disparities (1992, 6). This may equally apply to regional development goals and related policies adopted by different countries.

Hu, Wang and Kang (1996) have attempted to classify government regional policies into four models, i.e.:

- Tunisia model, where the government is both weak in capability and desire to redress regional disparities;
- Indian model, where the government is weak in desire, but has a strong capability to redress regional disparities;
- Yugoslavian model, where the government is weak in capability but strong in desire to redress regional disparities; and
- Indonesian model, where the government is both strong in desire and capability to redress regional disparities.

Their typology seems immature as it has left out many countries with long experiences in regional development policies and which do not fit into any of the four models. Besides, the models lack clarity as to how government desires and capabilities are defined and measured. The authors, nevertheless, have pointed to an area in regional development policy studies that is still not much covered.

Based on the existing literature, we may, however, generalize that 3 types of policy approaches exist, i.e.:

A. General Laizefaireism Approach

This refers to the general approach of countries such as the U.S., where governments have seldom interfered with market forces in regional economic development.

B. Cohesion and Equity Approach

This is found in some industrialized/developed countries such as Canada and the EU where government concern over regional disparities has become an important factor in macro policy formulation and where such policies are often based on fund transfer and stimulation of growth in backward regions.

C. Efficiency and Zonal Approach

Japan's adoption of a progressive zonal approach in 1960-1990, and its recent approach of 'urban regions (Economic Research Institute, State Planning Commission, 1996) as well as China's open door measures in 1978-1990 that favour the coastal region are examples of such an approach. They emphasize on efficiency and growth and treats equity and the spatial aspect of growth a lower parity.

After having broadly sketched the situation of studies on regional disparities, we now turn specifically to study on regional disparities in the PRC in the coming sections.

RECENT STUDIES ON REGIONAL DISPARITY IN CHINA

We have noted earlier that China's experience in regional development is different from most other countries, as has been revealed by many studies on regional disparity and regional development policies of the PRC in the past decade. We shall briefly review a selection of major publications to throw light on the main concerns on the Chinese experience already expressed as a background of this collection.

1. English publication

Linge and Forbe's edited book (1990) on the spatial economy of China contains ten chapters. It is led by a general discussion on PRC's regional disparities and regional policies and has thematic chapters on coastal China, seaports, transport, population increase and migration, and the two populations cities of Shanghai and Beijing. It points out the wide range and mixture of regional development policy goals of the country in different periods of time which include military objectives and politically-inspired regional policies, and the complicity of entanglement of local (provincial) development goals with national development goals.

Goodman and Segal's edited book (1994) has 12 chapters. It has a good coverage on the politics of regionalism, central-local relationship, regional trade and production, as well as the regional economies of Hong Kong-Guangdong, Fujian, North-China, Xinjiang, Yuanum and the Yangtze Delta. The focus of the book is on post-1978 development. In the introduction, Goodman remarked that 'The PRCs rapid economic development since 1978 has been based on decentralization and an explicit regional policy, in which some regions have been positively encouraged to become wealthy before others' (Goodman, 1994, 1).

Yang's 1997 book is the first English book devoted completely to a presentation and scrutiny of regional economic development and policies in the PRC. It has 8 chapters analysing regional economy development through the themes of competitive liberalization, resource, regional rivalries and market integration, the Special Economic Zones, regional dominance, besides examining the policy dilemma of the early 1990s and its possible re-orientation. However the data used are up to 1991 only. In the conclusion, Yang attached significance of the political power transition from Mao to Jiang in the country's future regional development re-orientation: 'In contrast to Mao and Deng, Jiang Zemin is compelled to be more even-handed in his effort to balance reform, development and stability. The debate on regional development policies thus reflects China's tradition beyond the politics of strongman role'. (142).

In 2000, Wei (2000) contributed an update book on regional development in China which covers the entire 1949-1995 period, though its stress is on post-reform (post-1978) China. Two of the seven chapters are devoted to a detailed intra-province study of Jiangsu. There is one short introductory and a concluding chapter. The remaining three chapters of substance present the patterns of regional development, and explain them in terms of the nation state, local state and globalization forces. Wei finds persistent regional inequality under Mao while confirms that post-1978 developments have brought dramatic changes to the role of the nation state, local state and have induced increased influence of global forces. During the post-1978 period, interprovincial inequality is said to have declined, as five of the coastal provinces that registered low level and slow pre-1978 economic development and growth have grown dramatically while the traditional leading provinces have stagnated. The chapters on intraprovincial inequality of Jiangsu have underlined the rural-urban gap in economic development in both the Mao and Post-1978 periods.

While there are many book chapters and journal articles on China's regional inequality and regional development, they will not be separately mentioned here. However, some of the relevant ones will be treated in next section dealing with the nature of the Chinese state and the situation of regional inequalities in the pre- and post-reform China.

2. Chinese publications

Among the Chinese publications, Lu's (1995) single-author book is the first comprehensive publication on regional development and spatial structure in China. Its nine chapters include three devoted to an extensive review of both Western and Chinese literature on regional development. In the main chapters he discussed the relationship between technological development, economic structure, resource, environment and spatial structure on regional development, using examples from China and other countries. Based on the unbalanced development concept, he put up a model of development centre and development axial to portrait China's past regional development. His model is said to be useful for China's future regional development policies. In chapter 3 he summarized China's post-1947 regional development and the related research by Chinese scholar. The latter is a useful general reference for the present collection and its summary table is therefore attached below:

Liu, Li and Xue (1994) edited a 20-chapter book dealing with a host of regional development issues of the PRC. Most of the data used are just up to the 1980s. The twenty chapters are of four types. Two deal with regional development strategy of the country. Four discuss on methodology, especially on regional development indicators. Nine chapters are devoted to the extent and situation of regional disparities and on analyses of the factors behind. The rest of the book discusses thematic topics such as the central-local fiscal issues, the tertiary sector, national income transfer and Third Line industries. The book in general supports the strategy of unbalanced development in the form of progressive zonal development/transition.

Hu, Wang and Kang (1996) first published their book on regional development in China in Chinese. It was later modified and published as Wang & Hu (1999) in English. The book contains 7 chapters and six annex. It reviews the changing regional disparities in China overtime through a variety of index systems, compares the physical, capital, consumption and administrative characteristics of the more developed and less developed regions, and provides explanations on China's widening disparities. Three chapters are devoted on a discussion on the ill-effects of widening disparities and proposes ways to redress them. The authors show displeasure on the popularity of the view of 'efficiency' priority in China, and strongly argue that rising inequality is not inevitable in post-reform China. They believe that rising inequality is also not a necessary trade off for economic development and that in China, the will of the nation-state and its capability matters much in the balancing between inequality and development. Therefore, the Chinese government should avoid serious imbalance in economic and political resources as their goal in regional development and they recommended to the Chinese government the following five targets under such a goal:

- poverty eradication
- equal opportunity
- balance between parity and efficiency
- democratization
- economic democratization

The Economic Research Institute of the State Planning Commission (1996) was engaged in a study on China's regional development in1995. Its report contains 7 chapters, the first three of which review theories and methodology in regional development experiences of the West and the PRC. The rest discuss China's choices in regional development strategy and estimates China's economic development potential in 2010. It then chooses a strategy based on the development of nine 'mega urban regions' and maps out detailed policy measures for the promotion of each of them. The 9-mega urban region strategy originated from a key team member of the Institute, Wang Jin, who had separately published his idea in detail in a book (Wang, 1996). The strategy is a replica of the Japanese approach, though the Chinese situation deviates significantly from the

Japanese. Wang's strategy assumes that each of the mega urban regions is more or less a closed system that possesses similar resource endowment. In fact they are not. Besides, the Chinese regions are more diverse in their internal dynamics, and are subject to much stronger central and foreign influences. There is also lack of attention to the effects of China's physical geography and the established transportation network on the formation and well-being of the regions.

Finally, Weng and Hsu (1996) produced a six-chapter book on the dynamics of China's regional development and their policy environment, in which they discussed the issues of social choice, unbalanced development, economic structure and future trends of regional disparities of the country.

The above brief surveys on the English and Chinese literature are by no means comprehensive. They have also excluded numerous book chapters and journal articles on related issues. Selected reference to such would be made in the following section that summarizes major views on the impact of the nature of the Chinese state on regional disparities as well as on whether regional disparities have enlarged in the PRC.

NATURE OF THE STATE AND REGIONAL DISPARITIES IN CHINA IN 1949-1995

a) Nature of the Chinese State

Many belief that socialist development policies have a natural built-in goal of equalization of regional development (e.g. Forbes & Thrift 1987; Sidaway and Simon, 1990). This is assumed to be achieved through population distribution and location of industries (Murphey, 1976; Simmons, 1979). Yet many have also noted the existence of marked regional inequalities in many socialist and former socialist countries (Koropeckyi, 1972; Fuchs & Demko, 1979; Spechler, 1979; Clark, 1983; Schiffer, 1989). Jia, Sachs & Warner (1996) relate this to the fact that within socialist countries, most channels of convergence common in market economies are cut off, and that their investment allocation are based on political and bureaucratic rather than market considerations. Thus, allocation of resources in these countries tends to concentrate on existing rather than new geographical locations. They therefore felt that convergence in socialist countries, if it happens, is somewhat accidental.

In the case of China, regional disparities are said to have resulted from both the physical geography of the country and the pre-1949 coastal bias in development, especially in industrialization (Cannon 1990). Such inheritance has little to do with the policies of the Chinese Communist Party (CCP). Besides, the CCP showed only occasional concern in the reduction of certain form of spatial inequality based on military strategy or Mao's notion of economic efficiency. In general, there was a strong prevalent view within China that the government should await the achievement of greater affluence of fully developed communism before dealing with regional disparities seriously. In reverse, the PRC had followed, until 1978, the Stalinist industrialization strategy and restrictive rural-urban migration policies which had aggravated rural-urban disparities (Perkins & Yusuf, 1984; Kueh, 1989; Riskin, 1984, 1987, 1990; Yang, 1990; Wu, 1987; Cole, 1987). In addition, Yang (1997) found explanation for such in the process of China's modern economic growth which was driven by foreign trade that favoured the coastal regions as well as a result of unbalanced growth in a 'Hirschman fashion', as China was still at an early stage of economic development.

However, not all have seen China in that light, i.e. that the Chinese government had not paid much attention to regional development and that regional disparities in China have increased since 1949. Lyons (1991) stressed that one of the most widely remarked elements of the Chinese model of development is the high priority assigned to regional objectives, i.e. to maintain a 'balance' in the distribution of productive capacity and equity in distribution of income, and hence consumption. One of the reasons for the more equitable regional development in China, according to Lyons, is that the provinces have exercised authority over a wide range of economic activities and that central government constraints on the contours of provincial plans had at times been loose and ineffective. Such a view is echoed by Goodman (1994) who finds that the CCP traditionally encourages a degree of local flexibility in national policy implementation and which makes regionalism a constant feature of the PRC. Linge and Forbes (1990) also acknowledged the wide range and mixed regional development goals of China over the years and the mediating effects of the constant struggle between the technocrats and professionals in central ministries and local officials in the regions.

As to the post-1978 turn in the role of the government and hence the modified nature of the state from one based on 'socialism' or even 'communism' at times, i.e. during the Cultural Revolution (1966-1976), to one of 'socialist-market economy', there is general agreement that the PRC as a whole has since paid more attention to regional development and have adopted new policies leading to increasing regionalism. Goodman (1994, 1) is a good example:

"The PRC's rapid economic development since 1979 has been based on decentralization and an explicit regional policy, in which some regions have been positively encouraged to become wealthy before others. The result is a highly differentiated economic geography and pattern of regionalism that has suggested to some a replay of the 'warlord era' of the 1920s, except that power is now based on economic rather than military might. "

He suggested that 3 key elements of the economic reform that started in 1978 have a great potential for increased localism/regionalism in the country, i.e.

- decentralization of economic management;
- development of more market-oriented and market-determined economy;
- determination to utilize economic and geographical comparative advantages on China's participation in the international division of labour.

In fact the 1978 reform implies three major changes that affects the country's regional development, i.e. (1) a movement from the pre-1978 political priority to economic priority in post-1978 development, (2) a change from an emphasis on regional self-sufficiency to regional comparative advantages; and (3) a change from self-reliance and central planning to open-door, encouragement of FDI and decentralization of planning powers and devolution of economic power to enterprises (Canon, 1996; Fan, 1995, 1997). These new directions are certainly grounds to support the fear of increasing regional disparities in the country.

Implicit in the new strategy is also the faith that the 'trickling down' theory in economic development would work. This of course, is being doubted, as China's regional policies for reducing spatial inequality are still relatively weak, and that powerful interest groups in the better off places have already been formed which would heighten polarization in the coastal areas and the negative backwash in lagging regions. Against the background of different national goals before and after 1978, it is therefore interesting to examine whether regional disparities have increased or decrease over-time before and after the 1978 reform.

b) Pre-1978 Regional Disparities

Many authors subscribe to the view that in pre-1978 or pre-reform China, there was no evidence of increased income disparities. Some even described the country as 'quite respectable' in its regional development achievement (Lardy, 1978, 1980; Lyons, 1991). Riskin (1987) and Lippit (1987) further believed that there was a narrowing of relative disparities, and praised Maoist China as having successfully avoided the widening of regional disparities and have differed markedly from the classical interverted-U curve frequently associated with the early stage of growth of an economy.

Yet, there are people who hold the view that Maoism had increased regional inequality (Friedman, 1987). Using a wider range of data other than income level, both Tsui (1993) and Seldon (1988) supported the view of a widening of interregional disparities in income, social conditions and cultural opportunities in Pre-reform China. Using more reliable data available to Chinese scholar in the mid 1990s, Hu, Wang and Kang (1996, 46-47) concluded that despite the implementation of the equilibrium development strategy, the total relative disparity in regional per capita national income of the PRC has increased overtime, i.e. from 65.6% during the 1st Five-year Plan (1953-57) to 74% in 1963-65, 81.3% in 1966-70, 94% in 1971-75 and further to 98.1% in 1976-80.

Thus the pre-1978 regional disparity pattern has not yet been clear.

c) Post-reform (1978) Regional Disparities

In spite of the coast-oriented regional strategy based on regional comparative advantages (Yang, 1990), or a disguised core-periphery concept, or the appearance of the 3 key elements that would increase regionalism (Goodman, 1994), there had been little increase in interregional inequality in China in the 1980s against a background of rapid overall economic growth for the country as a whole (Sloan, 1994, Jian, Sachs, Warner, 1996). After the reform of 1978, real income tended to converge much more among the Chinese provinces associated with marketization and openness. In particular, rural-urban inequality had narrowed, as agriculture and rural industries took advantage of the reform and a freer market, together with relaxed control over rural-urban migration (Oi, 1993; Tsui, 1991, 1993; Chen & Fleisher, 1996; Yang, 2000). Subscribing to the view that China's decade (1980s) of economic reform did not produce an ever widening gap between economic performance of the 'have' and 'have not' provinces, as measured by per capital net material product, Denny (1991) suggested further explanations:

(1) foreign lending had partially offset FDI in coastal locations;
(2) the central government had retained the ability to subsidize poor provinces; and
(3) a more natural pattern of regional economic development had replaced extreme and irrational equalitarianism in Mao's era.

Thus, despite claims that China's regional disparities had been widening at a 'remarkable' rate after the 1978 reform (Lakshmanan & Hua, 1987; Whyte, 1986; Aguignier, 1988), there exists general agreement that in the 1980s, convergence was the predominant trend. This is also supported by Hu, Wang, Kang (1996) who found that both provincial disparity in per capital GDP and intra-provincial disparity had declined in 1978-90.

Although little has been published on the development of the 1990s, it is already noted that widening regional disparities had resurged since 1990, particularly in the form of a widening income gap between coastal and interior provinces (Yang & Wei, 1996; Jian, Sachs, Warner, 1996; Fan, 1995, 1997; Chai, 1996). The 'ladder-step doctrine' was also doubted against the more popular concepts of sustainable development, fairness and political and social stability that had surfaced in China in the early 1990s. It is argued that interior provinces which have already lagged much behind coastal provinces in development should receive the same development emphasis such that they would have the ability to produce the raw materials and energy for supporting higher growth in the coastal region as well as to narrow the income gap between them and the latter. Or, alternatively, increasing frustration in the 'periphery' due to slow growth and widened disparity may push it to demand autonomy or may induce a prolong conflict between the periphery and the core.

Since the beginning of the 1990s, China has adopted a new regional strategy emphasizing more on the development of the Western and Interior regions, with more efforts aimed at eliminating poverty and to protect and cure the environment which had suffered during the first decade of reform. In addition, economic and technological cooperation between the coastal and interior provinces have also received high priority in national development policies. These are major new grounds which this collection will cover compared to the existing literature. In addition, the collection will also present more comprehensive reviews and analyses on China's regional pattern of development, and will evaluate the changing regional policies and their achievements, based on better data available in the mid 1990s as well as improved methods of analysis. The major content of the collection will be summarized in the next section.

ORGANISATION OF THE CHAPTERS

This collection has a total of 11 chapters. Following this introductory chapter, Sit and Lu's chapter will be a historical review of regional policies of the PRC in 1949-2000. Against the background of the 3 main policy goals of western countries, i.e. rapid growth, stable growth and balanced growth, the authors discussed China's varying policy goals. In the pre-1978 period, (1) the two predominant goals were to promote a fast transit into Communism and (2) military security. Since the 1980s, China's regional policy goals have become diversified, incorporating objectives such a growth, comparative advantage, environment and equity. The chapter also sets the geographical stage for regional disparity in the country in terms of the 3 natural and 3 topographical divisions and the administrative/management system of the PRC. It then reviews the four periods of regional policies, i.e. (1) pre-1972, (2) policy adjustment of 1973-78, (3) policy based on comparative advantages of 1979-82, and (4) the second policy readjustment of 1992 and beyond. The chapter ends with an evaluation of China's regional policies.

Chapter 3 by Liu assesses the 'total' regional development of China in 1990-96. Different from most studies based on a few economic indicators, the author measures regional development by a wide range of data, including variables on location, resource endowment, economic vitality, social development, living quality etc., besides the common yardsticks of per-capita GDP, income and other economic variables. Using this broader concept, Liu has compiled a set of indices and generated 7 types of regions. The latter are meant to be empirical guides for local and national governments in formulating regional development strategies.

The central-local relationship has been a hot topic in studies on regional development of China. Liu and Lin's chapter 4 focuses on the most important and sensitive of the central-local issues behind regional development problems in the PRC, i.e. the central-local financial relationship. It provides a terse review of the changing financial relationship, focusing on the tax division scheme that was introduced in the 1990s. It discusses problems after the introduction of the scheme and suggests policies on central-local financial relationship for redressing regional imbalance in future.

Chapter 5 of Pang and Loo addresses the question of what are the best ways to achieve regional cooperation in China. Their main effort is on analyzing major policies of regional economic and technological cooperation since the 8th Five-Year Plan. This provides them a unique angle for assessing the rationale and effectiveness of regional policies. In the conclusion, they note the existence of major constraints on the effectiveness of regional cooperation and regional development. These include the conflict in policy goals between the central and local governments, 'departmentalization' of the administration, and problems related to the banking and tax systems.

Chapter 6 takes another important angle in studying regional disparity in China, i.e. the transportation and communication infrastructure. Jim and Wang first present the characteristics of the infrastructure as it has evolved since 1990. They then examine the relationship between regional infrastructure provision and spatial disparities in economic growth. They have classified China's space into 4 regions according to infrastructure provision and note the existence of five corridors. They suggested that the areas of concern in enhancing the role of the transportation and communication infrastructure in regional development include the improvement in management, awareness and efforts in multinodalism, and development in logistics.

Regional disparity is not just an issue of economics. The regions of China also suffer from varying eco-environmental problems. Chapter 7 of Fu and Li analyses China's environmental policies, traces their implementation at the national and local levels, and discusses the results. There is a detail presentation of public-private investments in environmental management and ecological rehabilitation in various regions. It concludes that in order to solve the problems of the regional eco-environment, institution and financial reforms are urgently needed, so is more public participation.

Gu and Loo's Chapter 8 reviews food supply and agricultural production as a background for analyzing regional growth. It also assesses the national food grain production policy and its implementation in different regions. Finally, China's provinces are classified into five types in terms of agricultural potential and food production as a basis for regional agricultural and food production policies.

The physical geography of China, combined with its history of development, has given rise to many poor areas, most of which are in the Western and Interior regions. Liu Hui's Chapter 9 describes the standards used in defining poor areas and examines the characteristics of these areas. National and regional poverty relief policies have also been presented and evaluated. Liu stresses on the importance of developing the household and the individual in poverty relief. Poverty relief policies should also incline towards restructuring of the local economy and protecting the environment.

China's Open Policy since 1978 has taken an obvious 'zonal-spread' development and hence is regional biased. Sit and Cai (Chapter 10) look at the impact of foreign direct investment (FDI) and foreign trade (FT) on China's economic development. The chapter contains an analysis on the objectives, sectoral composition, source country and locational preferences of FDI into China, followed by an examination on FDI and FT's effect on regional disparities in recent years. It concludes with recommendations on how China should be more effectively and efficiently integrate with the world economy.

Fan and Lee's Chapter (Chapter 11) characterizes the spatial form of China's Open Policy, i.e. the Special Economic Zones (SEZs) and Open Cities (OCs) as special 'regional growth economies'. Their role in the regional economy of their respective province has been examined. Besides, the chapter discusses on the growth, economic strength and change of these areas in 1991-96, as well as examines their function, economic structure and spatial patterns. It ends with a discussion on the 'second industrial discovery' policies of the SEZs and OCs.

Concluding Remark

As the brief summaries of the Chapters reveal, the collection has cut deeper in topics already existed in the literature and has covered a wider range of issues important for our understanding of China's regional development. Most of the chapters show a strong bias towards detailed policy presentation and their evaluation. This too, is another strength of the collection. Blessed with more data and details on the respective policies, the collection can supplement existing literature and broadens the discussion on regional development in China towards more practical and informed grounds.

In its original form, the collection was published in Chinese (Lu & Sit, 1998), and was written in the format of the *World Development Report* of the World Bank. It was then targeted at Chinese officials at the national, provincial and municipal/district levels of the administrative hierarchy. Thus although it was a collaborated product of the Chinese Academy of Sciences and the Department of Geography, University of Hong Kong, the chapters are written basically in the style of 'official Chinese'. The present collection is a modified version more suited to the Western reader, although it has retained some of the 'official Chinese' style, a character that makes this collection unique in the English literature on regional development of the PRC.

ACKNOWLEDGEMENT

The editors thank the Hui Oi Chow Trust Fund of the University of Hong Kong for supporting the collaborative work involved, and Ms Meimei Wong and Mr Wong Ting Bor for typing the manuscripts and producing the line drawings.

REFERENCES

1. Adams, J. and Pigiliaru, F. (eds) 1999. *Economic Growth and Change: National and Regional Patterns of Convergence and Divergence*. Cheltenham: Edward Elgar.
2. Aguignier, P. 1988. Regional disparities since 1978. In Feuchtwang, S., Hussain, A. and Pairault, T. (eds), *Transforming China's Economy in the Eighties*, Vol. 1, 93-106.
3. Amin, A. and Thrift, N. (eds) *1994. Globalization, Institutions, and Regional Development in Europe*. Oxford: Oxford University Press.
4. Barro, R. J. 1991. "Economic Growth in a Cross-section of Countries". *Quarterly Journal of Economics* 407-43.
5. Barro, R. J. and Sala-i-Martin, X. 1991. *Convergence across states and regions. Brookings Papers of Economic Activity* 1: 107-82.
6. Barro, R. J. and Sala-i-Martin, X. 1992. Convergence. *Journal of Political Economy* 100: 223-51.
7. Barro, R. J. and Sala-i-Martin, X. 1995. *Economic Growth*. New York: McGraw-Hill.
8. Borts, G. H. 1960. The equalization of returns and regional economic growth. *American Economic Review* 50: 319-47, no. 3.
9. Borts, G. H. and Stein, J. L. 1964. *Economic Growth in a Free Market*. New York: Columbia University Press.
10. Boudeville, J. R. 1966. *Problems of Regional Economic Planning*. Edinburgh: Edinburgh University Press.
11. Bradshaw, M. 1991. *The Appalachian Regional Commission: Twenty-five Years of Government Policy*, p. 2. Lexington: University Press of Kentucky.
12. Cameron, G. C. 1974. Regional policy in the United Kingdom. In Hansen, N. M. (ed.), *Public Policy and Regional Economic Development: The Experience of Nine Western Countries*, p. 65. Cambridge, Mass.: Ballinger Publishing Co.
13. Cass, D. 1965. Optimum growth in an aggregative model of capital accumulation. *Review of Economic Studies* XXXII: 233-40.
14. Castells, M. 1977. *The Urban Question: A Marxist Approach*. London: Edward Arnold.
15. Cannon, T. 1990. Regions: spatial inequality and regional policy. In Cannon, T. and Jenkins, A. (eds), *The Geography of Contemporary China*, 28-60. New York: Routledge.
16. Chai, J. 1996. Divergent development and regional income gap in China. *Journal of Contemporary Asia* 26(1): 46-59.
17. Chen, J. and Fleisher, B. M. 1996. Regional income inequality and economic growth in China. *Journal of Comparative Economics* 22(2): 141-64.
18. Clark, C. 1983. Regional inequality in communist nations. In Nelson, D. N. (ed.), *Communism and the Politics of Inequalities*, 73-104. Lexington, MA: Lexington Books
19. Cole, J. P. 1987. Regional inequalities in the People's Republic of China. *Fijdschrift voor Economische en Sociale Geografie (Journal of Economic and Social Geography)* 78: 2011-13.
20. Courchene, T. and Melvin, J. 1988. A neoclassical approach to regional economics. In Higgins, B. and Savoie, D.J. (eds), *Regional Economic Development: Essays in Honour of Francois Perroux*. London and Boston: Unwin Hyman. See also Courchene, 1986. *Economic Management and the Division of Powers*, 169-92. Toronto: University of Toronto Press.

21. Crown, W. H. and Wheat, L. F. 1995. State per capita income convergence since 1950. *Journal of Regional Science* 35(4): 527-52.
22. Denny, D. I. 1991. Regional economic differences during the decade of reform. In Joint Economic Committee, Congress of the United States, *China's Economic Dilemmas in the 1990s*, 186-208. Washington, DC: US Government Printing Office.
23. Economic Research Institute, State Planning Commission, PRC. 1999. *Study on China's Regional Development Strategy*. Beijing: State Planning Commission (Chinese text).
24. Emmanuel, A. 1969. *L'echange inegal*. Trans. 1972. New York & London: Monthly Review Press.
25. Fan, C. C. 1995. Of belts and ladders: state policy and uneven regional development in post-Mao China. *Annals of the Association of American Geographers* 85: 421-49.
26. Fan, C. C. 1997. Uneven development and beyond: regional development theory in post-Mao China. *International Journal of Urban and Regional Research* 21: 620-39.
27. Forbes, D. and Thrift, N. 1987. *The Socialist Third World*, 53-97. New York: Basil Blackwell.
28. Frank, A. G. 1967. *Capitalism and Underdevelopment in Latin America*. New York: Monthly Review Press.
29. Friedman, E. 1987. Maoism and the liberation of the poor. *World Politics* 39: 411.
30. Friedmann, J. 1966. *Regional Development Policy*. Boston: MIT Press.
31. Friedmann, J. 1969. *A General Theory of Polarized Development*. Los Angeles: UCLA Press.
32. Fu Chen Lo and Salih, K. 1978. Growth poles and regional policy in open dualistic economies: Western theory and Asian reality. In Lo and Salih (eds), *Growth Pole Strategy and Regional Development: Alternative Approaches,* 243-70. Oxford: Pergamon Press.
33. Fuchs, R. J. and Demko, G. J. 1979. Geographic inequality under socialism. *Annals of the Association of American Geographers* 69(2): 304-18.
34. Goodman, D. S. G. and Segal, G. (eds) 1994. *China Deconstructs: Politics, Trade and Regionalism*. London: Routledge.
35. Gore, C. G. 1984. *Regions in Question: Space Development Theory and Regional Policy*. London: Methuen.
36. Hansen, N. (ed.) 1972. *Growth Centers in Regional Economic Development*. New York: The Free Press.
37. Hansen, N. 1988. Regional consequences of structural changes in the national and international division of labor. *International Regional Science Review* 22: 128, no.2.
38. Hansen, N. 1990. Regional economic development policies and programs in the US. Hansen, N., Higgins, B. and Savoie D. J. (eds), *Regional Policy in a Changing World*, p. 119. New York: Plenum Press.
39. Harrod, R. 1948. *Towards a Dynamic Economics: Some Recent Developments of Economic Theory and Their Application to Policy*, p. 86. London: Macmillan.
40. Harvey, D. 1982. *The Limits to Capital*. Oxford: Basil Blackwell.
41. Higgins, B. 1983. From growth poles to systems of interactions in space. *Growth and Change* 14, no. 4.
42. Higgins, B. and Savoie, D. J. 1997. *Regional Development Theories & Their Applications*. New Brunswick: Transaction Publishers.
43. Hirschman, A. O. 1958. *The Strategy of Economic Development*. New Haven: Yale University Press.
44. Hu, A., Wang, S. and Kang, X. 1995. *Report on Regional Disparities in China*. Shenyang: Liaoning People's Press (Chinese text).
45. Hu, Angang; Wang, Shaoguang; Kang, Xiaognang. 1996. *Regional Disparities in China*. Taipei: Chinese Association for Eurasian Studies (Chinese text).
46. Jian, T., Sachs, J. D. and Warner, A. M. 1996. Trends in regional inequality in China. *China Economic Review* 7(1): 1-21.

47. Koopmans, T. C. 1965. On the concept of optimal economic growth. In *The Econometric Approach to Development Planning*. Amsterdam: North-Holland.
48. Koropeckyi, I. S. 1972. Equalization of regional development in socialist countries: an empirical study. *Economic Development and Cultural Change* 21: 68-86.
49. Kueh, Y.Y. 1989. The Maoist legacy and China's new industrialization strategy. *China Quart.* 119: 420-447.
50. Kuznets, S. 1955. Economic growth and income equality. *American Economic Review* 45(1): 1-28.
51. Lakshmanan, T. R. and Hua, C. 1987. Regional disparities in China. *International Regional Science Review* 11(1): 97-105.
52. Lardy, N.R. 1978. *Economic Growth and Income Distribution in the People's Republic of China*. New York: Cambridge University Press.
53. Lardy, N.R. 1980. Regional growth and income distribution in China. In Denberger, R.F. (ed.), *China's Development Experience in Comparative Perspective*, 153-90. Cambridge: Harvard University Press.
54. Lawless, P. and Brown, F. 1986. *Urban Growth and Change in Britain*. London: Harper & Row.
55. Li, Si-ming. et al. (eds) 1996. *Issues in China's Regional Development*. Taipei: Taiwan National University and Hong Kong Baptist University (Chinese text).
56. Linge, G.J.R. and Forbes, D. K. 1990. The space economy of China. In Linge, G. J. R. and Forbes, D. K. (eds), *China's Spatial Economy*. New York: Oxford University Press.
57. Lippit, V. D. 1987. *The Economic Development of China*, 159-60. Armonk, N.Y.: Sharpe.
58. Liu, S. S., Li, Z., Xue, T. D. (eds) 1994. *Study on China's Regional Development*. Beijing: China Statistical Press (Chinese text).
59. Lu, D. 1995. *Regional Development and Spatial Structure*. Beijing: Kexue chubanshe (Science Press) (Chinese text).
60. Lu, D. and Sit, V. 1998. *Report on China's Regional Development*. Beijing: Science Press (Chinese text).
61. Lyons, T. P. 1991. Interprovincial disparities in China. *Economic Development and Cultural Change* 39(3): 471-506.
62. Martin, R. 1988. The political economy of Britain's north-south divide. *Transactions, Institute of British Geographers*, N. S. 13: 389-418.
63. McCallum, J. D. 1979. The development of British regional policy. In MacLennon, D. M and Parr, J. B. (eds), *Regional Policy Past Experience and New Directions*, p. 38. Oxford: Martin Robertson.
64. McCrone, G. 1969. *Regional Policy in Britain*, p. 114. London: Georges Allen & Unwin.
65. Mera, K. 1975. *Income Distribution and Regional Development*. Tokyo: University of Tokyo Press.
66. Mera, K. 1977. Population concentration and regional income disparities. In Hansen, N. (ed.), *Human Settlement Systems*, 155-76. Cambridge, MA: Ballinger Publishing Company.
67. Murphey, R. 1976. Chinese urbanization under Mao. In Berry, B. J. L. (ed.), *Urbanization and Counterurbanization*, 311-29. Beverly Hills: Sage.
68. Myrdal, G. 1957. *Economic Theory and Under-Developed Regions*. London: Gerald Duckworth.
69. North, D. 1996. *The Economic Growth of the US 1970-1860*. New York: Norton.
70. Ohlin, B.G. 1933. *Interregional and International Trade*. Cambridge: Harvard University Press.
71. Oi, J.C. 1993. Reform and urban bias in China. *J. Dev. Stud.* 29(4): 129-148.
72. Okun, A. M. 1975. *Equality and Efficiency: The Big Trade-off.* Washington, DC: Brookings Institution.
73. Paci, R. and Pigiliaru, F. 1999. "European Regional Growth: Do Sectors Matter?" in Adams and Pigiliaru (eds).

74. Parsons, D. W. 1986. *The Political Economy of British Regional Policy*, p. 51. London: Croom Helm.
75. Perkins, D. and Yusuf, S. 1984. *Rural Development in China*. Baltimore: Johns Hopkins University Press.
76. Perroux, F. 1950. Economic space: Theory and applications. *Quarterly Journal of Economics* 64: 89-104.
77. Perroux, F. 1955. A note on the notion of growth pole. *Economie Appliqu · e*. Nos. 1 and 2.
78. Richardson, H. W. 1980. Polarization reversal in developing countries. In *Papers of the Regional Science Association* 45: 67-85.
79. Riskin, C. 1984. *China's Political Economy: The Quest for Development Since 1949*. Oxford: Oxford University Press.
80. Riskin, C. 1987. *China's Political Economy*. Oxford: Oxford University Press.
81. Riskin, C. 1990. Food, poverty, and development strategy in the People's Republic of China. In Newman, L. F. (ed.), *Hunger in History: Food, Shortage, Poverty and Deprivation*, 331-52. Cambridge: Basil Blackwell.
82. Sachs, J. and Warner, A. 1995. Economic reform and the process of global integration. *Brookings Papers on Economic Activity* 1: 1-118.
83. Savoie, D. J. 1992. *Regional Economic Development: Canada's Search for Solutions*, 2nd ed. Toronto: University of Toronto Press.
84. Selden, M. 1988. *The Political Economy of Contemporary China*, 11-23. Armonk, N. Y.: Sharpe.
85. Schachter, G., Bevilacqua, C. and Schachter, L. 1999. "The EE's Regional Development Programmes: Allocation of funds by basic macro-economic indicators in Adams and Pigiliaru (eds).
86. Schiffer, J. R. 1989. *Soviet Regional Economic Policy*. London: Macmillan.
87. Sidaway, J. D. and Simon, D. 1990. Spatial policies and uneven development in the 'Marxist-Leninist' states of the Third World. In Simon, D. (ed.), *Third World Regional Development*. London: Paul Chapman.
88. Simmons, A. B. 1979. Slowing metropolitan city growth in Asia. *Population and Development Review* 5(1): 87-104.
89. Slater, D. 1975. Underdevelopment and spatial inequality. *Progress in Planning* 4: 97-167.
90. Sloan, J. 1994. Measures of inter-provincial inequality in China during the reform period, 1981-1990. Paper presented to the Midwest Conference on Asian Affairs. Department of Economics, University of Illinois.
91. Smith, D. 1975. Neoclassical growth models and regional growth in the US. *Journal of Regional Science* 15: 165-81.
92. Smith, D. 1989. North and south: Britain's economic, social and political divide. London: Penguin Books.
93. Soja, E. 1980. The socio-spatial dialectic. *Annals of the Association of American Geographers* 70(2): 207-25.
94. Solow, R. M. 1956. A contribution to the theory of economic growth. *Quarterly Journal of Economics* 70: 65-94.
95. Spechler, M. C. 1979. Regional developments in the USSR, 1958-78. In *Soviet Economy in a Time of Change*, U.S. Congress Joint Economic Committee, 1: 141-63. Washington, DC: Government Printing Office.
96. Thomas, D. 1978. United Kingdom. In Clout, H. D. (ed), *Regional Development in Western Europe*, 191-210. Chichester: Wiley.
97. Tsui, K. Y. 1991. China's regional inequality, 1952-1985. *Journal of Comparative Economics* 15: 1-21.
98. Tsui, K. Y. 1993. Decomposition of China's regional inequalities. *Journal of Comparative Economics* 17: 600-27.
99. Vernon, R. 1981. The product cycle hypothesis in a new international economy. *Economic Bulletin*, p. 256.

100. Wang, Jian. 1996. *Study on China's Regional Development.* Beijing: State Planning Commission (Chinese text).
101. Wang, Shaognang and Hu Angang. 1999. *The Political Economy of Uneven Development: The Case of China.* Armonk: M. E. Sharpe.
102. Wei, Yehua D. 2000. *Regional Development in China: States, Globalization and Inequality.* London: Routledge.
103. Weng Jun-yi and Hsu Hua. 1996. *Unbalanced Growth and Coordinated Development.* Beijing: China Development Press (Chinese text).
104. Whyte, M. K. 1986. Social trends in China: the triumph of inequality? In *Modernizing China*, (ed), Barnett, A. D. and Clough, R., 114-18. Boulder, Colo.: Westview.
105. Williamson, J. G. 1965. Regional inequality and the process of national development. *Economic Development and Cultural Change* 13(2): 3-83.
106. Wu, C. 1987. Chinese socialism and uneven development. In Forbes, D. and Thrift, N. (eds), *The Socialist Third World: Urban Development and Territorial Planning*, 53-97. New York: Basil Blackwell.
107. Yang, D. 1990. Patterns of China's regional development strategy. *China Quarterly* 122: 230-57.
108. Yang, D. 1997. *Beyond Beijing.* New York: Routledge.
109. Yang, D. L. and Wei, H. 1996. Rural enterprise development and regional policy in China. *Asian Perspective* 35: 1.

Chapter 2

CHINA'S REGIONAL DEVELOPMENT POLICIES: A REVIEW

LU Da-dao

Professor and Former Director, Institute of Geography,
Chinese Academy of Science

V.F.S. SIT

Professor, Department of Geography,
University of Hong Kong

1. PURPOSES OF REGIONAL POLICIES AND CHINA'S REGIONS

1.1 Regional Policies: Purposes and Approaches

Propelled by increasing regional imbalances in some developed industrialized countries since the 1960s, international research on regional development policies has been accelerated. In many of such countries, recession hit some regions, and there was rift imbalance in labour supply and demand, leading to uneven regional growth and serious social and political conflicts among regions. The 'problematic' regions so generated have become major concerns of both academics and governments. The Japanese Prime Minister Tanaka published his book *Restructuring the Islands* ( ) in the early 1970s and set an example for top leaders of nations to be concerned with regional policies. In addressing regional 'problems', most Western countries follow three basic approaches, i.e.: (1) to allocate funds from the central government to assist the development of lagging regions, (2) the central government invests in infrastructure and human resources in the lagging regions, and (3) to apply high tax and high fees and other restraining measures in the developed regions while promote growth in lagging regions through tax concessions, subsidies and other enhancement measures. The last approach has been dubbed as the 'stick' and 'carot' method for altering the economic cost-surface for attaining balanced growth. It has been practiced in the U.K. and a number of West European countries (see Chapter 1). All three approaches are predicted on the capability of the central government both in terms of its larger share in the total national revenue and its administrative centralization and political might.

Regional policies may be defined as policies negotiated between the central and local governments for regulating regional differences which have an impact on the socio-economic processes of the regions. In a country with a vast territory with spatially diverse geographies and resource endowment e.g. China, the regional economic structure is of paramount importance among the list of goals of regional policies. According to western scholars, there are three main purposes of regional policies, viz.: rapid growth; stable growth and balanced growth (Furst,

Klemmer & Zimmermann, 1976). For a given period of time, the actual policies adopted have to be considered against the basis of the development level, the phasing of development, as well as the socio-economic development targets of the country concerned.

In many Western countries regional policies development has undergone two stages. At first, regional policies were mainly targeted at reducing regional disparity in economic development. Recently, they are more inclined towards promoting technological development. In the 1980s regional policy study has become a sub-discipline within the sciences, with clearly defined contents which include the theoretical foundation of regional policies, purposes of regional policies and the related evaluation methodologies. Notable researchers have also emerged. They include the founder of regional science, W. Isard (1960) who has contributed significantly to its theoretical foundation; Robert Solow (1956) on the mechanism of regional growth, and P. Klemmer ( ) on the typology of regional policies and regional growth measures.

1.2 Purpose of Regional Polices of the PRC

After the founding of the PRC, China has launched a large scale exploitation of natural resources and put in much efforts in economic development. It has since established a huge and comprehensive economic structure with a widely distributed network of regional production system. Along with the huge stride achieved in economic development, the central government has attempted a number of times to promote regional development and to solve the problem of regional imbalance. It treated balanced regional development as a major strategic move in socio-economic development of the country. As a consequence, the Chinese government has learnt from experience on how to formulate regional policies, and have developed related theories.

Chairman Mao's work: *On the Ten Major Relationships* which was first published in 1956, is the proclamation of the country's regional policies in the pre-Opening & Reform period. After 1978, academic discussions on regional policies by Chinese scholars have been very active and are clearly focused on the 'East-West Issue' and on 'Efficiency and Fairness' in evaluating regional disparity, especially in proposing measures to promote growth and development of the Western region. More representative results of such deliberations include Chen Dongsheng's *Theory and Practice in Economic Distribution* (1989), Zhou Sulian, et. al.'s *China's Regional Production Structures* (1991), Yang Kaizhong's *Research on China's Regional Development* (1989), Hu Angang et. al.'s *China Regional Disparity Report* (1995), Wei Houkai's *New Patterns in Regional Economic Development* (1995), and Sun Shangqing et. al's *China's Regional Complimentary Growth Strategy* (1994).

Put simply, the purpose of China's regional policies differs substantially from the 3 basic principles of Western countries, i.e.: fast growth, stable growth and balanced growth. This is especially so during the Pre-1978 period. On top of the three approaches, the Chinese government stressed on the overriding considerations of: (1) the political requirement of a fast transition of the economy and society towards communism, and (2) military security of the PRC. Since the adoption of the Opening and Reform policies in 1978, both the CCP and the government have shifted their 'core' activities from politics to economics. China has thence turned away from the post-1949 closed-door mode of development into one based on opening to and integration with the world markets. As a result, regional development has also entered into a new phase. On 15 July 1979, the Central Committee of the CCP and the State Council jointly approved the requests of Guangdong and Fujian for special policies and flexible measures in promoting an export-oriented economy in the two provinces. Starting from 1980, these provinces established new economic institutions and adopted special policies in foreign trade, foreign exchange control, local budgeting, economic planning, and in material supply and pricing. At that time, few people realized the significance of such changes on the Chinese economy and their importance in its regional development. At present, most regions of the PRC are striving for growth and development based on their own resource endowment, locational attributes, as well as the varying degree and ways in opening and reform. Thus the old mode of regional development in China had gone, leading to diversity and increasing disparity in its regional economies.

1.3 Regional Divisions

There is no commonly accepted division of the Chinese territory into regions. Different researchers with different objectives have produced very different regions. The Chinese government has also used different regional topologies to pursue its varying objectives at different times. Thus most regional divisions used are time-bound. Besides reflecting the history of China's regional policies, they also represent varying interpretations of the basic geographical attributes of the Chinese provinces. Fortunately most these typologies are based on the 30 provincial administrative units, i.e. the provinces, autonomous regions, and the 3 centrally-administered municipalities. In the present volume, these province-level units are generally referred to as the provinces. Since March 1997, Chongqing has been separated from Sichuan to become a new centrally-administered municipality. However, in most of the present volume, it is still subsumed under Sichuan. In our analysis in the chapters that follow, we mainly follow the official regional divisions, though we also used a new North-Central-South division to throw light on some of the regional issues. Some of our analyses in later chapters have also generated new regional divisions to underline major current spatial processes.

In brief, we have stuck to and made use of the following regional divisions which have to be first defined here (see Fig. 2.1).

(1) *Coastal and inland areas*. This regional dichotomy was first used by Mao in 1956. The coastal areas refer to the East, or Eastern Region as defined in the next regional division, while the interior areas embrace the rest of the country.

(2) *The East-Central-West Division*. In the mid 1980s, the Chinese government divided the country into three broad regions. The East embraces 12 coastal provinces. The Central and West each include 9 provincial units. This regional division provides the basis for one of the major targets of the Second Five-Year Plan (SFYP), i.e. to reduce regional disparity of the three regions by shifting the focus of economic construction to the Central and West.

(3) *The 3rd Line/Front areas*. In the Cultural Revolution Period (1966-1976), the country was re-divided into 3 fronts. The 1st Front areas are the coastal areas. The 3rd Front areas are the core of the interior areas including Sichuan, Hubei, Shaanxi, Henan and Guizhou (see Fig. 2.1). The rest of China belongs to the 2nd Front. At that time, the government's investment and construction programmes were largely located within the 3rd Front.

(4) *The Six Economic Coordination Regions*. In 1958, the State Council proposed a more refined regional division for guiding regional development planning. It first took the form of the Seven Economic Coordination Regions. In 1961, they were modified to become the Six Economic Coordination Regions. They have become the basis for coordinating energy production, the construction of large scale machinery industries as well as serving as the frame for economic planning.

(5) *The Seven Large Regions*. When preparing the Eighth FYP, the State Council announced a new division of the country into Seven Large Regions as basis for regional policies in the EFYP. The same regional division was followed in the Ninth FYP. Many provinces also adopted it for their 2010 long-range plan.

(6) *The Three North-Central-South Regional Belts*. We find that existing official topologies are inadequate for an emerging North-south discrepancy in development – within China as a result of its post-1978 Opening and Reform policies. Thus, a tripartite division based on three horizontal, North-south belts has been used to analyze some of the regional issues in the coming chapters.

We have attempted to define and illustrate the above regional divisions by Figure 2.1. In the rest of the chapter we shall review the content and historical background of the various phases of regional policies implemented by the Chinese government and evaluate their success or failure.

2. Geographical Background and Development Setting

Within the 50-year history of the PRC, China has gone through the preparation, initial development and medium-level stage in industrialisation. At the same time, the world has experienced major changes in geo-economic and geo-political situations. In these 50 years, China's regional objectives have been a product both of its ideology as well as its response to the changing world situations. As the regions of China possess different geographical settings, they have been given different policy emphases from the central government and have thus tracked divergent paths in the development of the local economy in different historical periods of the last PRC. This section provides a brief account of these varying physical economic, and geo-political factors which have affected the formation of regions in China

2.1 The Three Major Natural Regions and Three Topographic Steps

China is among the world's most diverse and complicated geographical areas, a product both of latitudinal and non-latitudinal effects. Climate (latitudinal effect) and geomorphology (non-latitudinal effect) are these two critical natural factors leading to the natural geographical spatial variation of the country, which may be summarized by the three major natural and three topographical steps.

The three natural regions are: (1) The Eastern Monsoon Region (Figs. 2.2 and 2.3). It covers 45% of China and contains 96% of its population, 90% of its total agricultural and industrial output value. In the region are the country's major lowlands, i.e. the North-east Plain, North China Plain, Changjiang Delta and Pearl River Delta. The region's climate is dictated largely by the monsoon winds with a wet hot summer and cold dry winter. (2) The North-west Dry Region is part of the Euroasian Steppes. It covers about 30% of the country's territory, and is mainly of highlands of over 1,000 m. Its climate is dry and the region is generally deficient in water supply. The region has for less than 4% of the country's population and gross agricultural and industrial output. (3) The Qinghai-Tibetan Plateau is unique natural region of a world scale. Its of high altitude averaging above 4,000 m. The natural conditions there are harsh and they impose great difficulties on human activities. The region covers roughly 25% of China's territory but accommodates less than 1% of its population.

Figure 2.1 Types of Regional Divisions

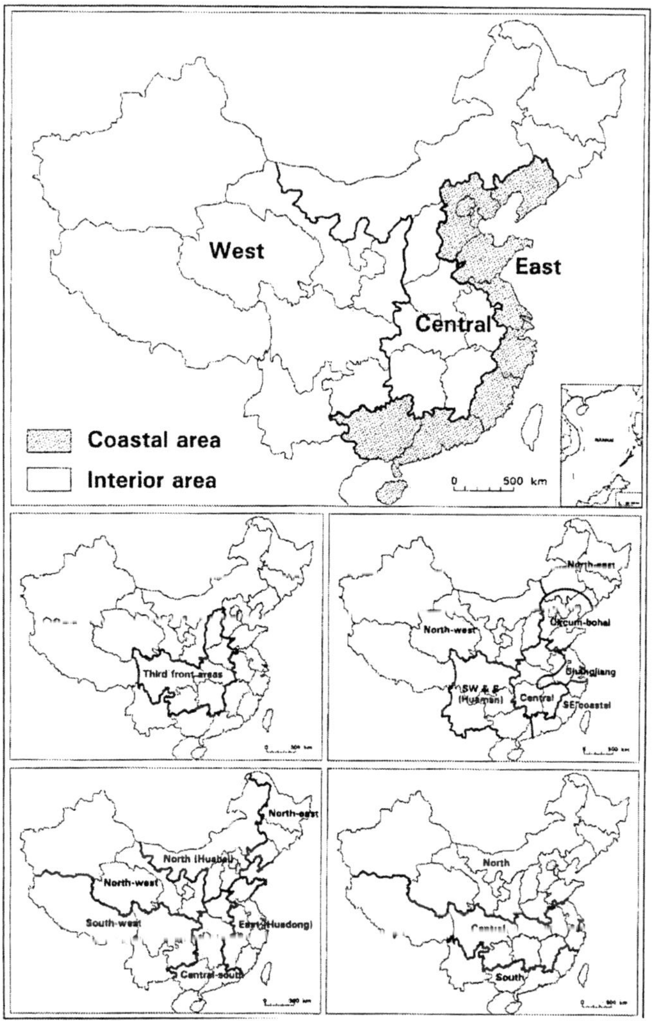

Fig. 2.2 The Three Natural Regions

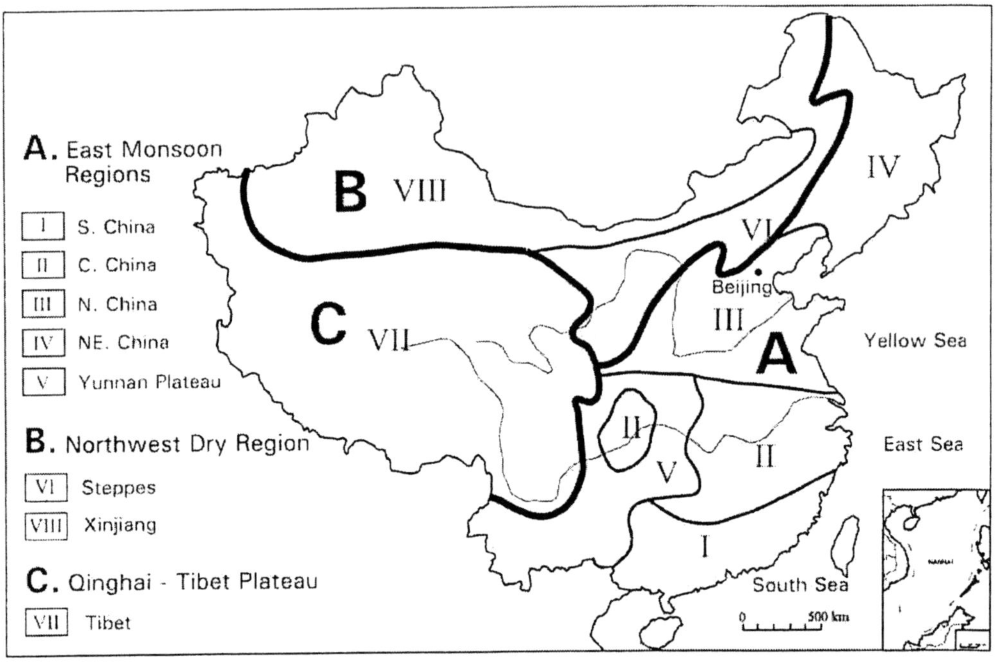

Fig. 2.3 China's Relief Regions and Three Topographic Steps

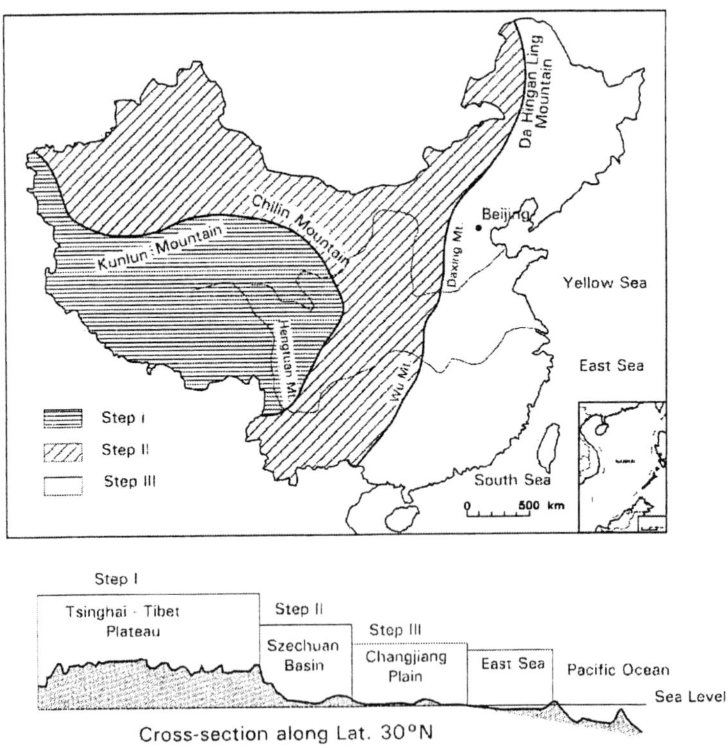

The Qinghai-Tibetan Plateau forms the first and highest topographical step. East of the Greater Xingan Mountains, Taihang Mountains, Funiu Mountains and all the way to the coastline lies the third and the lowest step which is where most of China's lowlands and low hills are found. In-between is the vast area of the second step.

The natural and topographical regions provide the physical framework for regional development and are important factors to be considered when adopting development strategies in regional development.

2.2 Effects of the Economic Management System: From Central-Planning to Market Forces

A nation's administrative system, particularly the way it manages the economy will have a major impact on regional development. China is a socialist country. The nature of the state sets a basic frame for its administration. Before 1979, China adopted the central planning system for managing the economy. "Under a highly centralised planned economic system and a strong sentiment of national unity, resource allocation of the country was a form of administrative behaviour. Basically the central government played the role of the investor. It planned and coordinated all investment projects and was responsible for their regional distribution. Products and raw materials were produced, supplied and distributed under one unified system of administrative allocation. Local revenue and expenditure were collectively controlled and balanced by the central government" (Chen, 1991). In addition, the local economies were also too weak to adopt and organise their own development strategies as their development targets were fully integrated (in most cases decided by the central government) into the national targets. Since 1980, local governments are given more and more decision making and budgetary autonomy. This new development, combined with the varying resources endowment and other factors of the regions, are generating increasing disparity in China's regional development.

"At the end of 1978, based on a detailed review of the country's economic development experience and the new international situation, the central government decided to adopt the open door policy as the country's fundamental national strategy" (People's Daily 11/7/1993). China has thence, pursued 15 years of opening and reform and has gradually established a model of development targets based on the market economy (Feng, 1984). Concurrent with increasing local administrative autonomy, local economies have charted significant increase in size and strength. They have by now become the main body in China's economic management system. Regional competition has appeared and become intense, with increasing disparity in socio-economic development in the regions. Under this new circumstance, the ability of the central government to manage development has also to some extent undermined. There should be a new regional development policy that takes into consideration overall national interests as well as the demands of regional economies and communities. Such a new strategy forms a critical challenge to the CCP and the central government.

Under central planning, most of the regions stressed on the development of heavy industry. The location of modern manufacturing then also tended to be raw material-biased. After the Opening and Reform in 1978 and the setting up of the modern market economy, forces of the market have exerted greater influence on economic development. However, market forces often represent short-term interests with little attention towards fairness and stability of the community. Thus, longer term government behavior, in the form of scientific regional policies for national security, stability, and more regional equity becomes, in the eyes of top decision markers of the country, a critical counter-weight to market forces. Thus regional policies assume a new meaning and significance as China further liberalises and de-volves power to local authorities.

2.3 Geo-Politics and Geo-Economics

In the last half-century, the geo-political and geo-economic situations relevant to China have experienced a number of major changes.

After the Bostain Declaration and Efta Agreement in 1945, the world was divided into two major camps. The Korean War in 1950 and the announcement of the USA on 27 June the same year to stop China's military campaign to liberate Taiwan, and its economic blockade on the PRC posed China imminent security threats. The fall-out of Sino-Soviet relationship in the late 1950s and the Sino-India border clashes in the early 1960s further reinforced China's fear of military attack and worsen the East-West geo-politcal situation.

The Shanghai Communique of 28 February, 1972 between China and USA marked a major turn as the two countries started a long process of normalization of relations. With this change of geo-politics, China embarked onto a new foreign relation strategy.

The Oil Crisis in 1973 promote globalization of production and international capital transfer. It had led to a down-turn in markets in industrialised countries in the West, while technological progress and demand situations in the labour market made it feasible as well as an added pressure on many firms in the West to globalised their production by transferring capital, technology and some of the production to foreign countries of lower costs and with more disciplined labour. In the late 1970s and 1980s, the Asian Newly Industrialised Economies of Hong Kong, Macau and Taiwan had also experienced high costs and labour shortage. They too demand outward expansion through trade and FDI.

The demise of the USSR and East European Bloc in late 1980s and early 1990s had brought to an end the Cold War. It boosted economic globalization and regionalisation one step further, and has led to the ascendance of economic issues over political and military strife in the international arena. It is therefore not a coincidence that FDI and foreign trade have increased significance in China. When China enters the World Trade Organisation (around 2001), it will mean another round of new economic and trade opportunities for the different regions of the country.

The above broad stroke geo-political and geo-economic settings have affected China's regional policies in the past half-century. They are to be borne in mind when evaluating the nature and implementation of China's regional policies.

3. Pre-1972 Regional Policies Under the Dictate of National Security

From 1949 to 1972, for more than two decades, China's regional policies were, to a large extent, subservient to national security based on the leadership's analysis of China's geo-political relationship with neighbouring countries and the possibility of major military conflicts that China might have with them. Chairman Mao's ultra-leftist ideological, such as his desire for a rapid transition into communism, had also significant impact. The latter, however, had more directly affected the national programmes such as the nationalization of the private sectors, rapid industrialization and the development of the Peoples' Communes.

We may further divide this period into two phases. Both subperiods are predominated by a quest for national security, but they show divergence in policy background and spatial implementation.

3.1 1949-1964

After the Civil War, there was a 3-year (1949-52) economic rehabilitation in which the central government's regional policy goal is "to correct the imbalanced productive capacity distribution pattern left by Old China", by shifting some existing coastal factories to inland raw material locations. In 1953, with the launching of the First Five-Year Plan such a regional shift was further stressed that "for the purpose of correcting the old unreasonable regional imbalance, new industrial bases have to be constructed. However, to provide the necessary pre-condition the Chinese government sanctioned the use, restructure and expansion of "existing industrial bases". Thus a 'regional balance' approach by strengthening the development of interior regions through industrial project had been firmly established.

In the three years of rehabilitation, the first priority was given to the heavy industry bases in the North-east, with those in East China and South-east taking a second priority. In the First Five-Year Plan period (FFYP), the country's major investment was predominated by 156 soviet aid-projects[1] and the 694 locally financed projects that exceed 10 million Yuan each. These laid the foundation of modern industrialisation of the PRC and its spatial pattern. In short, industrial construction in the FFYP shifted its focus from the coastal areas into the interior, as illustrated by Table 2.1 using the Six Economic Cooperation Regions at the time:

Table 2.1 Spatial Distribution of the Major Construction Projects in the First Five-Year Plan (1953-57)

Region	Project No.	% of Total
North-east	56	37.3
South-east	33	22.0
North	27	18.0
Central-south	18	12.0
South-west	11	7.3
East	5	3.4

Using the coastal-interior division, of the 156 major Soviet-aided projects, the coastal areas accounted for 32, or about 1/5. Of the 694 domestic projects, the coastal areas accounted for 32%. Thus the interior had the lion's share of these huge investment projects. As funds for the reconstruction, expansion and new construction in the coastal areas were inadequate, the potentials of old industrial bases had not been fully realised. Thus is particularly true for Shanghai, the Yangtze Delta and the coastal part of North China. Lacking large scale raw material producing plants and other complimentary projects, their growth rates had fallen to lower than that of the interior. For example, in 1952-57, China's industries grew at 15.5% annually, the rates for the interior and coastal areas were 17.8% and 14.4% respectively.

3.2 Strategical Shift to the 'Third Front' (1965-72)

In 1965, based on an unrealistic prediction of the central government that there would be imminent outbreak of war, China turned to two major policy decisions, i.e. a rush towards the 'Third Front' construction and, drastic implementation of a strategic shift emphasising on the setting up of independent industrial systems in the interior. The tenet for such moves was set by

[1] Two of them were double counted and four (including the Second Auto-works and the Second Tractor Factory) were not implement in the FFYP period. Thus the actual total is 150.

Chairman Mao in a speech given at a Conference of the Secretariat of the CCP in mid-August 1964. Mao said, "We must be prepared for military aggression of the imperialists. All new construction projects have to be located in the interior and we must put on hold all requests for new investment in the coastal areas". The National Construction Commission soon organised a 'national relocation conference' in August 1965 which concluded that in order to prepare the country for an early and major battle, projects of high-tech and the defense industry have to follow the principle of 'dispersed, into the hills and being hidden' in their location planning and that some even have to be actually 'caved-in' (constructed inside man-made caves). This principle was taken to be the main rule for economic construction for the rest of the period.

Thus in the Third FYP (1966-70), the focus of the nation's investment was moved from the North and the coast to the Third Front areas, with its core in Sichuan, Guizhou and the western parts of Henan, Hubei and Hunan. Panzhiha, Jiuquan and Chongqing were the new industrial construction foci. In the entire period, basic infrastructural investments in the interior accounted for 66.8% of the national total (Table 2.2).

Table 2.2 Distribution of Basic Infrastrural Investment in 1952-75 (%)

Time Period	Coastal*	Interior* subtotal	Third Front Areas
1st FYP	41.8	47.8	30.6
2nd FYP	42.3	53.9	36.9
Readjustment Period	39.4	58.0	38.2
3rd FYP	30.9	66.8	52.7
4th FYP	39.4	53.5	41.1
1952-75	40.0	55.0	40.0

* The two don't add up to 100. The balance is represented by purchase of related transport equipment which do not have a spatial breakdown.

In March 1973, the National Planning Conference proposed that during the Fourth FYP (1971-75), there should be a number of self-contained economic cooperative regions in the interior and each should have it's own character and be self-reliant. In sum, the interior was targeted to develop into a comprehensive economic structure with coordinated agricultural and industrial sectors for serving as the strategic hinterland of the country. Industries there pursued a spatial pattern of 'major dispersal, and local clustering' and should avoid major cities.

Investment by the state in the South-west and North-west regions which had a poor economic foundation accounted for only 16.9% of the country's total investments during the First FYP period. But it jumped to 35.1% in the Third FYP period. In the more developed East, the share of Shanghai receded to only 3.6%, ranking number 13 among the provincial units, while Tianjin just ranked 21. The share of the Third Front areas and coastal areas during the peak of 'Third Front Construction' may be gleaned from Table 2.3.

Table 2.3 The Share in National Total Investments, 1969-72

Third Front Areas		Selected Coastal Areas	
Sichuan	12.09	Liaoning	5.13
Hubei	7.38	Shanghai	2.38
Shaanxi	6.07	Tianjin	1.62
Henan	4.87	Jiangsu	2.40
Guizhou	4.22	Guangdong	3.44

4. First Regional Development Policy Readjustment, 1973-78

4.1 Background

A major regional policy readjustment took place in 1973-78 when the focus of economic development entered into a transition of shifting back towards the coastal areas. The major reason behind is the tremendous economic cost and difficulties encountered during the 'Third Front Construction'. The massive investment and huge human resources that were transferred to the interior provinces had not strengthened the economic base of these provinces. It was reported that 'the development of the interior was too slow, and the effort excessive' (Li, 1981). Combined with major and widespread mistakes in site selection, the strategy had led to tremendous waste and loss. Of the total fixed capital investment of ¥400 billion in 1953-80, only ¥250 billion are said to be 'effective' (operative). Among the latter, a proportion could not be efficiently operated as they lacked a suitable external environment and were beset by problems of bad location (Hsia, 1995).

A new turn in China's foreign relations also set in a new international environment for a major strategic change in China. In February 1972, China's relationship with the U.S. was normalised and its overall foreign relations with the rest of the world was also improving. At last, wars that would affect China directly were no longer imminent. International economic cooperation through equipment and technology import from advanced industrialised countries became an appropriate agenda in China's new state policy. While the long-drawn Cultural Revolution which had almost brought the country's economy to the brink of collapse had drawn the awareness of the leadership that economic development had to pay heed to efficiency, economic development and the consequent improved living standard for the people were also seen as important factors for maintaining social stability and sustainable development.

4.2 Major Projects Located in the East

In January 1973, the central government adopted a new strategy to 'enhance import of equipment and to expand economic exchange with other countries'. It thus opened a new phase of economic development emphasising the 'open door' approach. The pattern of economic development of the period shows a gradual shift of emphasis from the interior towards the coast where there have been more advanced socio-economic development.

Thus major new investment projects were mainly located in the more developed Central and East provinces. In the 1970s, a total of 47 comprehensive production plants were imported. 24 of these were located in the coastal areas. The Central and West received much less, i.e., 12 and 11 respectively. In the East, they were further concentrated in the central part of Liaoning (1 Petro-chemical, 1 Fertilizer Plant, 1 Electricity Plant), the Capital City Region (3 Petro-chemical, 2 Electricity Plant, 1 Fertilizer plant) the Changjiang Delta (4 Petro-Chemical, 2 Fertilizer Plant, 1 Steel Plant), East Shandong (1 Fertilizer, 1 Petro-chemical and 1 Tannery), and the Han River Plain (1 Steel Rolling Plant, 1 Fertilizer, and 1 Petro-chemical). In the West, the projects were mainly in the southern part of the Sichuan Basin. These locations possess a better environment for raw material and iron and steel industries to which the majority of the imported plants belong.

Another feature in location is that these new plants were either along the coast or along major rivers so that they are close to sea or fresh water supply. Some of them exploited water transport to increase the efficiency of investment, construction and operation of the projects. In 1973-80, of the 10 major oil refineries completed and the four under expansion, six are on the Changjiang banks (excluding those in Shanghai). The new projects represent a major improvement in China's economic location decision making after years of illogical and bitter experience.

In 1973, Premier Zhou pointed out that 'China should improve the ports within three years'. In the late 1970s, a large number of coastal ports witnessed major expansion. They include Dalian, Yingkou, Qinhuangdao, Tianjin, Ningbo, Xiamen, Guangzhou, and a number of new ports, e.g. Beilun. Improvement in rail transport was equally rift. In 1965-75, China had completed over 9,000 km of track, mainly in areas west of the Beijing-Guangzhou Line. The total rail mileage in the Western areas was increased from 3,900 km in 1949 to over 20,000 km in 1975 and the West accounted for 65-70% of the country's total investment in railway construction in the period. However, as 80% of the cargo volume growth happened in the eastern part of the country, the trunk lines in the East faced inadequate funds for new development and upgrading. In the late 1970s more efforts were spent in channeling coal export by rail from Shanxi to Bohai, and in linking the interior with the ports. At the same time, major improvement of existing tracks in the East was also launched.

In sum, the shifting back towards the East can be seen in terms of both new investments and expansion in major production projects, in imported facilities as well as in the development of the transport infrastructure.

5. Regional Policy Based on Comparative Advantages and Rapid Growth, 1979-1992

China experienced a major change in the late 1990s. The Third Plenary of the 11th Communist Party Conference marked a strategic shift of the national development path from emphasizing political and class struggle back to economic construction. This was immediately followed by a series of new policies and implementation measures soon known collectively as the Opening and Reform policies. In 1979, the central government announced that Guangdong and Fujian were granted the privileges 'to follow special policies and flexible management' marking the economic and institutional reforms that soon followed. In August 1980, the National People's Congress passed a legislation on the creation of Special Economic Zones (SEZ) in Guangdong and Fujian.

5.1 Coastal Development Strategy, 1981-86

As a consequence of the new philosophy, the Sixth FYP (1981-85), stressed on maximising existing economic bases in the coastal areas through opening policies, with the stated goal of 'exploiting their comparative advantages for enhancing economic development of the interior'. The major measures adopted for implementing the new policy include:

1. Enlarging the scope of opening

Since the creation of the 4 SEZs, in 1984, the central government further promoted opening by designating 14 Open Port Cities, i.e. Dalian, Qinhuangdao, Tianjin, Yiantai, Qingdao, Lianyungong, Nantong, Shanghai, Ningbo, Wenzhou, Fuzhou, Guangzhou, Zhanjiang and Beihai. Between 1985 and 1987, it further designated 5 Coastal Economic Open Areas, i.e. the Zhujiang Delta, Changjiang Delta, South Fujian Triangle, Shandong Peninsula and East-Liaoning Peninsula. These different categories of open areas represent China's attempt to foster a coastal open system of fast growing industrial cities and areas. They are the most developed areas of the nation with comparative edges in industry, agriculture, transport, communication, trade, technology and education.

2. Regional bias in policy measures and investment allocation

The regional favour towards the open areas can be charted by many new measures. The central government further strengthened investments in energy, high-tech industries, port and airport development, urban construction and the building of expressways and communication infrastructure in the open areas and open cities. Preferential treatment in revenue sharing, taxation, loans and investment had also been bestowed to these areas as will be listed and discussed in more details in later chapters.

In sum, in 1981-85, the fixed capital investment of the 11 coastal provinces increased from a 44% share of the national total in the Fifth FYP to 46%. Guangdong ranked first in the nation in the amount of fixed capital investment in 1984. Shanghai, being long neglected in the earlier periods, became number one in fixed capital investment in industries, taking a 9.6% share of the country's total. Making use of the opening policy, both Guangdong and Fujian had attracted a large amount of FDI, leading to hectic growth of their light industries. Shanghai, Southern-Jiangsu, Qingdao, Tianjin and Dalian also made use of such a policy bias to restructure and renew most of their industries in become more of high-tech oriented.

3. Increasing disparity

As a result of the new strategy, investment in the interior declined proportionately to the national total. During the height of the Third Front period (1965-67), Sichuan took 14.5% of the total fixed capital investment of the entire country. It dropped to 4.4% in 1984. In the Central, investment went mainly to energy and metallic minerals bases. In the West where the conditions are more favorable, such as the banks of the upper and middle reaches Huanghe, the southern and western parts of Sichuan and around Urumchi, there were investments in new and expanded projects.

Increasing regional disparity becomes evident. In the Sixth FYP period, China's GDP increased at an annual rate of over 9%, about twice the rate of the Fifth FYP. The pace of the East was much more hectic. The 14 open cities that accounted for 7.7% of the population of the country in 1984 increased their share in the national economy. Their industrial gross output value accounted for 23.1% of the nation's total; the per capita industrial output value was 3 times the national average and the fixed industrial capital asset was 12%, and contribution to remitted profits and tax was 26.2% the national total. Thus the East had drawn increasingly away from the Central and West due to faster rate of growth.

5.2 The Strategy of the Three Belts, 1986-1992

By the mid 1980s, China's development had made such significant progress that the dichotomy of the Coast and Interior had become too broad for more rational planning for the distribution of production. Thus in October 1982, Premier Zhao in a speech entitled "The New Industrial Revolution of the World and Our Challenges" proposed a spatial division of the country into the East, Central and West, which later had become an important regional concept of the State Planning Commission. Accordingly, in China's Seventh FYP (1986-90) and the Ten Year Plan (1986-1996) that followed, a new development strategy laying emphasis on the East and Central regions was used in the distribution of the nation's productive capacity. This is to say that while the coastal region continues to grow rapidly, an effort would be made to gradually shift the focus of growth into the Central region.

1. Regional Policy Goal

The Seventh FYP (1986-90) stated categorically that the level of development of the country showed a East-Central and East-West disparity. As such, the goal of development of the country was to respond to such a reality:

> "China's regional economic development should be based on the correct handling of relations between the three economic belts of the East, Central and West. From the Seventh FYP to the 1990s, we need to accelerate growth of the Eastern coastal areas,

while simultaneously put the focus of development of energy and raw material industries in the Central region. We should also make preparation for finally developing the West. In developing the East, coordinated exploitation of the West should be followed for mutual support and enhancement", and "We should encourage some regions, enterprises and individuals to get rich first".

In the early 1990s, although the country attempted to modify existing regional policies, the above-quoted "Three Belts" strategy remained in force. The "1987 Poverty Relief Program" adopted in 1987 was an important ancillary policy for the "Three Belts" strategy and is itself a major regional development goal of the nation in the last years of the 20th century.

2. Major Measures

In the coastal areas, while energy, iron and steel, petrochemical, machinery, automobile and ship-building industries were being enhanced, efforts were spent on developing electronics, consumer electrical products and communication products to meet market demands. Many of them had also registered significant increased in exports.

In the period, throughout the East, different types of economic and technology development zones, high-tech parks and SEZs had been fostered as growth centres. They had become Appendix bases for technological innovations. The list of these areas is appendixed to this chapter.

In the energy rich areas of the Northwest and Southwest, energy development was the focus of new growth. The principal development projects included the exploitation of natural gas in the Shaanxi-Gansu-Ningxia area, coal in Shaanxi-Shanxi-Inner Mongolia, and HEP in the upper reaches of Huanghe and Changjiang. Other than these, basic industries, like iron and steel and the refining of metallic minerals, were either newly built or expanded in the Central areas that have relevant resource endowment.

6. Second Major Regional Development Policy Readjustment (1992 and After)

6.1 Historical Background

In 1979-92, amid overall economic growth in the whole nation which averaged at 9% regional discrepancy in the rate of growth in the country had become obvious. Regional disparity has been demonstrated in a number of respects, i.e.

- between the provinces,
- between the Coast and the Interior,
- between the East, Central and West,
- between areas of national minorities and the predominant Han people,
- between urban and rural areas.

In 1978-90, China's National Income grew at 8.5% annually and the fastest growth was recorded in Zhejiang (11.9%), Guangdong (11.3%), Jiangsu (11.1%), Fujian (10.5%) and Shandong (9.9%). Yet the old industrial bases and less developed areas had grown much slower and usually lagged behind the national average by 2-4 percentage points. For example, the per capita National Income of Guizhou and Liaoning in 1978 were ¥192 and ¥502. In 1991 they were ¥1,817 and ¥1,933. In 1992, on average, the high income areas had a per capita income more than thrice that of the low income areas.

Policy and investment bias had been major reasons behind increasing regional disparity in the 1980s and early 1990s. The central government had thus criticised for regional bias in 'input', 'policy environment', 'price setting'and 'market competition'. At the same time, regional economic conflicts had led to a rise in local protectionalism. Increasing regional disparity had indeed started to pose new issues that may threaten social stability and overall development of the country.

In addition, high speed growth may not be sustainable indefinitely. The decade of hectic growth has brought serious environmental problems which are especially serious in areas that registered an extended period of growth in excess of 15% annually. Such areas, like the SW coastal areas, Changjiang Delta and some of the resource-based development areas in the Central, are currently faced with problems of ecological sustainability.

6.2 Goals

Due to the above-mentioned reasons, China did not set an explicit regional policy goal for the Eighth FYP period like what was done in the 1980s. Another major readjustment was underway. In 1992, the central leadership started to moot the idea of 'regional coordinated development'. It is a new strategy that focuses on compromises between the regions and are more acceptable by all of them. More details of the new concept was spelled out in 1996 in the Nineth FYP and the 2010 Long Term Plan. For the first time, these two documents adopted the principle of 'regional coordinated development' as the new guideline for the nation's socio-economic development:

> "Uphold the principle of regional coordinated development to gradually reduce regional disparity. . . more efforts are needed to stress the development of the interior. . . by applying measures that would reduce the trend of increasing disparity. . . Increase government efforts to reduce regional disparity".

The documents also request a careful balance between the comparative advantages of the developed areas and overall national planning, the relationship between the coastal and interior areas, and between the developed and less developed areas. The balancing act has to be one that promotes a rational spatial division of labour, and the complimentarity and coordinated development of the regions. Thus, the goal of the present stage of regional policies in China has been set.

6.3 Policy Measures

In 1992, the Chinese government adopted a policy of more or less complete opening to the outside and announced the opening of: 13 border cities, an addition of 5 port-cities along Changjiang, 4 provincial capitals in the border provinces and the capitals of other 11 interior provinces (see Figs. 2-3).

It also further expanded investment in major new projects in infrastructure, industrial plants and the creation of new agricultural bases into the Central and West to exploit their natural resources. The most notable investments include the energy base in Shaanxi-Shanxi-Inner Mongolia, and the cluster of coal-fired electricity plants on the banks of Huanghe, the Red River HEP, new oil exploitation and the development of 30 million mou of new cotton fields in Xinjiang. According to available data, of the 119 major projects under construction in 1996, the spatial distribution according to East-Central-West division is 46-41-22, i.e. the majority being in the Central and West regions. The balance, i.e., 10 projects, are cross-region in location.

In order to raise the proportion of total investment in the West out of the national total, the central government adopted a new practice of raising the proportion of investments in major construction projects of the West. Starting in 1996, it centralised some investments on infrastructure and resource utilisation. The latter was utilised by the central government for investing mainly in the West. In particular, the state had obviously used the new resources to favour growth of the Northwest (see Table 2.4).

Table 2.4 Source of Investment for Major Construction Projects (% ratio between central and local source)

	1988	1992	1993
Whole nation	1.32:1	0.79:1	0.67:1
Northwest	1.25:1	1.65:1	1.43:1

Poverty relief is another means used to reduce disparity as poverty was found mainly in the Interior. The central government annually allocates ¥10 billion to assist about 70 million poor residents, through infrastructure development, resources exploitation, education and the setting up of other economic activities. In addition, the National Comprehensive Agriculture Fund and the Township and Village Enterprises Fund also invested heavily in the Central and West. From 1994, China started the Tax Division Scheme (see Chapter 4). It will also help to increase the financial resources of the central government and hence its ability to support the less developed areas.

In the focal policy realm, the central government has raised the proportionate share of the Central and West in state policy loans and encouraged more foreign capital inflow there, e.g. by earmarking 60% of international syndicated loans and foreign government loans for projects in the less developed areas. In 1994, the State Development Bank was set up. The same year, it approved a loan amount of ¥33.5 billion to the Central and West, which is about 65% the total such loans in the year. The trend persisted in 1995 when ¥58.4 billion or 67% of all loans of the bank was granted to those regions. Solid measures like the decision to reduce policy lending for constructing coal- or oil-fired electricity plants in the coastal areas, while maintaining support for similar projects in the Interior, and the channeling of low or interest-free loans from foreign sources such as the Overseas Cooperation Fund of Japan and loans from the World Bank towards Interior projects.

From 1996 onwards, China took steps to reform taxation on import and export with an aim to gradually do away with regional tax preferences. This tends to reduce the biases towards the SEZs and Open Cities while forcing them to develop new competitive advantages. New measures for encouraging cooperation between the East and West have also been forged to enhance complimentary economic growth. This includes an attempt to guide resource processing and labour-intensive industries to shift towards the West and encourages capital in the East to invest into the West. Being mooted are measures for setting prices for resource-based products for enhancing their growth and development, hence the development of the Interior.

7. Evaluation of China's Regional Development Policies

To understand comprehensively and evaluate China's regional development strategies and policies since the founding of the PRC is a difficult task. It involves different perspectives on regional disparity and its intensification. Yet, it is significant for drawing lessons for future regional strategies and policies, something impinging on the nation's future growth. This is the main purpose of the present volume. For this introductory chapter, we attempt here a brief summary and evaluation of the major strategies and policies that have just been reviewed.

7.1 Goals of the Strategies

International experiences indicate that regional development strategies aim at three different goals: rapid growth, stable growth and balanced growth. For a country like China which is in the medium stage of industrialisation, it is difficult to achieve the latter two in the short and medium term. They are more relevant as long term regional development targets. Therefore, the more logical present goal for China is one of fast growth which to some extent is at conflict with the other two, because:

1. Fast growth by its nature is not balanced growth, especially so for a developing country with a vast territory. Economic historian J. Williamson (1965) had applied the inverted 'U' relationship between economic growth and regional imbalance in regional development research. From then on many follow-up studies proved his point. Recently N. Gradman ( ) of the University of New Jersey discovered that the disparity between the various states of the US had increased in the 1990s compared to the 1980s. He explained this as the result of faster growth of the decade after the Oil Crisis. Changes of the distribution patterns in population, transport and technology being other reasons.
2. Stable and balanced growth may be attainable simultaneously, but fast growth and balanced growth cannot be achieved at the same time.
3. In 1949-78, China's regional development strategy was dictated by national defense or national security considerations. It may be seen as a version of balanced growth though the major policy goal behind was not to attain balance in National Income in different regions.
4. China being a large developing country and is at its early, though vigorous, stage of industrialisation and urbanization, its economy is still predominated by the primary sector. Thus its priority should be the pursuance of fast growth. Its present stage of development therefore would show regional imbalance. It may be concluded that regional disparity is a logical by-product in pursuing fast growth.

At the same time, China is a socialist country. The nature of the state ideology requires that development should proceed within a 'socialist' framework which includes the continuing share of state-own enterprises or public-ownership in basic infrastructure and major economic sectors, and the maintenance of basic social equity and stability in the region. Therefore, although in periods when fast growth was the explicit goal there was some attention to socialist equity and stability.

7.2 The Two Major Strategic Readjustments

The two readjustments in 1973-78 and 1992 are transitional in nature. The first readjustment was propelled by economic reasons and was implemented mainly by way of imported equipment and turn-key projects. The second was a response to social demands for equity and sustainability. Its implementation took the form of policies.

7.3 Expanding Regional Disparity

A recent hot topic concerning China's regional development is how to view increasing regional disparity after the Opening and Reform in 1978. The 15th Communist Party Congress reached a major strategic conclusion that China is still 'in the early stage in building socialism'. It believed that the present socio-economic attributes of the country will have an 'irresistible' duration of one hundred years. The period will be characterised also by a temporal process which starts "from an imbalance in regional economic and social conditions, through even development to gradually reduced regional disparity". The process will also be evidenced by a "gradual reduction of disparity with the world's advanced economies and achieved the revival of Chinese people on the basis of socialism". We believe that regional disparity and its enlargement is one of the basic feature of the early stage of Chinese socialism.

In the 1970s economic globalisation had gained speed as advanced countries transfer in a in a large scale low value-added production and exported capital into the developing countries. The Opening and Reform of China and its related regional development policies were appropriate and timely responses to such a global development. It is not a 'regional biased' approach to development. In the process, the coastal areas had successfully exploited their comparative advantages and potentials. Some developed predominantly export-oriented economies. Such development was appropriate under the internal and external conditions of the country at the time.

After the Opening and Reform, regional disparity has not been fully reflected in the tripartite division of the country into the East, Central and West, nor is it mainly an issue of the areas of concentration of the Han Chinese and the national minorities. The largest disparity exists between the rural and urban areas. Chapter 3 will address more on such issues.

Besides the influences of the three Natural Regions and three Topographical Regions, the existing economic foundation and social and historical attributes also affect divergent development of China's regions. In the short and medium term, the appropriate regional strategy should be one that focuses on "relieving the tendency of increasing disparity", and "moving towards reducing disparity" (Ninth FYP and 2010 Plan). Efforts must be made to firmly establish the system for the operation of the Chinese socialist market economy as a sure way to strength various regional economies. 'To reduce disparity' is not an appropriate goal at the present. Another major reason behind is that the central government do not have the financial capability yet. In 1995, of the total invested fixed capital of about ¥2,000 billion, only ¥400 billion or 20% came from the central government. Thus, without faster growth of central revenue, a meaningful 'strategic transfer' may not be achieved in the medium term.

Under the present system of the socialist market economy, the support of superior administrative units remains important in the growth of a local area. Yet lateral administrative cooperation are playing an increasing role in promoting growth, as partnership with the right locale, or regional economic cooperation for capital, market, raw material, equipment, technology, human resources and information is an important means for growth based on the principle of regional complementarity. This works for both the developed and less developed regions. A national policy for promoting and guiding regional cooperation and for setting up measures for regional economic management across provincial borders will have a lot to gain, particularly in relaxing regional disparity in the process of enhancing growth. This is yet a major area for the Central and local governments to take action and will be detailed in chapter 8. Of course, to crown all, the central government's financial capability and decision power on key infrastructural projects and sectoral policies have to be much enhanced. To find a suitable balancing point in such matters between centralisation and localisation that fits the reality of the present stage of Chinese socialist market economy is not an easy job, though it is of critical significance.

REFERENCES

1. Chen, Dongsheng. 1989. *Theory and Practice in Economic Distribution*. (Chinese text).
2. Chen, Dongsheng. 1991. Issues in regional economies. *National Land Development Reconstruction*. Vol. 1, no. 1, (Chinese text).
3. Feng, Weizhung. (ed.) 1984. *Major Economic Events of the Peoples Republic of China*. Beijing: Social Science Press (Chinese text).
4. Furst, D., Klemmer, P. and Zimmermann, K. 1976. *Regionale Wirtschaftspolitik*. Duesseldorf: Werner Verlag.
5. Gradman, N.

6. Hsia, Senmin. *et al.* 1995. Policy suggestion regarding production readjustment in China's East, Central and Western regions. *National Land Development and Reconstruction*. Vol. 1 (Chinese text).
7. Hu, Angang. *et al.* 1995. *China Regional Disparity Report*. Shenyang: Liaoning People's Press (Chinese text).
8. Isard, W. 1960. *Methods of Regional Analysis: An Introduction to Regional Science*. Cambridge: MIT Press.
9. Klemmer, P.
10. Li, Jebao. 1981. *Infrastructure, Environment and Regional Comparative Advantage*. Beijing: People's Press (Chinese text).
11. *People's Daily*. 1993.11.17. Decisions of the CCP Central Committee on the System of Socialist Market Economy (Chinese text).
12. Solow, R.M. 1956. A contribution to the theory of economic growth. *Quarterly Journal of Economics*, 70: 65-94.
13. Sun, Shangqing. *et al. China's Regional Complimentary Growth Strategy*. (Chinese text).
14. Tanaka, *Restructuring the Islands*.
15. Wei, Houkai. 1995. *New Patterns in Regional Economic Development*. Kunming: Yunnan People's Press (Chinese text).
16. Williamson, J.G. 1965. Regional inequality and the process of national development: a description of patterns. *Economic Development and Cultural Change* 13(3).
17. Yang, Kaizhong. 1989. *Research On China's Regional Development*. Beijing: Ocean Press (Chinese text).
18. Zhou, Sulian. *et al.* 1991. *China's Regional Production Structures*. (Chinese text).

Chapter 3

ASSESSING TOTAL REGIONAL DEVELOPMENT IN CHINA

LIU Weidong
Associate Professor, Institute of Geography,
Chinese Academy of Science

1. IMPORTANCE OF REGIONAL DISPARITY IN DEVELOPMENT

In the past two decades, China as a whole has achieved unprecedented and world-known economic growth, lifting her national strength and people's living standard to a new level. At the same time, since the different regions of her vast territory show different economic vitality, regional development disparity has become one of the hot issues that have drawn much concern from the government, academic circles and international organisations. In recent years, the regional development gap has been the major complaint from the deputies and officials of the Central and Western regions during the annual National Peoples' Congress. Quite a few international organisations and world famous scholars have also listed regional disparity as one of the uncertainties which the Chinese government has to deal with properly. For example, in its report on China published in 1993 the US-based Morgan Stanley Corp. (Group) cautioned that increasing regional disparity and urban-rural differences are dangerous for China's future development. Of course, the debate goes much deeper and wider than these general concerns (for example, Lu, 1995, -96; Hu, 1995; Chen, et al, 1996; Yang, 1993; Wei, 1992; Liu, 1994; Zhao, 1994, -95; Li, 1996; Xue, 1994; Cui, 1992, -94; Cannon, 1990; Maruya, 1997; etc.). Several authors, represented by Hu (1995), indicate that recent regional disparity has been so serious that the gap should be reduced immediately; while other authors like Lu (1996) argue that it is not only impossible but also unnecessary to reduce the absolute differences in the near future and a reasonable and necessary step to be taken now is to slow down the increasing rate of the gap. No matter which viewpoint they hold, most stress that the central government should pay sufficient attention to the issue and take suitable and definite countermeasures.

In fact, in its Ninth Five Year Plan (FYP) the Chinese central government has taken regional disparity as an important issue and has changed its regional policy from one emphasising coastal priority towards a coordinated regional development strategy (see Chapter 2). However, these recent policies are based mainly on the recognition of regional disparities according to a simple regionalization concept of the three regions, i.e. the East, Central and West regions, which is already not able to reflect the real situation. The three-region division did facilitate the coastal priority regional policy of the 1980s and had played an important role in promoting rapid national economic growth there. Since the mid-1990s such a regional division is evidently inadequate as the development of regions has become much diversified. For example, in the East /coastal region, there exist both "growth stars" like Guangdong, Zhejiang and Jiangsu and "problem regions" like Liaoning. Similarly, the middle income provinces, such as Jilin, Heilongjiang, Xinjiang and Hubei, are not facing the same development problems. Cannon (1992) points out that China is an enormously intricate patchwork of social and economic variation, not easily subsumed into a few large regions. Cui (1992), too, provides an empirical evidence that the disparity between the three regions can explain only 50 per cent of the total regional differences of China. Thus intra-region difference is in fact no less than inter-region difference, and it becomes necessary and important for policy making and implementation to analyse and assess recent economic development of each province and to classify them into proper development types.

This chapter will firstly review the historical process of regional disparity in China since the 1950s, and it will then use a set of indice to analyse and evaluate economic and social development of the provinces in 1990-1996. Lastly it will try to regionalize the socio-economic development in China based on provincial data. It hopes to offer an empirical guide for the central and local governments in making regional policies and in choosing a development strategy. The basic spatial unit used for analysis will be the provinces due to the availability of data although a smaller unit may reveal more information.

2. Post-1949 Regional Development Disparity

It is essential for assessing current regional disparity to have a proper understanding of its historical evolution which can be glanced from a large literature on inter-province income disparity in China (e.g. Yang, K., 1988; Wei, H., 1992; Yang, W., 1992; Liu, S., 1994; Zhao, S. B., 1994; etc.). However such studies are limited to the late 1980s or early 1990s. Besides, when analysing the standard deviation (*SD*) of income per capita, many of these studies have not transferred the current price to fixed price. In this section we will review such important aspects of regional development by examining the historical changes in inter-province disparity of 'national income (*NI*) per capita (1952-77)' and 'gross domestic product (*GDP*) per capita (1978-95)' as well as 'consumption per capita (1952-96)'. As Beijing, Tianjin and Shanghai are much different from other provincial units in area and urbanisation level, i.e. their difference with other provinces may be seen as a urban-rural difference, it may not be reasonable to take them as a provincial unit and compare them with others. Thus, in the following analysis, they have been excluded unless it is particularly pointed out.

2.1 Inter-province Disparity in Economic Development

2.1.1 The inter-province gap of economic development has been increasing since the 1950s. It increased slowly before the Opening and Reform in 1978 but very rapidly in the 1980s and -90s.

From 1952 to 1977, the *SD* of per capita *NI* increased by 86 per cent and its change was very smooth except during the *Great Leap Forward* (GLF) period (Fig. 3.1), which indicates that inter-province economic disparity had not altered much before the *Opening and Reform*. The smooth curve of regional development disparity is usually explained by the so-called "balanced regional development strategy" although the strategy actually has much more to do with national defence purpose due to the then geo-political environment (see Chapter 2). While not with deliberate intention, the policy did arrest regional disparity at the cost of a low growth rate for the entire national economy. The average growth rate of *NI* of China was just 6.3 per cent during the same period. If Beijing, Tianjin and Shanghai are included, the *SD* of provincial per capita *NI* increased quickly in the 1970s (Fig. 3.1), but this mainly reflects the increase in the urban-rural gap.

Fig. 3.1 Standard Deviation of Provincial Per Capita NI (1952-77)/GDP (1978-96) in China (*SD* of per capita *NI* is in 1952 fixed price, and that of *GDP* is in 1978 fixed price.)

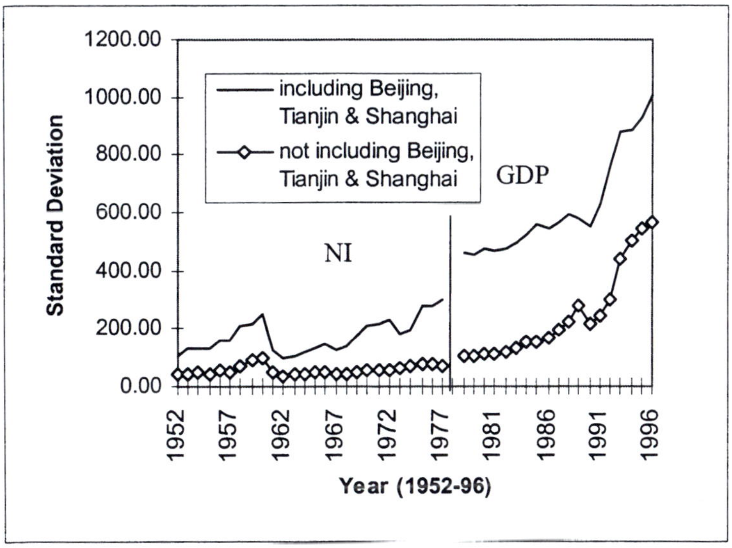

After the *Opening and Reform*, the decentralisation of planning power and coastal priority strategy had released development potentials of the coastal provinces, particularly the southeast coastal regions and the Yangtze River delta. At the same time, the *SD* of provincial per capita GDP had increased quickly, from 103.6 in 1978 to 219.0 in 1990. In this period, five of the seven fastest growing provinces were within the coastal region, including Guangdong, Zhejiang, Jiangsu, Fujian and Shandong (Appendix 3.1). The other two were Xinjiang and Yunnan in the West. Before the 1980s most of these five coastal provinces had a record of per capita *NI* lower than the country's average. The descending provinces were traditional industry bases, like Shanghai, Tianjin, Liaoning and Heilongjiang, and several "third line" provinces. They are mainly in the coastal region but not concentrated in the Central and West.

Deng Xiaoping's tour to southern China in 1992 encouraged a new surge of economic development in China and a wave of huge investment, including foreign direct investment (FDI), which was mainly concentrated in the coastal regions. From 1990 to 1996, China recorded a high annual GDP growth rate of 11.2 per cent. But it also registered a sharp increase of *SD* of per capita *GDP*, from 219.0 to 567.6. All coastal provinces except Beijing, Tianjin, Shanghai and Liaoning had an annual GDP growth rate of 20 per cent higher than the national average while the growth rates of most western provinces were lower than the average.

Thus the inter-province development disparity is rapidly enlarging. The *SD* of per capita *NI*/*GDP* had first doubled in 26 years. It then took 12 years to get another doubling. Recently this happened only in 5 years. The accelerated increase of regional disparity can be attributed to two reasons, i.e. rapid economic growth of all provinces and the difference in growth rate between provinces. The first reason implies that even if all provinces have grown at the same rate, the *SD* of per capita GDP will still increase. The second reason may strengthen the increasing tendency. The sharp increase of regional disparity may not be satisfactorily explained according to only one of these two reasons. No matter which is the principal reason, absolute regional disparity is enlarging rapidly.

2.1.2 The relative inter-province disparity of economic development has been much fluctuated, but shows a decreasing tendency in a long term, indicating that increasing regional disparity is mainly due to faster growth of former low-income provinces.

Figure 3.2 shows that before the *Opening and Reform* in 1978 the coefficient of variation (*CV*) of per capita *NI* in China had experienced three major fluctuations, i.e. during the *GLF*, late 1960s and mid-1970s. Since 1978, however, the indicator has been decreasing, except for a little rise between 1992 and 1994. The *CV* in 1996 (0.67) is less than half of its highest level (1.42) in 1960. If measured by the extreme gap of per capita GDP, inter-province economic disparity has been quite stable with the gap between 4 and 5 in the last four decades (Fig. 3.3).

Fig. 3.2 Coefficient of Variation of Provincial Per Capita NI (1953-77)/GDP (1978-96) in China

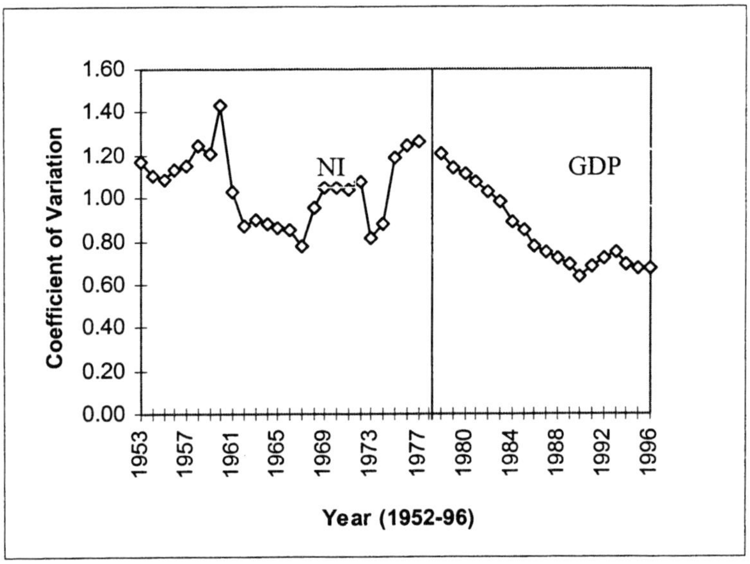

The decreasing *CV* of per capita *GDP* seems reverse to regional disparity to that *SD* has shown. This antithesis is the direct reason that while some authors argue that China's regional disparity is reducing while others believe it is enlarging. Past researches have not given a reasonable explanation to such statistic results. Most authors have avoided such a contradiction by choosing only one of the two indices, i.e. either the *CV* or *SD*, as a sole criterion. Actually, a mathematics treatment can relate the two to provide a laudable conclusion. The *CV* mainly reflects the extent of difference between each unit's growth rate and the distribution feature of the units' growth rate. For example, if all high-income regions grow faster than average and all low-income regions grow slower than average, then the *CV* will increase. Conversely, it will decrease. The actual changes are of course more complex, but the statistical analysis still make sense in understanding more about the results of the *CV* and *SD*. The decreasing *CV* of per capita *GDP* indicates that the most rapidly growing provinces are not all high-income provinces, at least some of them are former low-income provinces. The rise of *CV* between 1992 and 1994 shows that the higher the income level, the quicker the GDP growth rate in the period. Thus it can be inferred from the statistical fact of decreasing *CV* that rapid enlargement of absolute regional disparity since the 1980s has mainly been caused by the quicker growth of some former low-income provinces.

Fig. 3.3 Inter-province Extreme Gap of Per Capita NI (1953-77)/GDP (1978-96) in China

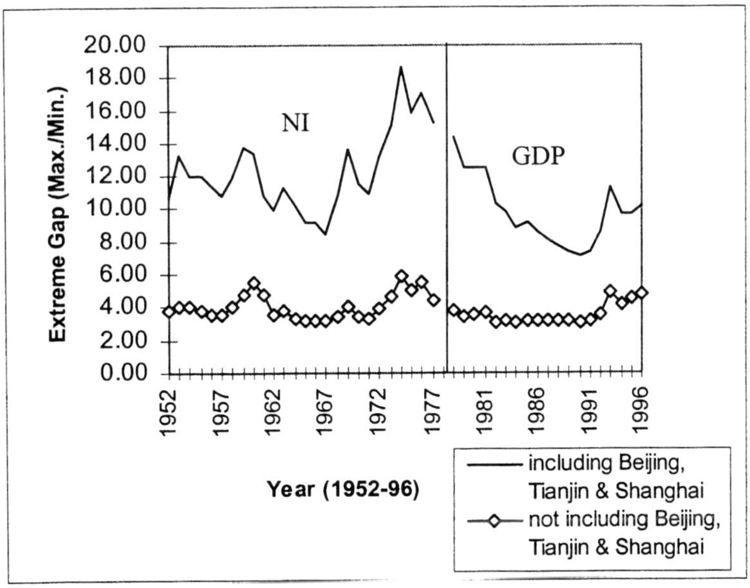

2.1.3 In quite a long period the regional economic disparity in China had mainly been a north-south gap, and it became an east-west gap only since the 1980s.

Before 1949, modern industries in China were concentrated mainly in a few coastal treat-port cities and central Liaoning, and they were predominated by foreign interests. In spite of this inheritance, in the early 1950s inter-provincial economic disparity in terms of per capita *NI* showed a north-south gap (Fig. 3.4-a). In 1952, most high-income provinces were in northern China. Beijing, Tianjin, Liaoning, Jilin, Heilongjiang, Xinjiang and Inner Mongolia had a record of per capita NI 25 per cent higher than national average while only one southern provincial unit, i.e. Shanghai, had such an income level. Two reasons can be attributed to this spatial pattern. One is that agriculture was a major contributor to *NI* at that time. The other is the fact that the central government used most of its resources to build the northeast region during the recovery period of 1950-52.

During the first and second *FYP* period (1953-62), China adopted a regional development strategy that stressed industrial construction in the inland areas (see Chapter 2). The principal analyses on the spatial result of such a strategy usually focuses on the shift of industrial construction towards the inland areas and the economic disparity thereafter between the coastal and inland regions. Although this shift is part of the truth, the biggest economic gap then actually existed between the north and south, not between the coastal and inland regions. From 1953 to 1962, about 42 per cent of China's total capital construction went to the coastal region, quite proportional to its population share (40 per cent). This means that, if evaluated by per capita capital construction, the coastal region was actually not in a down-played position though the investment there had declined compared to that before 1949. The key point, however, is where the inland investments had gone to. They were concentrated in northern China. The Northeast, North and Northwest China (usually shortened as the Three Norths) accounted for 58 per cent and 46 per cent, respectively, of the total investment in the first and second FYP periods while their total population was only 26 per cent of the country's total. Thus, in terms of per capita capital investment, regional disparity at that time was indeed a big north-south gap, leading to the persistence of north-south differences in per capita *NI* throughout the 1950s and even 1960s. In 1962, nearly all provinces with per capita *NI* 25 per cent higher than the national average were in northern China (Figure 3.4-b). The only exception was still Shanghai. This spatial distribution pattern is explainable by the then international environment, i.e. the antagonism between the capitalist camp centred around the US and the communist camp led by the former USSR. The northern regions were considered a safety belt since they are near USSR, the close ally of China.

In the third and fourth *FYP* period (1965-75), as a result of the "third line" strategy (see Chapter 2), huge capital investments had gone to peripheral provinces like Sichuan, Guizhou, Shaanxi, Gansu, Qinghai, Hunan, etc. However, due to serious external diseconomies caused by the adoption of uneconomic locational principles, i.e. '*regional dispersal and situational concentration*', '*far away from big cities*' and '*hidden in mountains or caves whenever possible*', these investments achieved little in increasing the per capita *NI* of the 'third front' provinces. By 1978 most of the high-income provinces were still in the northern regions (Figure 3.4-c) though the north-south gap had become less distinct. A clear change then was the rise of Jiangsu and Zhejiang which were already high-income provinces in 1978. After the *Opening and Reform*, the coastal priority policy has promoted rapid growth of the coastal region and led to the emergence and prominence of a new east-west gap (Figure 3.4-d). It should be acknowledged, however, that the most rapidly rising provinces in the last two decades are mainly those 'cold' regions before the *Opening and Reform*.

Figure 3.4 Spatial Distribution of Provincial Per Capita NI / GDP in China

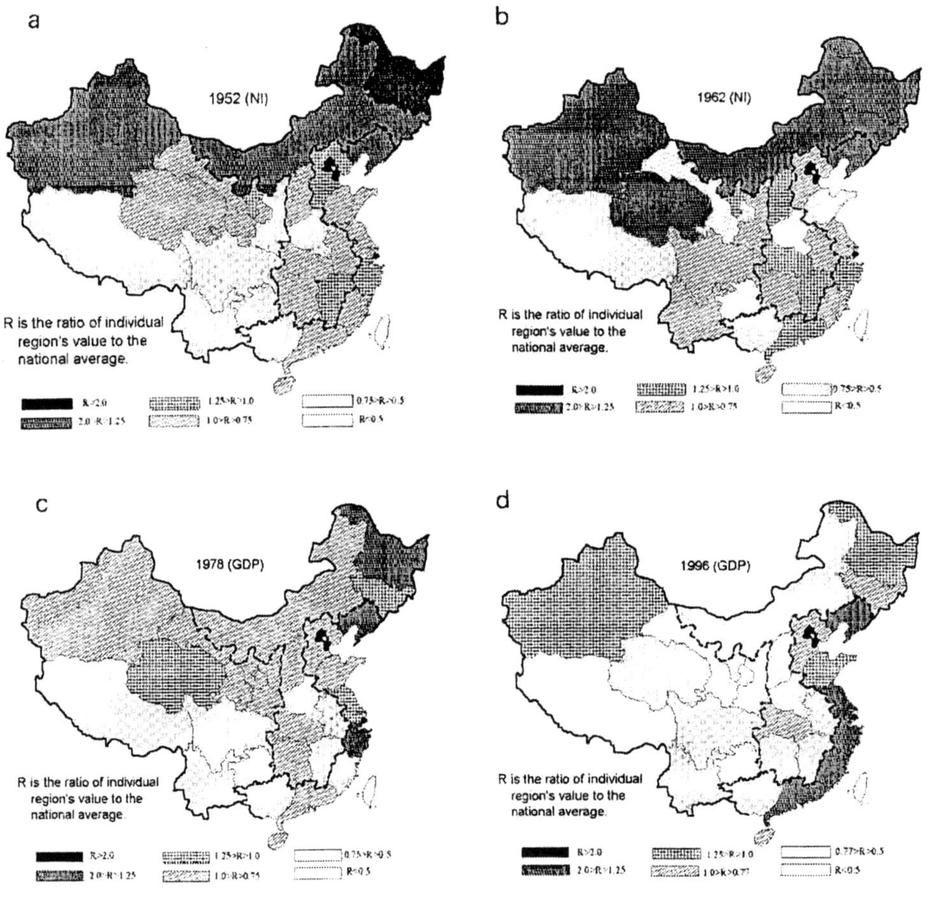

2.2 Inter-Province Disparity of Consumption Level

Under the central planning system before the *Opening and Reform*, the central government derived its revenue mainly from comparatively developed provinces. For example, Shanghai once contributed nearly one-third of the central revenue. Thus regional disparity of per capita *NI* was not equated with the difference in people's livelihood. Even after the *Opening and Reform* when the financial system has been changed towards a sharing mechanism between the central and local governments (Chapter 4), these two kinds of disparity still do not conform because the financing system is still not a standard one in which the revenue turned over to the state by an individual province is proportional to its per capita GDP. Therefore, it is helpful for an overall understanding of regional disparity to look into changes of the inter-province consumption gap. Since complete time series data for Zhejiang, Anhui, Qinghai and Guangxi are not available, the following *SD* and *CV* of per capita consumption can only offer an approximate view into the regional livelihood disparity.

2.2.1 The inter-province consumption disparity shares the same tendency of change as that of per capita NI/GDP, but it is much smaller than the latter.

From 1952 to 1978, the *SD* of per capita consumption increased by only 50 per cent (Fig. 3.5), much less than that of per capita *NI* in the same period, which indicates that during the highly centralised economy period the central government obtained much more revenue from the comparatively more developed provinces than other provinces. Since 1978, inter-provincial livelihood gap has been rising rapidly. The *SD* of per capita consumption in 1995 was five times that in 1978, increasing at nearly the same rate as *SD* of per capita *GDP*. In particular, in the four years of 1991-95, the index had doubled. One reason behind is that the central government had adopted a generous financial policy towards the rapid developing provinces, closely related to the strategy to encourage rapid growth of the coastal regions. However, the inter-province livelihood gap was much less than the economic gap in both period (before and after the *Opening and Reform*). In 1952, the *SD* of per capita consumption was 57 per cent of that of the latter while it was 36 per cent in 1978 and 34 per cent in 1995.

Fig. 3.5 Standard Deviation of Provincial Per Capita Consumption (1952-95) in China (*SD* in the period of 1952-78 is in 1952 fixed price, and that after 1978 in 1978 fixed price)

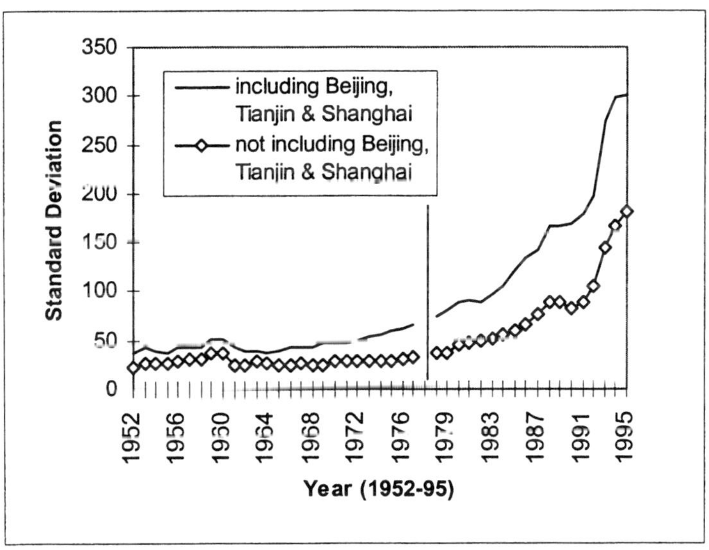

Fig. 3.6 Coefficient of Variation of Provincial Per Capita Consumption (1952-96) in China

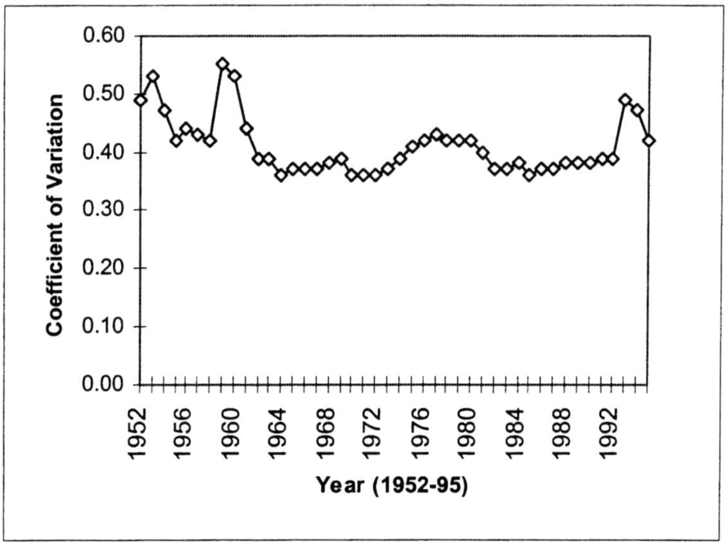

Fig. 3.7 Inter-province Extreme Gap of Per Capita Consumption (1978-96) in China

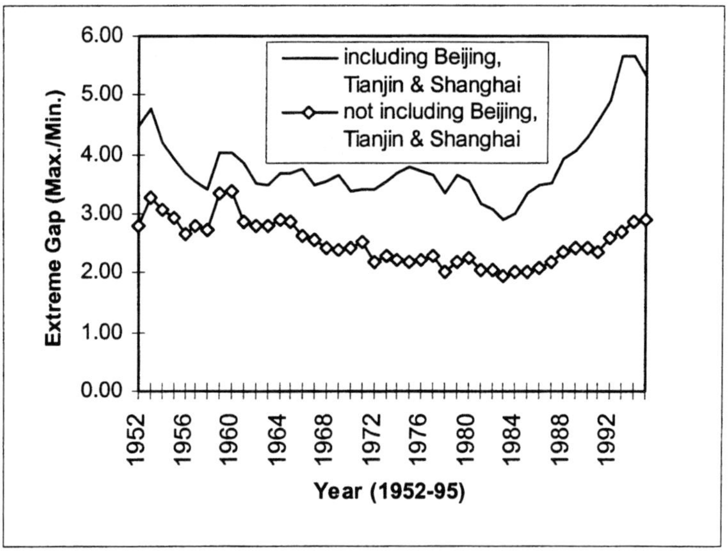

2.2 2 The relative gap of consumption level has been quite steady if compared with that of per capita NI/GDP, but the extreme gap has been increasing since the mid-1980s.

In the last four decades the *CV* of per capita consumption has not changed much (Fig. 3.6), fluctuating mainly between 0.4 and 0.5, which implies that the overall inter-province difference of livelihood level has been nearly constant. However, the gap between provinces of the highest and lowest per capita consumption has been rising in the last decade (Fig. 3.7) though the rise is actually less if inter-province difference of commodity prices is taken into account, i.e. the developed provinces usually have a higher price index. Besides, comparing figure 3.2 to figure 3.4, inter-province consumption gap measured by the *CV* is also smaller than per capita *GDP* gap.

2.3 The Nature of Regional Economic Disparity

In the 1990s the inter-province disparity of economic development and people's consumption level have been increasing very rapidly. Such an increase is due to the faster growth of several former low-income provinces like Guangdong, Jiangsu, Zhejiang, Fujian and Shandong. Before the Opening and Reform, none of these provinces were of very high per capita *NI*. Another reason is that all provinces have been developing rapidly, leading to the fast increase of per capita *GDP*/consumption. Under these circumstances, the *SD* would still increase quickly even if all provinces were growing at the same rate. On the other hand, relative regional disparity (*CV*) has shown a decreasing tendency as the inter-province consumption gap is much less than the per capita *GDP* gap.

The recent 'east-west' disparity should be understood rationally. In the early 1950s, while the coastal regions had much better infrastructure and more industrial bases, the regional disparity was a 'north-south' gap in terms of per capita *NI* due to the then geo-political environment, i.e. the antagonism between the capitalist and communist camps. The "balanced regional development strategy" helped to maintain such a spatial structure at the cost of a low increase rate of the whole national economy. Since the1980s, under an increasing tendency of globalization, the comparative advantages of the coastal regions have been brought into play, making the 'east-west' gap a focus of public attention. However, the change in regional disparity cannot be attributed wholly to the coastal priority strategy. It is fair and rational to understand it as a result of the government's correct step in conforming with the world's tendency of globalization and to make full use of available development opportunities. Indeed, in the formation of the 'east-west' gap, market forces and the entry of *FDI* have played a key role.

The "start off effect" is an important factor behind the increasing regional disparity. Since *Liberation*, China's economic development has seen periodic fluctuations of 'lax control' and "tight control", i.e. periods of high-speed growth interwoven with periods of retrenchment of capital investment. Such fluctuations are particularly evident after the *Opening and Reform*. Normally, when the government is going to relax control over capital investment, a few provinces with better economic bases and stronger market consciousness are the first to increase investment and achieve a new round of fast growth, while others lag behind in one year or so. When the retrenchment comes, however, all provinces will be affected nearly at the same time. Sometimes "agile" provinces may also find ways to get behind the central government's tight control during the retrenchment period and acquire faster growth than other provinces. This kind of "start off effect" is a reason leading to the increase in the regional economic gap. The empirical study in this chapter offers some evidence, that is there exists a close correlation between the increase rate of the *SD* of per capita *GDP* and the growth rate of the national GDP in China. In 1979-95, the correlation coefficient between the two is 0.66, indicating that when the national *GDP* increases rapidly, the *SD* of provincial per capita *GDP* also increases fast, and the same is true conversely.

3. Regional Development, Resources and Disparity: 1990-96

1990-96 is a special period for China. It is characterised by an ever-fast economic growth and equally rapid increase of regional disparity. It represents a complete economic growth cycle. The last retrenchment began in 1989 and continued until 1992 when a new round of vast capital investment started. The upsurge in investment reached its peak in 1993 and new retrenchment policies were put into effect in the year. By 1996 the government announced that it had successfully cooled down the over-heated economic growth without a new round of economic 'depression'. As discussed above, regional economic disparity and consumption disparity have risen very fast in the period. In the following analyses, we shall assess regional development and disparity in 1990-96 by applying a set of specifically designed indices for making a horizontal comparison (inter-region) on regional development, and to regionalize the provinces on that base. The indices used in this part are in relative value, i.e. the ratio of provincial value to the national value (represented by *R* hereafter). *R* does not reflect an absolute level of an index.

3.1 The Factor of Location

Location is a concept which is very popularly used but also very difficult to quantify. For a large region like a provincial unit, its locational conditions depend heavily upon national macro policies and the geo-political/geo-economic environment. For example, during the period of the anti-Japanese war or the 'Third front' construction, inland peripheral provinces like Sichuan were a good location for industrial construction, while coastal regions possessed evident locational advantages in economic development soon after the open-door policy was adopted in 1978. Thus under the presupposition of a steady geo-political environment and of the open-door policy, the locational condition of an individual province is mainly decided by its relative spatial position to the major sea ports and economic core regions, which are key factors affecting individual region's access to foreign capital, technology and information and its participation in the international division of labour (See Lu, 1995). In recent years, a border location (here meaning land border) has been used by many regions as a locational advantage. If its neighbour is a developed country, a border location can indeed be as important as, or more important than, a coastal location. For example, the Mexico border region has developed very quickly since the 1980s. In China, perhaps Shenzhen is also such a case. However, in most cases a border location in China does not seem to be very effective in economic development though it does show some advantage to develop limited border trade.

Based on the shortest railway distance from an individual province's capital to the nearest hub harbour and relative distances to Hong Kong, Shanghai and Beijing (including Tianjin), an assessment on provincial locational conditions has been made as shown on Figure 3.8. A hub harbour is defined following the Ministry of Transportation (see Chapter 6 for a list of them). Hong Kong, Shanghai and Beijing are given a weight of 0.5, 0.3 and 0.2 respectively to show their different importance of being an economic hub. According to this evaluation, Shanghai, Guangdong, Fujian, Jiangsu and Zhejiang are of best locations in terms of economic development. Beijing, Tianjin, Shandong and Hebei are the second layer provinces. Among the coastal regions, Hainan, Liaoning and Guangxi do not possess very good locations, but are still better than the inland provinces. In the Central region, the four provinces along the Yangtze River and Henan have a better location. The north-west and south-west provinces are at an unfavourable location. However, the construction of the railway from Nanning to Kunming has improved the locational conditions of Yunnan, Guizhou and Sichuan. Thus, generally speaking, from the viewpoint of economic globalization, the spatial distribution of provincial locational conditions conforms with the division of the East, Central and West regions, but there exist big differences in the Central region.

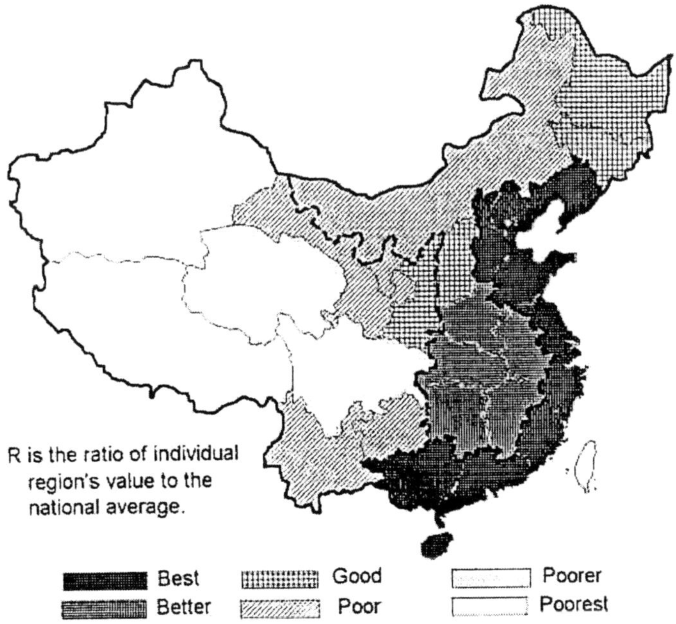

Figure 3.8 Locational Conditions of Provinces in China (1995)

3.2 The Factor of Natural Resource Endowment

In China, as the economic development level and degree of internationalisation rise, the role of natural resources in regional development is decreasing. Yet, natural resources still plays a basic and important role in sustainable regional development. Natural resources endowment of an individual province is here evaluated in terms of four indices, i.e. mineral resources, agricultural resources, water resources and ocean resources. Beijing and Tianjin will be included into Hebei province, and Shanghai into Jiangsu province.

Mineral Resources

In the late 1980s, the Geo-Sciences Board of the Chinese Academy of Sciences (CAS) entrusted a group of geologists to evaluate the potential market value of the country's 45 major mineral resources. As the evaluation does not take into account the different role of different resources and their combinations in current economic development, it is not suitable to use the results directly to assess provincial mineral resources endowment. Indeed, only a few important mineral resources, such as coal, iron, petroleum, copper, aluminium, etc. are critical to China's economic development. Here we have combined the work of the Geo-Sciences Board of CAS with the evaluation on the importance of key mineral resources and their combinations in each province, made in *Industrial Geography of China* (Li & Lu, eds., 1990), to assess an individual province's mineral resources endowment in terms of per capita possession. The result of this weighed and combined evaluation shows that Shanxi and Inner Mongolia are the best provinces in mineral resources endowment and Sichuan is second. Liaoning, Xinjiang, Qinghai, Ningxia and Yunnan are also among the superior provinces while others have no advantage. If only energy resources is used for evaluation, then Shanxi, Inner Mongolia, Ningxia, Xinjiang and Tibet are superior, and Shaanxi, Yunnan, Qinghai and Guizhou are second. Tibet is rich in hydropower potentials but has many difficulties in exploring these potentials.

Agricultural Resources

Agricultural resources include light, heat, precipitation and land resources. The first three are climatic resources. Based on data from *Manual of China's Natural Resources* compiled by the Commission for Integrated Survey of Natural Resources, CAS, an assessment on the climatic conditions of each province's capital is first made. [1] Then a weighed sum with per capita arable land is worked out to be the evaluation of provincial agricultural resources. The result shows that the top five provinces are Heilongjiang, Inner Mongolia, Xinjiang, Guangdong and Guangxi while Jilin, Fujian, Jiangxi, Ningxia and Gansu also have a good ranking. What is different from conventional impression is that Inner Mongolia, Ningxia and Gansu have good agricultural resources. This may only be partly truth, as land quality is not attached much importance in this assessment.

Water Resources

The spatial distribution of water resources in China is far from a good match with her patterns of population and arable land. The northern part of the country (north to the Yangtze River valley), where two-fifth of the nation's total population and three-fifth of the total arable land are located, owns only one-fifth of the total water resources. An assessment based on the ratio of each province's per capita water resources to the national average, modified by the regional degree of aridity[2], indicates that the Beijing-Tianjin-Hebei region, Shanxi, Henan, Liaoning, Jiangsu, Shandong and Ningxia are unfavourable in terms of water resources. Except Jiangsu which can transfer water from the Yangtze River, most of these provinces are in serious shortage. Xinjiang and Qinghai in the northwest are in fact rich in water resources in terms of per capita amount.

Ocean Resources

The ocean offers an extended space for economic development. For example, ocean fishing and breeding industry now provides more than half of the total aquatic product in China. A number of coastal provinces, like Liaoning and Shandong, have ambitiously planned to develop an ocean industry as big as their land economy. Thus it is necessary to bring ocean resources into our assessment. The ocean contains many resources, such as beach for breeding, tidal energy, bio-resources, crude oil in the ocean floor, etc. However, due to the availability of data, we take into account here just per capita area of beach for breeding and the amount of known crude oil reserves. Guangdong (including Hainan) and Liaoning take a lead in ocean resources, and Shandong and Fujian are second. Hebei, Jiangsu and Zhejiang are in the third class while Guangxi has poor ocean resources among the coastal provinces.

Comprehensive Evaluation of Natural Resources

By giving mineral, agricultural, water and ocean resources a weight of 0.6, 0.2, 0.1 and 0.1 respectively, a weighted sum of each province comes out as a comprehensive evaluation of its natural resources endowment (see Fig. 3.9 and Appendix 3.2). Mineral resources are given a high weight because China is still in the middle phase of industrialisation with heavy and chemical industries being the focus of industrial development. Shanxi and Inner Mongolia are best in natural resources, and Sichuan, Yunnan, Xinjiang, Ningxia, Qinghai and Liaoning are second. Heilongjiang, Guangdong and Fujian are also fairly good. Other provinces shows little advantage. Henan, Hubei, Jiangsu, Jilin and Hunan are among the poorest provinces.

---

[1] Here we use a provincial capital to represent its province in assessing the climatic condition.

[2] The coefficient is the ratio of evaporation to precipitation.

Figure 3.9 Evaluation on Provincial Natural Resources Endowment in China

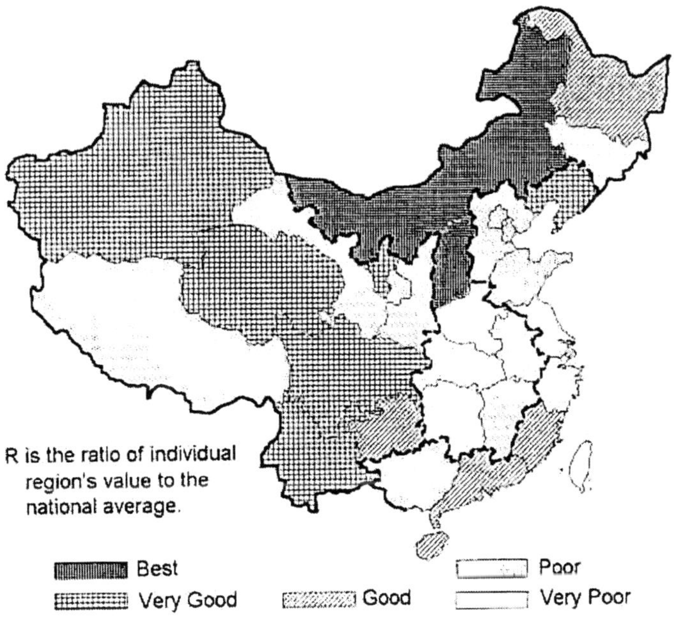

3.3 Region By Economic Development Level

GDP/GNP is a major indicator of economic development level in modern economic accounting. For a developing country, the industrial structure and internationalisation degree are also meaningful indicators. *GNP* is more appropriate than *GDP* in evaluating a region's economic development as it is the net regional income. However, since *GNP* is not available in many provinces, *GDP* is used here. For provinces in which Sino-foreign joint-ventures account for a high proportion of industrial output, the value of the two indicators may be a little different. For example, in Guangdong where Sino-foreign joint ventures and foreign owned enterprises account for nearly two-fifth of the total industrial output, *GDP* may over-estimate its actual net income. But for most provinces, the value of *GDP* is quite similar to that of *GNP*.

Per Capita GDP

In 1990-96, the distribution structure of provincial per capita GDP has changed greatly. By *R* in terms of per capita GDP, the country is divided into six income groups, i.e. the highest, high, upper-middle, lower-middle, low and lowest groups. The change shows the following features: (1) The number of mid-level income provinces declines very quickly while that of high and low income provinces increases, indicating a rising polarisation of the provincial economic development level. In 1990, the mid-level groups had 14 provinces but it had only 9 provinces in 1996. Zhejiang, Jiangsu and Fujian have moved up to the high-level group, and Qinghai and Ningxia have gone down to the low group. (2) The "east-west gap" is getting more and more evident. In 1990, seven of the nine coastal provinces (not including Beijing, Tianjin and Shanghai) were in the mid-level group and five of them had a per capita GDP lower than the national average. In 1996, however, five of the nine provinces ranked in the high income group, including Zhejiang, Guangdong, Jiangsu, Fujian and Liaoning. At the same time, the provinces low in per capita GDP were all in the north-west and south-west regions. (3) Most northern provinces are going down in terms of *R* of per capita GDP, including Beijing, Liaoning, Inner Mongolia and north-west provinces except Xinjiang; (4) The gap between the centrally-administrated cities (Beijing, Tianjin and Shanghai) and the top province in per capita GDP is reducing. In 1990 the per capita GDP of Tianjin was 40 per cent higher than that of Liaoning, the top among the provinces in the year, while it was only 21 per cent higher than Zhejiang, the top province in 1996.

Another eye-catching feature is the failure of Hainan's bubble economy. In 1990 the province was in the lower-middle income group with a *R* value of 0.9. Four years later it went up to the high-income group. But two years later, in 1996, it fell down to its original group.

Industrial Structure

There is no distinguishable difference between the ratio of non-agricultural sectors in *GDP* of each province, except Tibet and the three centrally-administered cities. They range from 70 to 80 per cent. The ratio has a strong positive correlation to per capita *GDP* with a coefficient of 0.82. However, among provinces in the same income group, Shanxi, Qinghai, Ningxia, Shaanxi and Yunnan have a higher proportion of non-agricultural activities in their economy while Hainan and Fujian have a lower proportion. Besides, in recent years, the non-agricultural ratio in the three northeast provinces has increased very slowly, by 1 to 2 percentage from 1990 to 1996, sharply in contrast to the rapid increase of the national average, of 7 percentage in the same period. Such a slow progress is closely related to the stagnancy of state-owned enterprises there as the region has the most important clusters of large and mid-sized state-owned enterprises. For example, to solve the increasing serious problem of unemployment, Heilongjiang provincial government encouraged the unemployed to make their living by engaging in agriculture.

Degree of Economic Internationalisation

The degree of internationalisation, in terms of the ratio of the import and export values to *GDP*, shows a big regional disparity, i.e. with an east-west gap for the whole country and a north-south gap for the coastal region. As a result of the "front-shop and back-factory" industrial development model of Hong Kong (Sit, 1997), Guangdong is now the champion in the economic internationalisation in China. The indicator of the province in 1990 was 53.1 per cent, a little higher than Shanghai (of 47.1 per cent). But it rose to 153 per cent in 1996, leaving other provinces far behind, including the three centrally-administered cities (of 77 to 81 per cent). Among the other provinces, only Fujian and Hainan were above the national level. Hainan, however, is an import-oriented type with its import value nearly four times the export value. It should be acknowledged that the high degree of economic internationalisation of the whole country is mainly due to Guangdong's high value. In 1996 the province accounted for 38.6 per cent of the total export and import of the country. If it is excluded, the indicator will go down to 24 per cent; then quite a few provinces will be above the national average. Besides, provinces in the low-income group have a nearly identical indicator of around 10 per cent. Zhejiang and Jiangsu shows strong domestic market orientation as their economic internationalisation degree was only 29 and 31 per cent respectively.

Comprehensive Evaluation of Regional Economic Development

A comprehensive evaluation on the provincial economic development level is made through a weighted sum of the above three indicators (see Fig. 3.11 and Appendix 3.3). Based on the result of 1996, the provincial units are categorised into seven groups. From the most developed down, the first group consists of the three centrally-administered cities (R>2.0); the second group is composed of Guangdong and Zhejiang (R=1.79 and 1.57); the third group includes Jiangsu, Fujian and Liaoning (1.41 to 1.31); the fourth group includes Shandong and Heilongjiang (R=1.15 and 1.04); the fifth group contains Xinjiang, Hainan, Hebei, Hubei and Jilin (0.97 to 0.86); and the sixth group (0.73 to 0.61) includes all other provinces except Guizhou, Gansu and Tibet which form the seventh group (R<0.5). Compared to the category in 1990, Fujian, Zhejiang, Jiangsu, Shandong, Hebei, Guangxi and Henan have moved up in rank while Gansu, Qinghai, Ningxia,

Shaanxi, Tibet and Guizhou have gone down a lot. Beijing, Liaoning, Jilin, Inner Mongolia and Shanxi have shown a decrease in the *R* value though their ranks have not changed. Thus the spatial distribution of the provincial economic development level in 1996 has shown an evident 'east-west' gap, and in the coastal region a reversed 'north-south' gap, i.e. south high and north low, is emerging (Fig. 3.10).

Figure 3.10 Provincial Economic Development Level in China (1990 and 1996)

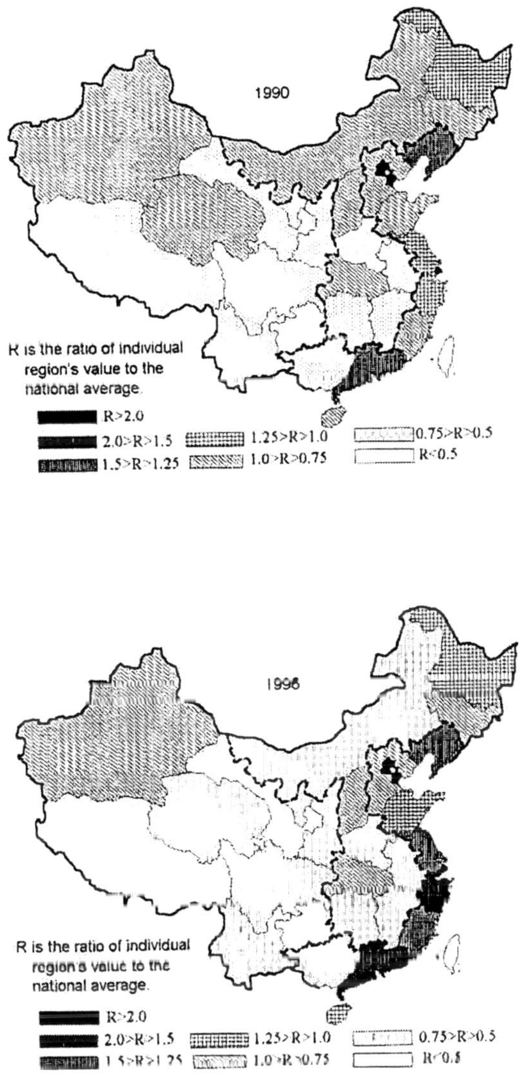

3.4 Regional Economic Vitality

Regional economic vitality is reflected not only by the growth rate but also by the growth quality. In the following part, the *GDP* growth rate, investment ratio, investment effectiveness, economic operation quality, ratio of *FDI* to the total capital investment, and non-state-owned economy ratio are employed to evaluate the economic vitality of the provinces.

Economic Growth Rate

The spatial distribution of *GDP* growth rate in 1990-96 shows a distinguishable east-west difference. In the coastal region, except the three centrally-administered cities and Liaoning, there is a fast *GDP* growth rate of more than 25 per cent higher the national average. A few provinces in the Central region, like Henan, Anhui, Jiangxi and Hubei, also have quite a rapid growth rate of 10 to 20 per cent higher than the national average while most Central and West provinces are under the national average. Qinghai, Heilongjiang, Ningxia and Guizhou are among the slowest provinces, and their growth rates are less than half of that of Fujian, the fastest growing province.

However, in the last two years, regional disparity in growth rate has been reducing, mainly due to the retrenchment policy. In 1995-96 the fastest growing provinces were not exclusively in the coastal region, as Henan, Anhui, Hubei and Jiangxi had achieved rapid economic growth as fast as those growth "stars" in the coastal region. Besides, Tianjin and Shanghai has recently ranked among the fastest growing regions for the first time after the *Opening and Reform*. These new and old growth "stars" form a "*T-shaped*" distribution pattern along the coast and the Yangtze River (Fig. 3.11). On the other hand, two coastal provinces, i.e. Hainan and Liaoning, had become problem regions while the whole coastal region still maintained an economic boom. In 1995, Hainan and Liaoning grew at a rate of 4.3 and 7.1 per cent respectively, and in 1996 they had improved to 5.2 and 8.5 per cent respectively. Such growth rates seem reasonable internationally, but are evidently too slow compared to other Chinese regions. It is not surprising that Liaoning has a slow growth as the province has suffered from an outdated industrial structure since the 1980s. In the case of Hainan which was a growth "star" in 1990-94, the drop in economic growth rate is a little unexpected though the province's investment had shown signs of a "bubble economy" in its booming period. Most of its investment had gone to real estates and services, leading to instant economic growth without much support by the growth of manufacturing. Such growth has little sustainability, particularly for a developing region.

In 1996, the first year of the ninth FYP, regional economic growth showed a new tendency, that is, the former most rapidly growing provinces, except Fujian, were becoming slow in growth while Hebei, Anhui, Jiangxi, Henan, Hubei, Hunan, Shanghai and Tianjin turn out to be new '*growth stars*'. Besides, in 1996, several former problem provinces, like Shanxi, Inner Mongolia, Gansu, Ningxia, Heilongjiang and Shanxi, had achieved more rapid economic growth than the national level.

Figure 3.11 Provincial GDP Growth Rate in China (1990-94 / 1995-96)

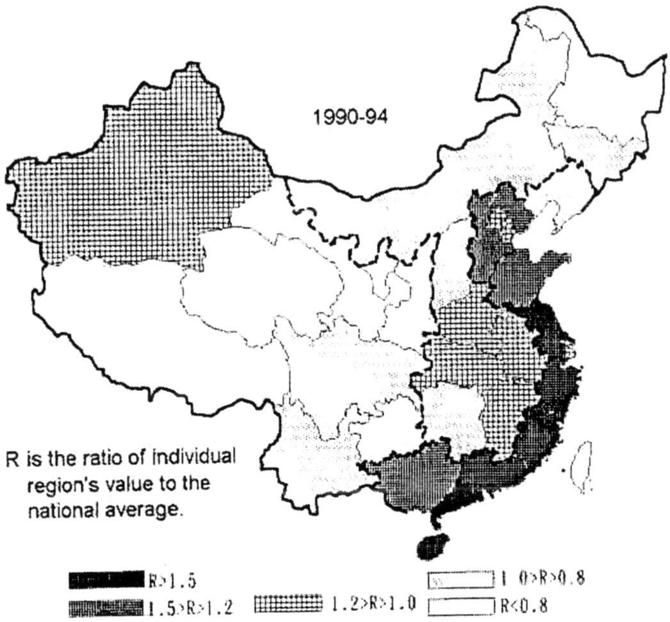

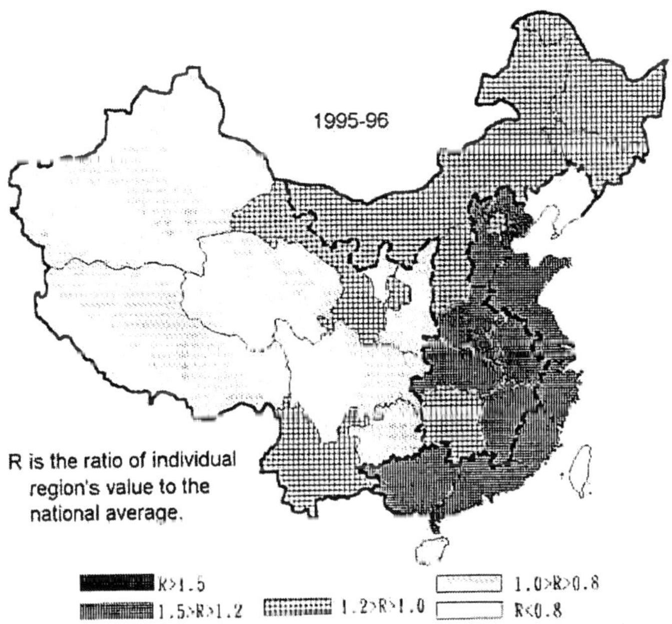

Investment Ratio and Effectiveness

Investment ratio is the percentage of the total capital investment in *GDP*, and investment effectiveness is the ratio of the net increase of *GDP* to the total capital investment in the preceding year. They reflect the quantity and quality of capital investment. In the last two decades, rapid economic growth in China as a whole has been mainly due to a high investment ratio. For individual regions, however, the GDP growth rate has not been closely correlated to the investment ratio but the investment effectiveness.

In 1995-96, the provincial investment ratio differed greatly from each other, but they were spatially not in the form of an East-West difference. The three centrally-administered cities, Hainan, Xinjiang, Guangdong, Qinghai and Ningxia had a very high investment ratio of more than 35 per cent while in Jiangxi, Anhui, Hunan, Guangxi and Shandong it was as low as 25 per cent or less. Other provinces were around 30 per cent. In the same period, the investment effectiveness also showed big difference among the provinces, but with a negative correlation to the investment ratio. The best provinces included Jiangxi, Anhui, Hunan, Henan, Guangxi, Hubei, Hebei, Fujian and Shandong which were 50 per cent above the national level. Hainan and Beijing were among the lowest provinces in the indicator. It should be acknowledged that the investment effectiveness as calculated here not only is a result of the investment structure but is also closely related to the construction speed and length. Thus the indicator itself cannot be independently used to diagnose the investment structure.

Economic Operation Quality

The economic operation quality is an integrated indicator reflecting the management condition of economic entities. Due to data availability, it is evaluated by agriculture labour productivity and a synthesised industrial index which combines the ratio of added-value to the gross output value, ratio of pre-tax profits to total assets, ratio of after-tax profits to total cost, overall labour productivity, turnover of circulating capital, etc.

In 1995-96, the synthesised industrial index showed a great regional difference between the Coastal, Central and Western regions. All provincial units in the coastal region, except Hainan, Guangxi and Liaoning, were above the national level. In the Central region, Shanxi, Henan, Anhui, Hubei and Hunan were close to the national average while other provinces were much lower. In the Western region, all except Yunnan were 20 per cent below the national average. Yunnan, Shanghai and Heilongjiang were the best in the indicator. Yunnan was among the best due to the high proportion of the high-profit tobacco industry in its industry structure, and Heilongjiang due to its strong petroleum industry. Other industries in these two provinces were actually not very good in making profit. Qinghai, Shaanxi and Xinjiang were among the lowest provinces. Qinghai was only 17 per cent of the national average. Due to the stagnancy of large and medium-sized state-owned enterprises, Liaoning had a bad record with a synthesised industrial index of 25 per cent lower than the national level although it is traditionally a very important industrial base in China.

The provincial agriculture labour-productivity is evidently polarised, in the form both of an East-West and North-South gap. The coastal provinces (except Guangxi) and several northern provinces like Heilongjiang, Jilin, Inner Mongolia and Xinjiang are much higher than the national level while other provinces except Hubei are much lower than the average. The three centrally-administered cities, Heilongjiang, Liaoning and Xinjiang are the best whose indicators are two times the average while Yunnan and Guizhou, the lowest ones, are only half the average. Besides, Shaanxi, Sichuan and Qinghai also have low labour-productivity in agriculture, 60 per cent of the national average.

Ownership Related Economic Vitality

In recent years, stagnant state-owned enterprises have strongly affected the economic vitality of provinces with a high proportion of such enterprises. Thus before the central government finds an effective way to turn around these state-owned enterprises, the proportion of non-state-owned economy can be taken as an important indicator of regional economic vitality. In the following analysis, the share of *FDI* in total capital investment and the proportion of industrial output of non-state-owned enterprises in total industrial output are employed to evaluate the ownership related element in regional economic vitality.

Since *FDI* has mainly concentrated in several coastal provinces, the share of *FDI* in total capital investment shows distinguishable regional difference between the Coastal, Central and Western regions. In 1995-96, indicators for Guangdong, Hainan, Fujian and Tianjin are as high as 30 per cent or more, more than twice the national average. Jiangsu and Shandong are also much higher than the average, that is 23.3 per cent and 16.5 per cent respectively. Provinces in the Central are around half of the national average. In most western provinces like Xinjiang, Qinghai, Ningxia, Yunnan, Guizhou and Tibet, however, *FDI* played an insignificant role in economic development. In these provinces, the share of *FDI* in total capital investment was less than 2 per cent. Nonetheless, *FDI* location has changed a little recently. In 1996, total *FDI* by contract decreased by 20 per cent for the whole country. The sharpest reduction happened in Guangdong and Hainan, which had decreased by 40 per cent and 90 per cent respectively. Shanghai, Tianjin, Zhejiang, Shandong and Liaoning maintained the previous year's amount or had increased slightly. In contrast, Xinjiang, Inner Mongolia, Shanxi, Henan, Ningxia, Shaanxi and Guizhou recorded an increasing *FDI* by 10 per cent or more than the previous year. *FDI* in Shanxi and Inner Mongolia had even increased by 430 per cent and 102 per cent respectively. Such shifts indicate some change in regional competitiveness. It could be expected that increasing *FDI* will add some growth energy into several provinces in the Central and Western regions.

The ratio of non-state-owned industry also show a clear East-West gap. In 1995-96, non-state-owned enterprises accounted for more than 70 per cent of total industrial output in Zhejiang, Guangdong, Jiangsu and Fujian; and around 60 per cent in Shanghai, Tianjin and Shandong. However, in other coastal provinces like Liaoning, Beijing, Hebei and Guangxi, the state-owned enterprise are still the major players in their industrial system, accounting for 50-60 per cent of total industrial output. In several rapidly growing provinces in the Central region, i.e. Hubei, Anhui and Henan, the ratio of non-state-owned industry has increased fast, approaching half of the total in terms of output value in 1996. While in other Central and Western provinces the state-owned industry still accounted for 60-70 per cent or more of total industrial output. The non-state-owned industry ratio correlates to some degree with provincial *GDP* growth rate (the coefficient is 0.6), an outcome of previous economic vitality, and somewhat a sign of future vitality as well.

Regional Economic Vitality in Summary

By giving a weight to the *GDP* growth rate, investment ratio, investment effectiveness, economic operation quality, *FDI* ratio and non-state-owned industry ratio of 0.5, 0.1, 0.1, 0.1, 0.1 and 0.1 respectively, and summing them, we obtain a final and comprehensive evaluation of regional economic vitality in 1995-96 (see Fig. 3.12 and Appendix 3-4). The best provincial units in economic vitality are Fujian, Tianjin, Shanghai, Guangdong, Jiangsu, Zhejiang and Shandong (R>1.2). The second includes Hebei, Hubei, Henan, Anhui, Jiangxi, Guangxi, Beijing and Hunan (R=1.2-1.0). While other provinces have a R value of around 0.8 except Qinghai which is the worst in economic vitality with a R value of 0.6. Overall, regional economic vitality shows a disparity both as an East-West and South-North gap. In each of the Eastern, Central and Western regions, southern provinces show stronger economic vitality than northern provinces.

Compared with the situation in 1990-95, economic vitality of the coastal provinces has come down a little in 1996 while that of the Central and Western provinces has been strengthened much (R increased by 15-20 per cent), indicating that the gap is reducing. Although GDP growth rates of Hubei, Anhui, Jiangxi, Hunan and Henan in the Central region have caught up with that of coastal provinces in recent years and some are even higher than most coastal provinces in 1996, these provinces are still inferior to coastal provinces in terms of comprehensive economic vitality. Besides, Hainan, among the best provinces in economic vitality in 1990-94, had fallen to much below the national average in 1995-96 with its R value falling between 1.43 to 0.87. On the contrary, Tianjin has recovered from relative stagnancy and acquired strong economic vitality with its R value increased to 1.44 in 1996. Different from the big changes in these two provinces, Qinghai has kept its position as a problem region since the late 1980s.

Figure 3.12 Provincial Economic Vitality in China (1995-96)

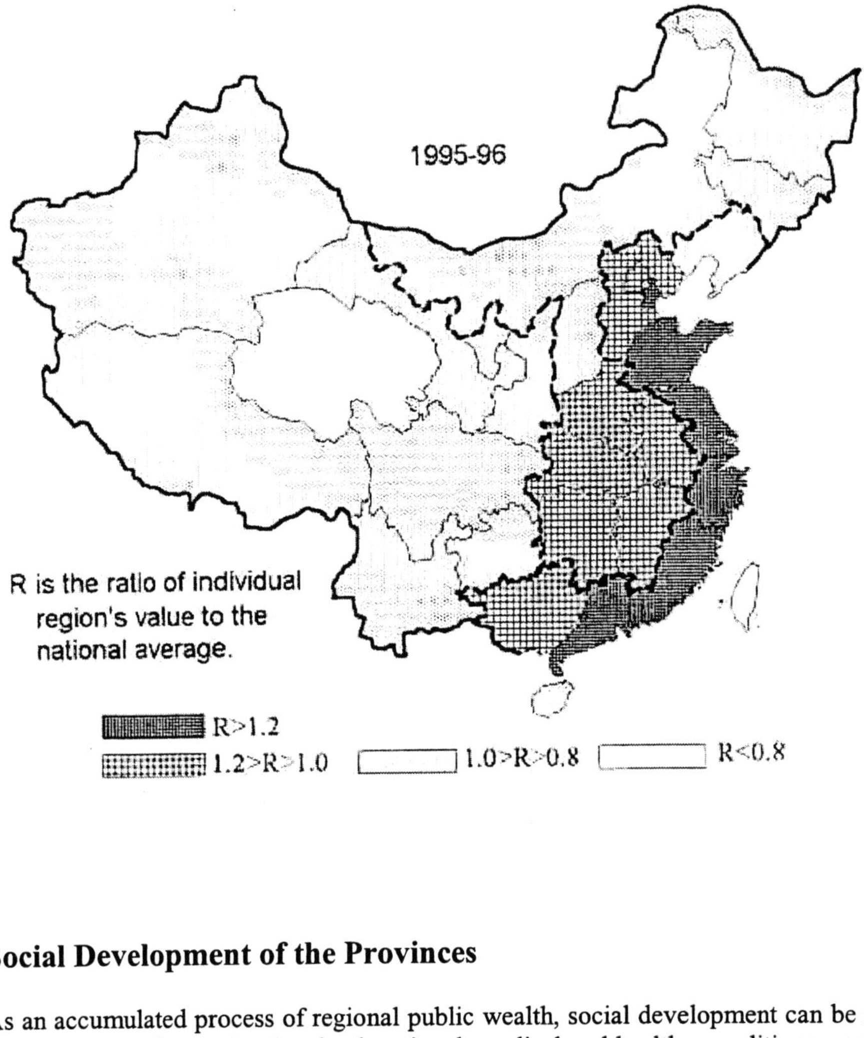

3.5 Social Development of the Provinces

As an accumulated process of regional public wealth, social development can be reflected by such indicators as urbanisation level, educational, medical and healthy conditions, social security, community service, population quality, etc. In short, social development is a direct result of economic development and also a support for further economic development. However, as it is heavily depended upon the distribution structure of *GNP*, social development may not be synchronised with economic development in the short term. That is, a comparatively high economic development in a region sometimes does not necessarily mean a high level of social development. We shall examine the key elements of social development one by one and then summarise them in a comprehensive statistics to demonstrate regional disparity in social development in China.

Urbanisation Level

The definition of China's urban population is ambiguous and inconsistent. In the fourth census in 1990, urban population is defined as: (1) population of subordinated districts for cities having districts and population of neighbourhoods for cities without districts; and (2) the non-agricultural population of towns subordinated to either cities or counties. Such a definition makes it difficult to compare the urbanisation level between provinces since whether to set up districts under cities is normally a local decision (Zhou, 1995, 55). As an indicator of social development, the urbanisation ratio should reflect the population which can access urban infrastructures daily. Thus urban population is here simply defined as the population of cities (not including counties subordinated to cities) and the non-agriculture population of towns.

Regional difference in the urbanisation level shows gaps between the Eastern, Central and Western regions. But the differences are not very marked. The highest urbanisation is found in the three centrally-administered cities, Liaoning, Jilin, Guangdong and Heilongjiang (R>1.4). Zhejiang, Jiangsu, Shandong and Hubei are second with R value between 1.2 and 1.4. Other provinces except Hainan (R=1.03) are all below the national average. Tibet, Qinghai, Yunnan and Guizhou have the lowest urbanisation ratio (<20%). Besides, Fujian's urbanisation level (R=0.99) is still a little lower than the national average though the province is very high in per capita *GDP*. The urbanisation ratios of Heilongjiang, Hubei and Jilin are much higher than that of other provinces in the same income group. It needs to be pointed out, however, that the high urbanisation level in Heilongjiang results from the huge amount of "non-agriculture" population actually engaged in agricultural activities in state farms and state forestry centres. Jilin is similar to Heilongjiang in this.

Educational and Medical & Healthy Supplies

The educational condition is evaluated by the number of schools, colleges and universities and the number of teachers in theses educational institutions per 10,000 people. The basic medical and healthy condition is examined by the number of doctors and medical technicians and the number of hospital beds per 10,000 people.

The educational condition shares a similar spatial distribution with the medical and health conditions. Their provincial disparity is not significant and shows a somewhat North-South gap. In 1995-96, provinces with these two indicators are much higher than the national level include the three centrally-administered cities, Liaoning, Jilin, Heilongjiang, Inner Mongolia, Xinjiang, Shanxi, Qinghai, Hubei, Ningxia and Shaanxi. Most of them are northern provinces. The growth 'star' provinces like Guangdong, Zhejiang, Fujian and Shandong have values lower than the national average though their per capita *GDP* is much higher than the average. Such a reverse spatial pattern to that of economic development is due to two reasons. One is that the improvement of educational and medical & healthy conditions depends mainly upon government inputs. In the last decade, the government's decreasing capability to derive revenues from rapid economic growth had obviously affected its inputs in these public sectors. As a result, provinces which were major construction areas before the *Opening and Reform* still show some superiority in public infrastructures. The other reason is closely related to the development strategy of local governments which gives priority to instant economic growth.

Social Security and Community Service

As social security and community service appeared quite recently in China, systematic data on them are still not available. Per capita amount of social security fund in the rural area is employed to represent provincial social security level because there is still no openly published data on urban social welfare. Community service is evaluated by the amount of community service facilities per 10,000 of non-agriculture population.

In 1995-96, there exist big differences in social security and community service between provinces, but the difference does not show a clear North-South or East-West gaps. The highest group in the indicators (R>2.0) includes Shanghai, Beijing, Tianjin and Hunan. Hunan ranks in this group due to its high level of rural social security. The per capita amount of rural social security fund in the province is 4.5 times the national average. Other provinces much above the national level are Guangdong, Xinjiang, Jilin, Jiangxi and Tibet. Among them, Jilin is a case of high-level urban community services while it is comparatively low in rural social security. The other four provinces are just to the contrary. Some economically developed provinces like Tianjin, Jiangsu, Liaoning, Fujian and Shandong have a low level of social security and community service (R<0.65). Fujian is among the lowest five provinces. The other four are Inner Mongolia, Hainan, Guangxi and Qinghai. The three lowest provinces in per capita *GDP*, i.e. Guizhou, Gansu and Tibet, have a comparatively high R value in the indicator. Tibet is much higher than the national average while Gansu is a little below the average and Guizhou is about two-third of the average.

Population Quality

Evaluated by the ratio of population with college education or above, population quality shows an obvious provincial disparity, but not an East-West gap. The three centrally-administered cities possess the highest population quality, and Beijing is the champion with its R value 4.6 times the national average. Ningxia, Liaoning, Xinjiang, Jilin and Shaanxi are also above the national level while the rest of the provinces are all below the average. Yunnan, Guizhou, Tibet and Guangxi have the lowest population quality. It is quite unexpected that Guangdong and Fujian are among the lowest provinces in the indicator. Among the seven highest provinces in per capita *GDP* (except the three centrally-administered cities), only Liaoning has a population quality higher than the national average. Thus the regional economic development is not correlated with population quality. Such a phenomenon is worth pondering. Either there is something wrong with the educational system or these economically developed provinces may have problems caused by low population quality to sustain their future economic development.

Comprehensive Evaluation of Social Development

By giving urbanisation ratio, educational supply, medical & healthy supply, population quality and social security and community service a weight of 0.3, 0.2, 0.2, 0.2 and 0.1 respectively, a weighted sum of comprehensive evaluation of regional social development is resulted for each province (see Fig. 3.13 and Appendix 3.5). Social security & community service are given a weight of 0.1 not because they are unimportant in social development but because available data are less than satisfactory. Regional disparity in social development is quite different from that of economic development. It does not show an East-West gap, and is much lower than regional economic disparity if we do not include the three centrally-administered cities. Beijing, Shanghai and Tianjin are the best in social development (R>1.6) due to their overwhelming urban nature. Liaoning and Jilin are second with R values between 1.4 and 1.35. Xinjiang (R=1.2), Heilongjiang (1.18), Hunan (1.08), Ningxia (1.05) and Guangdong (1.0) are also above the national level. It should be noticed that Hunan has a high R value mainly due to its very high rank in the rural social security evaluation. Other provinces are all below the national level. Guangxi (R=0.57), Guizhou (0.58), Yunnan (0.61), Anhui (0.66) and Sichuan (0.69) are the lowest in overall social development. Among the growth 'star' provinces of the last decade (see economic vitality evaluation), only Guangdong achieves national average in social development, while others are all below. Such a dualism in development not only indicates that these provinces have not paid enough attention to social development but also argues that social development cannot be achieved instantly by an economic "great leap forward".

Figure 3.13 Provincial Social Development Level in China (1995-96)

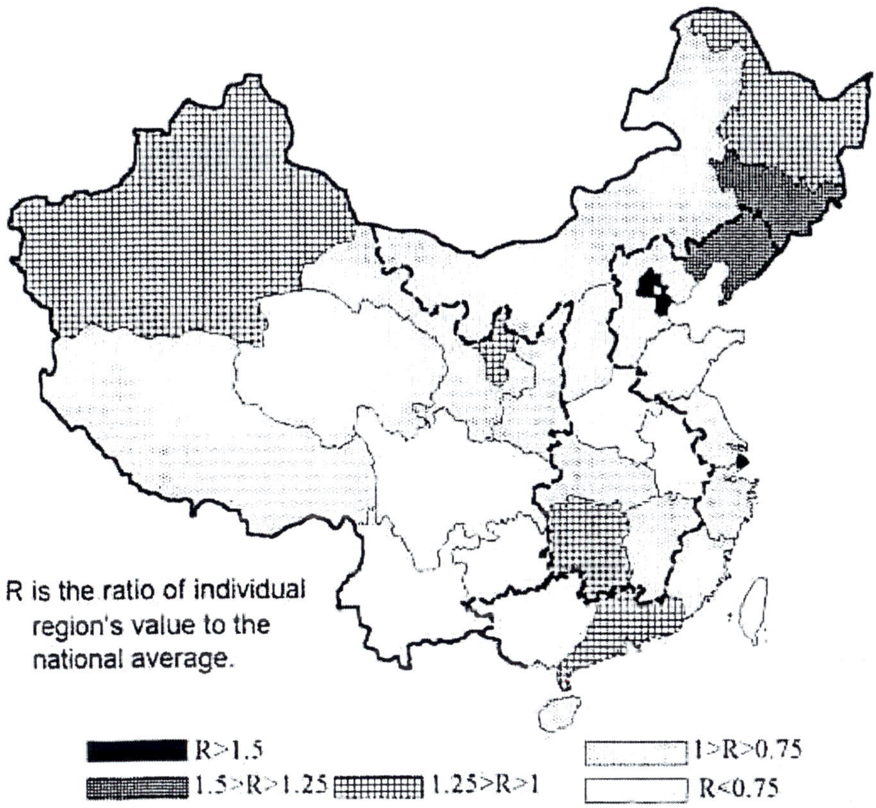

3.6 Regional Difference in the Living Quality

Living quality depends not only on per capita consumption but also such factors as the amount of private assets, surrounding natural environment, etc. Since per capita consumption does not reflect the expenditure structure and may be undervalued or overvalued as a result of regional differences in commodity prices, *Engel Coefficient* (i.e. the ratio of food expenditure to total domestic expenditure) and after-consumption remains of income are used to modify the indicator so as to make it more accurate in representing actual living quality. Although per capita living space and per capita household electricity consumption are quite important in evaluating living quality, these two indicators have to be given up due to lack of systematic data.

Consumption Level

If evaluated only by per capita consumption, Shanghai, Beijing, Tianjin, Guangdong and Liaoning have the highest living standard ($R>1.25$). Zhejiang, Fujian, Heilongjiang and Jiangsu are also much above the national average ($1.24>R>1.06$). Provinces with much lower per capita consumption than average include Yunnan, Guizhou, Sichuan and Tibet in the South-west, Shaanxi, Gansu, Ningxia and Qinghai in the North-west, Henan, Anhui, Jiangxi and Shanxi in the Central region, and Hebei and Guangxi in the coastal region ($R<0.75$). Per capita consumption in

Tibet, Guizhou and Gansu is only half of the national average. Thus regional disparity in living standard in terms of per capita consumption is not a very evident East-West gap.

A combined evaluation by consuming expenditure, *Engel Coefficient* and after-consumption remains of income shows that provincial disparity in urban living standard is quite narrow while rural living standard differs greatly from province to province. The extreme gap of the former is only 1 time but that of the latter is as high as 4 times, indicating great disparity in rural living standard. The latter is the major factor behind regional disparity in living quality. Beijing, Shanghai, Zhejiang and Guangdong have the highest urban living standard (R>1.25) while other provinces do not differ much from each other (1.1>R>0.8). It is worth mentioning that Tianjin, as a centrally-administered city, has an urban living standard even lower than Guangdong and Zhejiang. In rural living standard, the three centrally-administered cities, together with Zhejiang, Guangdong and Jiangsu, are the champions. Liaoning, Fujian, Heilongjiang, Shandong and Hebei are also much above the national average. Yunnan, Guizhou, Sichuan, Shaanxi, Gansu, Ningxia, and Qinghai have the lowest rural living standard (R<0.75). All of them are western provinces.

Liaoning, Tianjin, Heilongjiang, Hebei, Jiangsu and Fujian share one common feature, that is they had a comparatively low urban living standard but a high rural living standard. The first three provinces are the traditional industrial bases with a large amount of state-owned large and medium-sized enterprises which adversely affects income of the urban population there. The last three provinces possess such a feature because their township enterprises and rural economy are much more vital than their urban economy. Besides, Tibet, Gansu, Yunnan, Sichuan, Hunan and Guizhou have the biggest urban-rural difference, i.e. the rural living standard in these provinces is very much lower than urban living standard.

Environment Quality

Good natural environment is one of the important aspects of a high living quality. Such a view is not only the concern of wealthy countries, but of all human beings. Although developing countries generally attach less importance to environmental quality than economic development, the indicator for the former cannot be neglected when evaluating the peoples' living quality. It is not intended here to evaluate the quality of the natural environment of each region, but to look into negative environmental changes caused by economic development in terms of the discharge of wastewater, waste gas and solid industrial wastes. The environmental effect of wastewater discharge is evaluated by the ratio of wastewater discharged to total runoff, and that of waste gas and solid wastes by the amount of waste discharged per unit area. Such an evaluation indicates only a comparative environment quality of each region, not its actual situation.

Regional environment quality shows a reverse spatial pattern to economic development level, i.e. the western region is, as a whole, much better than the Eastern region. Tibet and Qinghai are the best (R>1.2), still maintaining an almost natural environment. Yunnan, Guizhou, Sichuan, Inner Mongolia, Gansu, Xinjiang, Guangxi, Hainan and Fujian are second, and are much better than the national level (1.19>R>0.1). Zhejiang, Jilin, Hunan, Heilongjiang and Shaanxi are a little above the national level while Guangdong, Jiangsu, Hubei, Anhui and Jiangxi are a little below the average. Shandong, Hebei, Shanxi and Henan have a low environment quality (0.88>R>0.79), but not among the worst ones which are Ningxia, Liaoning and the three centrally-administered cities. Beijing, Shanghai and Tianjin rank so due to the urban nature, i.e. high density of population and economic activities. Most of the low quality provinces, like Shanxi, Hebei, Shandong and Henan, have a high proportion of heavy industry but are endowed with a comparatively small amount of natural water resources. Being the growth 'star' in the last decade, Guangdong, Zhejiang, Fujian and Jiangsu are in fact not very bad in their environment quality.

Comprehensive Evaluation on Regional Living Quality

If environment quality is not taken into account, provincial disparity in living standard shows a clear East-West gap which is, however, smaller than the economic development disparity. Shanghai is the champion in living standard (R>2.0), and Beijing, Tianjin, Zhejiang and Guangdong are second in order (1.49>R>1.34). Jiangsu, Fujian, Liaoning and Heilongjiang are also above the national average (1.22>*R*>1.08). Other provinces are all below average though Shandong, Xinjiang, Hainan, Jilin, Hebei and Hubei are quite near the average. The south-west provinces and north-west provinces (except Xinjiang) have the lowest living standard. If the three centrally-administered cities are not included, the gap in provincial living standard (between Guangdong and Gansu) is 2.76 times while that in economic development is as high as 4.59 times (between Guangdong and Guizhou). Guizhou and Tibet rank higher in living standard than in economic development level, indicating a large amount of inward financial subsides from the central government in these two provinces.

A combined evaluation with environmental quality being given a weight of 0.2 reveals that provincial disparity of living quality shows evidently a gap between the south-east coastal provinces and other provinces, and the gap is much narrower without a dimension of environmental quality (see Fig. 3.14 and appendix 3.6). Shanghai still ranks first in living quality (*R*=1.8) and Guangdong and Zhejiang are second (1.38>*R*>1.36), while Beijing, Tianjin, Jiangsu, Fujian and Heilongjiang are third (1.2>*R*>1.08). Other provinces are below the national average but are very close among themselves. Xinjiang, Hainan, Liaoning, Shandong, Jilin, Hubei and Hebei have a living quality 10 per cent or less lower than the average, and Guangxi, Inner Mongolia, Hunan, Qinghai, Anhui, Jiangxi and Sichuan 10 to 20 per cent lower than the average. The extreme gap is still between Guangdong and Gansu, but it is only 2.1 times.

Figure 3.14 Provincial Living Quality in China (1995-96)

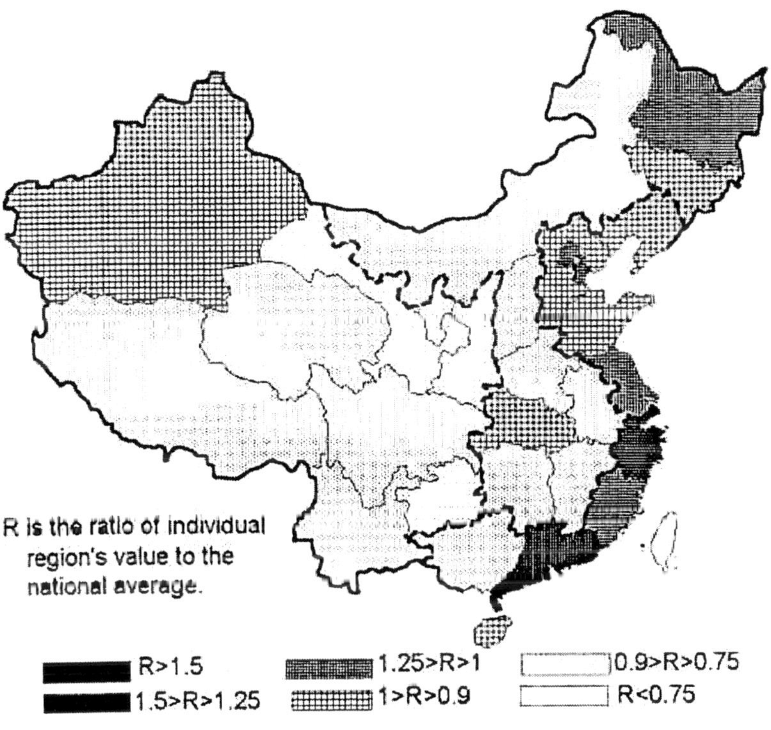

3.7 Regional Variation in Infrastructure Development

Improvement of the infrastructure is not only the result of economic development but also the foundation for further socio-economic development. A well-established infrastructure is an important sign of a developed society. Infrastructure includes many items. For a regional unit like a province, transportation, communication, energy supply and water supply can be chosen to represent the overall regional infrastructure.

Transportation Facilities

In China railway and highway are the major transportation modes. In the following evaluation, the density of railways and highways is a major indicator, supplemented by the total fixed assets of water transportation and harbour enterprises. In the past, evaluation on regional railway and highway condition was generally based on area density (length per area unit). Since it does not take into account the effect of population density, i.e. areas with a high population density normally need more transportation facilities given a fixed economic development level, it is not accurate enough. To improve this, area density and person density (per capita length) are combined through a weighted sum method. For the highways, trunk roads are given different weights as they play different roles in the highway system. It should be noticed that the transportation evaluation here is focused on the internal transportation of each province. External transportation will be dealt with in Chapter 6.

Regional transportation facilities show a clear gap between the Eastern, Central and Western regions. Most provinces in the Central region possess similar levels of railway and highways to the Eastern provinces, while the Western provinces, particularly the south-west provinces, are much lagged behind. The three centrally-administered cities have the best transportation facilities. Hainan, Liaoning, Shandong and Hebei are second. Hebei benefits a lot from being next to Beijing, hub of the national transportation system. Zhejiang, Guangdong, Jiangsu, Jilin, Shanxi, Henan, Anhui and Hubei are also above the national average. Yunnan, Guizhou, Sichuan, Tibet and Guangxi have the poorest transportation. Inner Mongolia, Qinghai, Heilongjiang and Gansu rank in the lower group. Other provinces are on the average.

Communication Facilities

Regional communication condition is not decided only by local facilities, but by local connections to external systems also. Restricted by availability of data, the telephone exchange capacity, telephone sets installed and long-distance call capacity are chosen here as indicators to evaluate provincial communication conditions.

Regional communication condition shows a clear East-West gap and is closely related to the regional economic development level. The three centrally-administered cities lead in communication facility supply, with Guangdong and Liaoning being the second. Zhejiang, Jiangsu, Fujian and Hainan rank in the third while other provinces are below national average except Jinlin and Heilongjiang which are a little better than the average. Again, south-west provinces (Sichuan, Guizhou, Yunnan and Tibet) and Guangxi have the poorest communication condition just as they do in transportation condition. Henan and Anhui also belong to this lowest group. Some of the provinces with superior transportation facilities, like Hebei, Shandong, Shanxi and Henan, rank low in communication condition.

Energy Supply

Energy supply includes coal, petroleum, electricity, etc. To evaluate regional infrastructure, electricity supply is a key indicator. In the following evaluation, electricity production per capita and per unit of industrial output (which is modified by the percentage of heavy industry) are chosen to represent provincial energy supply as an infrastructure facility. It should be emphasised that there have been less and less independent provincial electricity supply systems, i.e. provinces are getting more and more merged into regional electricity supply systems and are moving towards an integrated national system. A province's electricity production may not be equal to its electricity supply. Thus, strictly speaking, the following evaluation can only illustrate the extent to which local electricity production meets local demand in relation to local socio-economic development level and its comparison with the national level.

Electricity supply shows a complicated regional disparity, which is by and large a reverse of the East-West gap in regional economic development disparity, i.e. the Western provinces are a little better while the Eastern provinces are generally poor. In relation to local socio-economic development level, Shanxi, Ningxia, Guizhou and Tibet have the best electricity supply, and Qinghai, Gansu, Yunnan, Fujian, Hebei, Hainan, Shanghai and Shaanxi rank in the second level. Guangdong, Xinjiang, Jilin, Guangxi and Henan are also above the national average whereas other provinces are poor in local electricity supply. However, since Beijing and Tianjin are supplied with electricity by regional grids and Beijing has gone to Inner Mongolia to establish its own power plants with the collaboration of local government there to alleviate its shortage of electricity supply, these two cities are actually much better in electricity supply than the evaluation result has indicated. Hebei is just an opposite case, i.e. a large amount of its electricity production has to go to Beijing and Tianjin, resulting in lower local electricity supply than the evaluation has suggested. Shanxi and Inner Mongolia are similar cases.

Water Supply

The water supply condition is evaluated by the amount of water supply per capita modified by the local aridity index. The number of cities reported to be in want of water supply is also taken into account in the evaluation. Regional disparity in water supply condition shows both an East-West and South-North gap. Hainan, Guangdong, Jiangsu, Jiangxi and Guangxi have the best water supply capacity. Shanghai, Zhejiang, Hubei and Hunan are second. Fujian and Ningxia are a little above the national average. Shanxi, Shaanxi, Tibet, Tianjin and Gansu are the poorest in water supply. Yunnan, Guizhou, Liaoning, Hebei, Shandong and Henan rank in the poorer group.

Compared to the water resources endowment, Hebei (including Beijing and Tianjin), Liaoning, Shandong, Henan and Shanxi belong to the category of provinces with strong limitation of growth imposed by natural water resources, while Tibet, Yunnan, Guizhou, Sichuan, Xinjiang, Qinghai, Jilin and Heilongjiang have a low water supply capacity due mainly to the lack of water conservancy works or difficulty in building such works. Jiangsu and Ningxia benefit much from channelling water from the Yangtze River and Yellow River respectively and have a high water supply capacity though their per capita water resources is quite low.

Comprehensive Evaluation of Regional Infrastructure Development

Through a weighted sum of the evaluation result of transportation, communication, electricity supply and water supply with each individual indicator having a weight of 0.3, 0.3, 0.2 and 0.2 respectively, a combined evaluation of regional infrastructure is made. The final result is shown in Figure 3.15 and appendix 3.7. Regional disparity in infrastructures supports an east-central-west gap. Beijing, Shanghai and Tianjin have the best infrastructure, and Guangdong, Liaoning, Hainan, Zhejiang, Jiangsu, Shandong and Hebei rank second. Fujian, Shanxi, Ningxia and Henan also have an infrastructure condition better than the national average. Other provinces are poor in infrastructure. Among them Sichuan and Tibet rank the poorest. The five provinces in the Central region which show strong economic vitality in 1995-96 (i.e. Hunan, Hubei, Jiangxi, Anhui and Henan) do not have a good infrastructure condition except Henan. Shanxi ranks high in infrastructure evaluation due to the development of the energy base there.

Figure 3.15 Provincial Infrastructure Condition in China (1995-96)

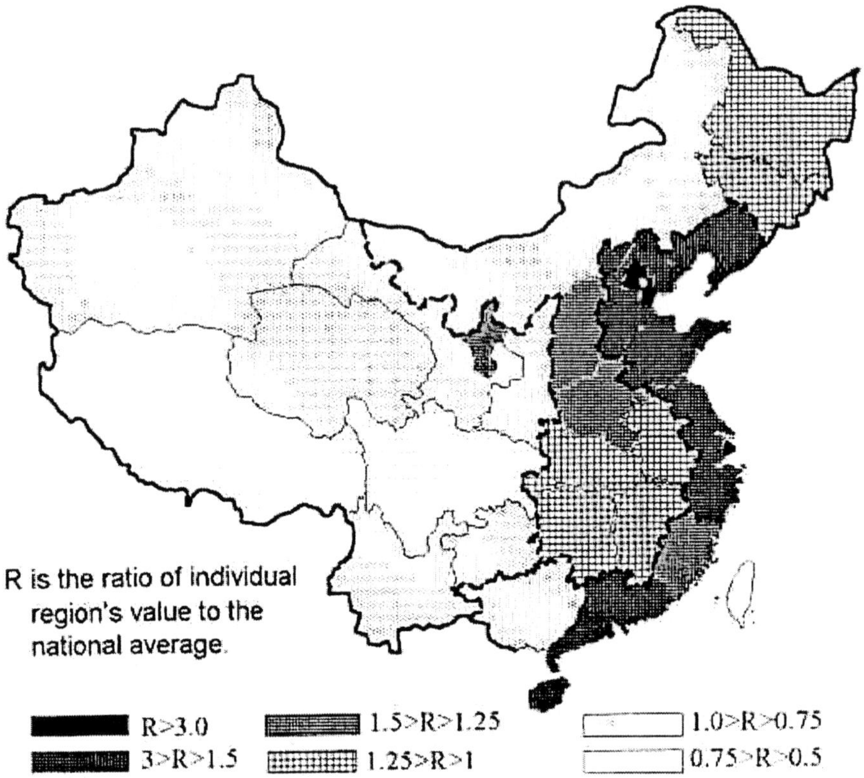

4. Regionalization of China's Socio-Economic Development

4.1

Based on the above analysis of economic development, social development, living quality and infrastructure, the current socio-economic development of China's provinces can be categorised into seven groups (Table 3.1 and Figure 3.16). That is[3]:

- ***developed city regions***, including Beijing, Shanghai and Tianjin;
- ***developed provinces***, including Guangdong, Zhejiang, Liaoning and Jiangsu;
- ***upper provinces***, including Fujian, Shandong, Hainan, Jilin and Heilongjiang;
- ***upper-middle provinces***, including Hebei, Xinjiang and Hubei;
- ***lower-middle provinces***, including Shanxi, Hunan and Ningxia;
- ***lower provinces***, including Henan, Inner Mongolia, Jiangxi, Anhui, Qinghai, Shaanxi and Guangxi;

[3] Here "developed" is a relative concept in relation to other regions in China, not regions in the world.

- ***underdeveloped provinces***, including Sichuan, Yunnan, Guizhou, Gansu and Tibet.

As shown in Table 3.1, the rank order of the comprehensive evaluation on socio-economic development is by and large correlated to that of per capita *GDP* though several provinces rank quite differently in the two. Fujian ranks among the upper provinces only because of its relatively low level of social development, though its per capita GDP is among the highest in 1996 (R=1.44). Hainan is among the upper provinces mainly due to the high level of its infrastructure while it has a per capita *GDP* just close to the national average (R=0.97). A similar case is Jilin. Its high level of social development and infrastructure leads to its relatively high rank in the comprehensive evaluation. The same reason makes Shanxi, Hunan and Ningxia rank among the lower-middle provinces. Sichuan, Guizhou, Yunnan, Gansu and Tibet have a low rank in each of the individual indicators and are the most under-developed regions.

Figure 3.16 Spatial Distribution of Provincial Socio-economic Development in China

A: Developed Cities; B: Developed Provinces; C: Upper Provinces; D: Upper-middle Provinces; E. Lower-middle Provinces; F. Lower Provinces; G. Underdeveloped Provinces

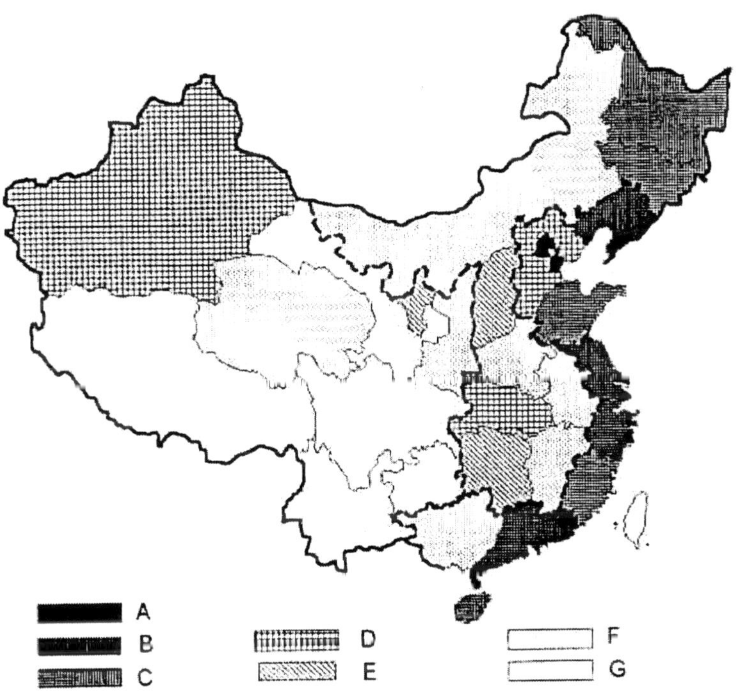

Table 3.1 Comprehensive Evaluation of Regional Socio-economic Development in China

Indicators	Combined Evaluation	Econ. Development	Social Development	Living Quality	Infrastructure Condition	Econ. Vitality	Resource Condition	Location Condition
Year	**1995-96**	**1996**	**1995**	**1995**	**1995**	**1995-96**		
Type of Value	R	R	R	R	R	R	4 Grades	4 Grades
Weight		0.25	0.25	0.25	0.25			
Highest Income:								
Shanghai	2.85	3.27	2.11	1.80	4.23	1.47	0.63	3.48
Beijing	2.54	2.17	2.74	1.20	4.05	1.19	0.91	3.2
Tianjin	2.03	2.01	1.69	1.12	3.30	1.46	0.91	3.2
High Income:								
Zhejiang	1.36	1.57	0.98	1.36	1.53	1.23	0.76	3.35
Guangdong	1.52	1.79	1.00	1.38	1.89	1.33	1.16	3.4
Jiangsu	1.25	1.41	0.93	1.16	1.50	1.31	0.63	3.4
Fujian	1.16	1.39	0.77	1.16	1.31	1.52	1.04	3.4
Liaoning	1.35	1.31	1.35	0.96	1.79	0.86	1.35	3.0
Upper-Middle Income:								
Shandong	1.13	1.15	0.90	0.93	1.55	1.24	0.97	3.18
Heilongjiang	1.08	1.04	1.18	1.08	1.02	1.07	1.17	2.10
Xinjiang	1.01	0.96	1.19	0.99	0.90	0.89	1.43	0.75
Lower-Middle Income:								
Hainan	1.18	0.97	0.89	0.99	1.85	0.86	1.16	3.08
Hebei	1.03	0.90	0.78	0.90	1.54	1.17	0.91	3.15
Hubei	0.99	0.86	0.99	0.92	1.20	1.18	0.59	2.75
Jilin	1.11	0.86	1.39	0.96	1.21	0.96	0.61	2.15
Low Income:								
Inner Mongolia	0.83	0.73	0.92	0.84	0.81	0.92	2.61	1.93
Shanxi	0.95	0.75	0.98	0.78	1.29	0.96	2.44	2.53
Guangxi	0.77	0.71	0.56	0.89	0.90	1.12	0.97	3.08
Hunan	0.93	0.70	1.08	0.87	1.06	1.03	0.74	2.83

Henan	0.86	0.69	0.71	0.76	1.29	1.20	0.58	2.68
Anhui	0.82	0.67	0.68	0.83	1.09	1.18	0.82	2.83
Qinghai	0.81	0.67	0.78	0.82	0.95	0.66	1.47	1.0
Ningxia	0.96	0.66	1.05	0.70	1.44	0.83	1.50	1.5
Sichuan	0.69	0.65	0.71	0.80	0.60	0.87	1.86	1.2
Yunnan	0.70	0.66	0.61	0.77	0.76	1.11	1.51	1.63
Jiangxi	0.84	0.62	0.83	0.84	1.06	1.16	0.81	2.9
Shaanxi	0.81	0.61	0.98	0.75	0.91	0.84	0.79	2.18
Lowest Income:								
Tibet			0.64	0.76	0.67		0.75	0.5
Gansu	0.71	0.50	0.79	0.66	0.87	0.91	0.81	1.25
Guizhou	0.63	0.39	0.58	0.73	0.80	0.83	1.05	1.95

*R is the ratio of individual indicator's actual value to the national average. The income group is divided based on per capita *GDP* in 1996. Econ. - Economic.

4.2

Taking into account other indicators like economic vitality, China can be regionalized into four major types of development regions (Fig. 3.17). Some of them can be further subdivided into sub-regions to produce a total of 10 sub-types. These divisions indicate the influence of both the natural resource endowment, accumulated human and infrastructure assets as well as their respective spatial advantages in the Opening and Reform period on China's regional development. They are useful for understanding the regional dynamics as well as the prospects of regional development in the country in the future.

4.2.1 Ascending Region

(1) *The Ascending Developed City Regions* includes Shanghai and Tianjin. These two centrally-administered cities have a very high level of per capita *GDP*, economic development, social development, living standard and infrastructures, and have shown strong economic vitality in recent years. In 1995-96, Shanghai and Tianjin rank in the top group in economic vitality and growth rate though they had a *GDP* growth rate of only the national average in 1990-94. Shanghai keeps far ahead in per capita *GDP* while Tianjin shows a tendency to surpass Beijing to be the second.

(2) *The Rapidly Ascending Developed and Upper Provinces* includes Guangdong, Zhejiang, Jiangsu, Fujian and Shandong. These provinces have been the growth 'stars' of the last decade. At the same time their living standard and infrastructures have improved quickly. In the eighth FYP, they had a *GDP* growth rate 50 per cent faster than the national average (except Shandong, 36 per cent faster than the average) whereas in 1996 they fell behind the top group in growth rate except Fujian. Now these newly developed provinces have ranked among the top six provinces in per capita *GDP*, but had a social development level only about the national average.

(3) *The Ascending Middle Provinces* includes Hebei and Hubei. These two provinces have been developing fast since the 1990s. In the eighth FYP, Hebei and Hubei achieved a *GDP* growth rate of 24 per cent and 9 per cent faster than the national average, but in 1996 they all rank among the fastest growing provinces. Besides, they are also among the areas with the best investment effectiveness in recent years, showing great development potentials.

(4) *The Ascending Lower Provinces* includes Anhui, Henan, Jiangxi and Guangxi. These provinces are among the problem provinces in the 1980s but have shown increasing vitality in economic development since the 1990s. In the eighth FYP, they all had a growth rate of 10 per cent higher than the national average and Guangxi was even in the fastest growing group. In 1995-96, they moved up to the fastest growing group except Guangxi and are also among the best provinces in investment effectiveness. However, the overall socio-economic development level in these provinces is still very low, and they have a long way before entering the relatively developed group.

4.2 2 Stable Region

(5) *The Stable Growth Provinces* includes Beijing and Xinjiang. Since the *Opening and Reform*, these two regions have basically maintained a growth rate of around the national average. In 1996 their growth rates are lower than the national average. Beijing still ranks second in per capita GDP but its superiority has been going down from a *R* value of 2.82 in 1990 to 2.3 in 1996. At the same time, Xinjiang's *R* value in per capita *GDP* has increased slightly though its rank among the provinces has not changed. The investment ratio of these two is quite high (R=1.9 and 1.21 respectively for Beijing and Xinjiang) but the investment effectiveness is far from satisfactory, which may be a reason leading to their relatively slow growth.

4.2 3 Descending Region

(6) *The Continuously Descending Provinces* includes Liaoning and Qinghai. Once ranked in the top group in economic development, they have suffered from declining economic vitality since the 1980s. In 1978-90, the annual growth rate of Liaoning was only 6.9 per cent and that of Qinghai was even lower (5.9 per cent), while in the same period the national average growth rate was as high as 8.4 per cent. In 1990-96 their growth rates were a little higher than the previous period with 9.7 per cent for Liaoning and 7.6 per cent for Qinghai, but were still much lower than the national average (11.2 per cent). Such a relatively slow growth has resulted in the rapid descent of their rank among provinces in per capita *GDP*. In 1995-96, these two provinces did not show any sign of recovery. They share some common features in their economy, i.e. both have a natural-resources-based industrial structure and a high proportion of state-owned enterprises, which may be the important factor behind their descent.

(7) The *Babble Economy Region* includes only Hainan. From 1984 when it was established as a new province to 1994 when the retrenchment policy was put into effect, Hainan had been a growth 'star' in China. In 1990-94, its annual growth rate was 50 per cent higher than the national average. At the same time its per capita GDP rank improved from the eleventh to the sixth among the provinces. Such a rapid growth did promote the overall development of its socio-economic system. However, since 1995 its economic growth rate had dramatically gone down to the lowest among the provinces, i.e. only 4.3 per cent and 5.2 per cent, respectively, in 1995 and 1996. In 1996 its rank in per capita GDP fell down to the ninth among the provinces. The improper investment structure may be responsible for its sharp descent. Most investment in the province went to such sectors as real estate and infrastructure with little in manufacturing.

4.2 4 Regenerating Region

(8) *The Recovering Upper Provinces* includes Jilin and Heilongjiang. Before 1995, these two provinces were the descending states, which form the so-called "North-east Phenomena". Actually, Heilongjiang was much worse than Liaoning in terms of economic growth. It was only a little better than Qinghai which had the lowest growth rate. Jilin was also descending though it was not so bad as Heilongjiang. However, in recent years there appeared a recovery tendency in these two provinces. In 1995 Heilongjiang and Jilin achieved economic growth at nearly the same rate as the national

average. In 1996 they even surpassed the national average in growth rate. Whether such recovery can be kept is still an uncertainty.

(9) *The Regenerating Lower and Underdeveloped Provinces* includes Hunan, Shanxi, Inner Mongolia, Gansu and Yunnan. These five provinces were by and large very poor and had growth rates much lower than the national average in 1990-94. In 1995-96, however, they all had achieved economic growth faster than the national average, indicating a regenerating tendency. Among them, Hunan, Shanxi and Inner Mongolia are much evident in their recovering growth vitality. Hunan's growth rate is much higher than the national level (R=1.16). Shanxi and Inner Mongolia have attracted more FDI (by contract) than previous years except that they have higher growth rates than the national level. Whether the regenerating tendency of Gansu and Yunnan can be sustainable needs further evidence.

4.2.5 Relatively Stagnant Region

(10) *The Relatively Stagnant Underdeveloped Provinces* includes Sichuan, Ningxia, Shaanxi, Guizhou and Tibet. These provinces have since the 1980s been much slower in economic growth than the national average and have shown little sign of regeneration in recent years. They are also among the lowest group in per capita *GDP* and other indicators like living standard, infrastructure and social development. In 1995-96 they recorded the slowest growth rates and weakest economic vitality except the three descending provinces discussed above. Their location and infrastructure are also poor. In 1996 Sichuan, Shaanxi and Ningxia achieved a little better economic growth than pervious years, which was still not enough to indicate there was regeneration.

Figure 3.17 Regionalization of Regional Development in China (1995-96)

AD: Ascending Developing Regions; AM: Ascending Middle-level Regions AL: Ascending Lower-level Regions, SR: Stable Regions; DR: Descending Regions; BER: Babble Economy Regions; RU: Regenerating Upper-level Regions; RL: Regenerating Lower-level Regions; RSL: Regenerating Stagnant Regions

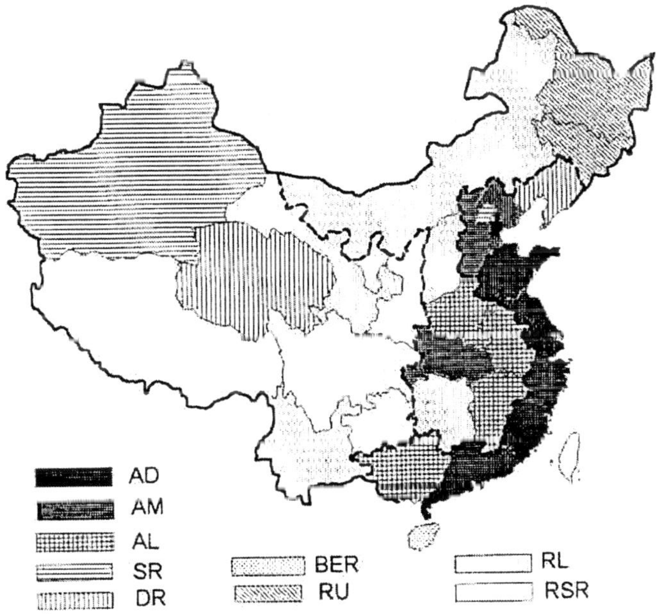

5. CONCLUSION

Regional development in China shows complicated and multi-dimensional features. To understand this is important for government decision making. In this chapter, we have applied a more comprehensive concept on regional socio-economic disparity and development level, as well as a new and more sophisticated quantification methodology. Our analysis have brought out a number of significant results which we summarise here.

Firstly, absolute regional disparity is enlarging rapidly while relative disparity shows a reducing trend. Besides, regional development had gone through a long period of "North-South" disparity before it turned to an East-West gap. Although regional disparity is increasing, it is not as serious as commonly believed. The fact that relative disparity is reducing indicates that since all the provinces are growing very rapidly, absolute disparity can be reduced only if most lower development level provinces have a much higher growth rate than the other provinces, i.e. most of the latter have such a descent as Liaoning. In fact, even in the "balanced regional development" period before the *Opening and Reform*, such reduction of regional disparity had not been achieved. Thus, slowing down the enlarging gap of regional disparity is probably the only rational policy. Besides, the change of regional disparity from a North-South gap before the *Opening and Reform* to an East-West gap in the 1990s is a result of the change in the major factors behind economic development, i.e. from geo-political concerns to market forces. In China, at present and most likely in the future, market forces and increasing economic globalization have and will play a determinant role in the formation of the east-west gap of regional development.

Table 3.2 Distribution of Provincial Development Levels and States in China (1995-96)

Type of Development State	Eastern Region	Central Region	Western Region
Ascending	A: Shanghai, Tianjin, Guangdong, Zhejiang & Jiangsu; B: Fujian & Shandong; C: Hebei; D: Guangxi	C: Hubei; D: Henan, Anhui & Jiangxi	
Stable	A: Beijing		C: Xinjiang
Descending	A: Liaoning; B: Hainan		D: Qinghai
Regenerating		B: Heilongjiang & Jilin; C: Hunan; D: Shanxi & Inner Mongolia	E: Yunnan & Gansu
Relatively Stagnant			D: Shaanxi & Ningxia; E: Sichuan, Guizhou & Tibet

* A -Developed Provinces, B -Upper Provinces, C -Middle Provinces, D -Lower Provinces, E -Underdeveloped Provinces.

Secondly, regional disparity may not be reflected by the simple division of China into the Eastern, Central and Western regions, nor by the simple indicator of per capita *GDP*. As shown in Table 3.2, there are four levels and three states of development in the Eastern region, three levels and two states in the Central region, and three levels and four states in the Western region. This is not to deny the existence of the East, Central and Western regions from a macro viewpoint, but to indicate that there are indeed big differences in any of these macro regions. Regional disparity within the coastal region is almost as wide as that of national disparity. It will be inadequate if regional policy is still made and implemented based on the tripartite spatial division. Besides, per capita *GDP* does not reflect the overall socio-economic development level, and is not equal to living quality as well. From a historical viewpoint, per capita *GDP* represents only a region's capacity to produce wealth at a given time, and does not necessarily reflect the accumulated regional wealth. Provinces which were more "glorious" in the past generally have a high social development level. Thus, evaluated by more comprehensive indicators, regional disparity in China is not as wide as it is shown by per capita *GDP*.

Thirdly, in 1995-96 a number of 'lower-middle' and 'lower' provinces had achieved economic growth as fast as or faster than the coastal growth 'stars', and several underdeveloped provinces showed a regenerating tendency. This may foretell some new growth 'stars' in the next stage of economic development. If so, relative regional disparity will continue to decrease. In 1996, Guangdong, Zhejiang, Jiangsu and Shandong had descended to the second group in economic growth rate while Hebei, Hubei, Henan, Anhui and Jiangxi moved up to the top group. If such a change in the distribution of growth rate is to maintain, then new growth 'stars' in the next stage will be those rapidly ascending middle provinces. This may imply a new spatial structure in China's future regional development.

In short, diversification in regional development level and status is a necessary result of the rapid economic growth as experienced by China in the 1990s. In a considerably long time in future, the enlargement of absolute regional disparity with the gradual reduction of relative disparity in China will be unavoidable. To understand such a fact will be helpful for regional policy making.

REFERENCES

1. Bureau of Territory Planning and Regional Development, State Planning Commission of China. 1994. *China Regional Economic Development Report*. Beijing: China Environmental Science Press (Chinese text).
2. Cannon, T. and Jenkins, A. (eds) 1990. *The Geography of Contemporary China: The Impact of Deng Xiaoping's Decade*. London: Routledge.
3. Chen, Dongsheng and Wei, Houkai. *et al.* 1996. *Roads to Economic Rise of the West Region of China*. Shanghai: Shanghai Far-east Press (Chinese text).
4. Commission for Integrated Survey of Natural Resources, the Chinese Academy of Sciences. 1991. *Manual of Natural Resources of China*. Beijing: Science Press (Chinese text).
5. Geo-Science Board of the Chinese Academy of Sciences. 1993. *Proceeding of the Potentials of China's Natural Resources and Countermeasures*. Beijing: Beijing Press (Chinese text).
6. Hu, Angang. 1995. *Challenging China*. Taiwan: News & Culture Enterprise Co. Ltd. (Chinese text).
7. Hu, Angang and Kang, Xiaoguang. 1995. *China Regional Disparity Report*. Shenyang: Liaoning People's Press (Chinese text).
8. Hu, Zhaoliang. 1992. The impact of the reform and opening policies on the changes of regional economic structure. In *Yunnan Geographical Environment Research*, No.2 (Chinese text).

9. Jin, Bo and Chen, Liying. (eds) 1996. *Breakthrough Between China Mainland and Taiwan: Regional Analysis of China's Industrial Economy*. Beijing: Economic Management Press (Chinese text).
10. Li, Siming. *et al.* (eds) 1996. *A Multi-dimensional Study on China's Regional Development.* published jointly by the Centre for Population Study, the National University of Taiwan and Li Siqi Institute of East-West Academic Exchange, the Baptist University of Hong Kong (Chinese text).
11. Li, Wenyan and Lu, Dadao. *et al.* (eds) 1990. *Industrial Geography of China*. Beijing: Science Press (Chinese text).
12. Liu, Shucheng; Li, Qiang and Xue, Tiandong. (eds) 1994. Study on China's Regional Economic Development. Beijing: China Statistic Press (Chinese text).
13. Liu, Weidong. 1997. Evaluation of the disparity of regional economic development in China: historical changes. In *Economic Geography* 17(2): 28-32 (Chinese text).
14. Lu, Dadao. 1995. *Regional Development and Its Spatial Structure*. Beijing: Science Press (Chinese text).
15. Lu, Dadao. 1996. Several problems faced by China in its sustainable development in the 21st century. In *China Soft Science* 55(7): 38 (Chinese text).
16. Maruya, Toyojiro. (ed.) 1997. *Regional Economic Development and Policy in China*. Tokyo: Institute of Development Economies, ASEDP NO.43.
17. Sit, V.F.S. 1997. The industry and industrial geography in Hong Kong. In *Economic Geography* 17(3) (Chinese text).
18. State Statistic Bureau of China. 1996. *China Development Report: The Period of the Eight Five-Year-Plan*. Beijing: China Statistic Press (Chinese text).
19. Wei, Houkai. 1992. The pattern of changes in inter-regional income disparity. In *Economic Research,* No. 1 (Chinese text).
20. Wei, Houkai. 1995. *New Patterns of Regional Economic Development of China*. Kunming: Yunnan People's Press (Chinese text).
21. Williamson, J.G. 1965. Regional inequality and the process of national development: a description of patterns. In *Economic Development and Culture Chang* 13(4).
22. Woo, Tun-oy. 1996. Regional economic development and disparities. In Brosseau, Maurice *et al.* (eds.) *China Review (1996).* Hong Kong: The Chinese University Press.
23. Yang, Dali. 1990. Patterns of China's regional development strategy. *China Quarterly* 122(2): 230-57.
24. Yang, Kaizhong. 1993. *Towards Spatial Integration*. Chengdu: Sichuan People's Press (Chinese text).
25. Zhao, Simon Xiaobin. 1994. Reforms and regional inequality in China: 1953-89. *China Report* 30(3): 331-43.
26. Zhou, Yixing. 1995. *Urban Geography*. Beijing: Commercial Press (Chinese text).

Appendix 3.1 Provincial Economic Growth Rate During Different Period in China

	Growth Rate (NI)	R		Growth Rate (NI)	R		Growth Rate (GDP)	R	Growth Rate (GDP)
Year	**1952-78**			**1978-90**			**1990-96**		**1996**
Ascending:									
(1)Beijing	12.54	1.99	(1)Zhejiang	11.94	1.42	(1)Fujian	18.08	1.62	14.8
(2)SHH	8.66	1.37	(2)GD	11.29	1.35	(2)Zhejiang	17.51	1.57	12.7
(3)Qinghai	7.74	1.23	(3)Jiangsu	11.06	1.32	(3)GD	17.37	1.56	11
(4)Tianjin	7.36	1.17	(4)Fujian	10.49	1.25	(4)Jiangsu	16.66	1.49	12.1
(5)Liaoning	7.29	1.15	(5)SHD	9.91	1.18	(5)Hainan	15.43	1.38	5.2
(6)Shaanxi	6.98	1.11	(6)Xinjiang	9.88	1.18	(6)Guangxi	15.21	1.36	10.6
			(7)Yunnan	9.33	1.11	(7)SHD	15	1.34	12.2
						(8)Hebei	14.08	1.26	13.5
						(9)Anhui	13.56	1.22	13.2
						(10)Jiangxi	13.38	1.20	13.4
						(11)Henan	12.81	1.15	13.9
						(12)SHH	12.71	1.14	13
						(13)Hubei	12.66	1.13	13.2
Stable:									
(7)HLJ	6.33	1	(8)Henan	8.99	1.07	(14)Tianjin	11.87	1.06	14.3
(8)Ningxia	6.2	0.98	(9)Anhui	8.89	1.06	(15)Beijing	11.06	1.00	9.1
(9)Yunnan	5.96	0.94	(10)Hubei	8.89	1.06	(16)Xinjiang	11.03	0.99	9.1
(10)Xinjiang	5.78	0.91	(11)Guizhou	8.31	0.99	(17)Hunan	11.01	0.99	12.6
(11)Gansu	5.77	0.91	(12)Ningxia	8.21	0.98	(18)Jilin	10.72	0.96	11.1
(12)SHD	5.73	0.91	(13)Beijing	8.16	0.97	(19)Sichuan	10.63	0.95	9.8
(13)Zhejiang	5.72	0.91	(14)Jilin	8.14	0.97				
(14)InnerM.	5.69	0.9	(15)InnerM.	8.11	0.97				
			(16)Shanxi	8.09	0.97				
			(17)Jiangxi	7.96	0.95				
			(18)Sichuan	7.87	0.94				
			(19)Hebei	7.87	0.94				
			(20)Shaanxi	7.76	0.93				
Descending:									
(15)Jiangsu	5.56	0.88	(21)Gansu	7.45	0.89	(20)Yunnan	9.92	0.89	10
(16)Jilin	5.52	0.87	(22)Hunan	7.23	0.86	(21)Shanxi	9.91	0.89	10.9
(17)Fujian	5.42	0.86	(23)SHH	7.16	0.85	(22)InnerM.	9.89	0.89	12.4
(18)Henan	5.36	0.85	(24)Guangxi	7.16	0.85	(23)Liaoning	9.69	0.87	8.5
(19)Hebei	5.33	0.84	(25)Liaoning	6.89	0.82	(24)Gansu	9.64	0.86	11

(20)GD	5.32	0.84	(26)Tianjin	6.86	0.82	(25)Shaanxi	9.36	0.84	10.2
(21)Hunan	5.19	0.82	(27)HLJ	6.23	0.74	(26)Guizhou	8.41	0.75	8.8
(22)Shanxi	5.17	0.82	(28)Qinghai	5.9	0.7	(27)Ningxia	8.32	0.75	10.5
(23)Hubei	5.12	0.81				(28)HLJ	8.19	0.73	10.6
(24)Sichuan	5.01	0.79				(29)Qinghai	7.6	0.68	8.6
(25)Guangxi	4.99	0.79							
(26)Guizhou	4.61	0.73							
(27)Jiangxi	4.6	0.73							
(28)Anhui	3.56	0.56							
Average	6.32	1	Average	8.38	1	Average	11.16	1.00	9.7

* SHH: Shanghai; GD: Guangdong; SHD: Shandong; HLJ: Heilongjiang;and Inner M.: Inner Mongolia.

Appendix 3.2 Evaluation On Provincial Natural Resources Condition in China

Indicator	Overall Natural Resources	Mineral Resources	Agricultural Resources	Ocean Resources	Water Resources
Type of Value	**4 Grades**	**4 Grades**	**4 Grades**	**4 Grades**	**4 Grades**
Weight		0.6	0.2	0.1	0.1
High-Income:					
Zhejiang	0.76	0.00	1.91	2	1.72
Guangdong	1.16	0.08	2.34	4	2.44
Jiangsu	0.63	0.03	1.89	2	0.38
Fujian	1.04	0.02	2.10	3	3.08
Liaoning	1.35	0.91	1.67	4	0.73
Upper-Middle:					
Shandong	0.97	0.46	1.83	3	0.32
Heilongjiang	1.17	0.74	2.76	0	1.75
Xinjiang	1.43	0.92	2.37	0	4.00
Lower-Middle:					
Hebei	0.91	0.48	1.95	2	0.29
Hubei	0.59	0.06	2.04	0	1.44
Jilin	0.61	0.10	2.10	0	1.26
Lower-Income:					
Inner Mongolia	2.61	3.16	2.64	0	1.87
Shanxi	2.44	3.36	1.95	0	0.40
Guangxi	0.97	0.09	2.35	1	3.53
Hunan	0.74	0.21	2.04	0	2.12
Henan	0.58	0.31	1.80	0	0.38
Anhui	0.82	0.55	1.97	0	0.95

Qinghai	1.47	1.27	1.54	0	4.00
Ningxia	1.50	1.74	2.22	0	0.17
Sichuan	1.86	2.11	1.82	0	2.32
Yunnan	1.51	1.19	1.97	0	4.00
Jiangxi	0.81	0.12	2.22	0	2.99
Shaanxi	0.79	0.50	1.92	0	1.07
Lowest-Income					
Tibet	0.75	0.15	1.30	0	4.00
Gansu	0.81	0.48	2.13	0	0.97
Guizhou	1.05	0.71	1.87	0	2.53

* Beijing and Tianjin are included in Hebei, Shanghai in Jiangsu, and Hainan in Guangdong.

Appendix 3.3 Evaluation On Provincial Economic Development Level in China

Indicator	Overall Econ. Development Level	Per Capita GDP	Share of Non-Agriculture in GDP	Dependency on Export & Import	Overall Econ. Development Level
Year	**1996**	**1996**	**1996**	**1996**	**1990**
Weight		0.80	0.10	0.10	
Highest Income:					
Shanghai	3.28	3.66	1.22	2.25	3.01
Beijing	2.17	2.3	1.18	2.14	2.44
Tianjin	2.01	2.1	1.17	2.14	1.90
High Income:					
Zhejiang	1.57	1.73	1.06	0.81	1.13
Guangdong	1.80	1.58	1.06	4.25	1.41
Jiangsu	1.41	1.52	1.05	0.86	1.10
Fujian	1.39	1.44	0.98	1.42	0.95
Liaoning	1.31	1.38	1.07	1.00	1.39
Upper-Middle:					
Shandong	1.15	1.23	1.00	0.69	0.91
Heilongjiang	1.04	1.12	1.00	0.44	1.01
Xinjiang	0.96	1.05	0.90	0.28	0.92
Lower-Middle:					
Hainan	0.97	0.97	0.79	1.14	0.94
Hebei	0.90	0.96	1.00	0.28	0.79
Hubei	0.86	0.92	0.95	0.25	0.82
Jilin	0.86	0.91	0.90	0.44	0.93

Lower Income:					
Inner Mongolia	0.73	0.77	0.86	0.25	0.76
Shanxi	0.75	0.76	1.06	0.39	0.80
Guangxi	0.71	0.74	0.87	0.31	0.57
Hunan	0.70	0.74	0.88	0.17	0.67
Henan	0.69	0.72	0.93	0.17	0.61
Anhui	0.67	0.69	0.90	0.25	0.62
Qinghai	0.67	0.68	0.98	0.28	0.85
Ningxia	0.66	0.67	0.97	0.28	0.75
Sichuan	0.65	0.67	0.90	0.22	0.63
Yunnan	0.66	0.66	0.96	0.33	0.63
Jiangxi	0.62	0.64	0.88	0.19	0.64
Shaanxi	0.61	0.6	0.97	0.36	0.68
Lowest Income:					
Tibet					0.64
Gansu	0.50	0.49	0.89	0.22	0.62
Guizhou	0.39	0.36	0.79	0.19	0.48

* The income group is divided based on per capita GDP in 1996.

Appendix 3.4 Evaluation on Provincial Economic Vitality in China

Indicator	Econ. Vitality	GDP Growth Rate	Invest. Ratio	Invest. Effective.	Industry Growth Quality	Agri. Productivity	Share of FDI in Total Investment	Share of Non-State-Owned Industry
Year	**1995-96**	**1995-96**	**1995-96**	**1994-95**	**1995-96**	**1995-96**	**1995-96**	**1995-96**
Type of Value		**R**	**R**	**R**	**R**	**R**	**R**	**R**
Weight		0.50	0.10	0.10	0.07	0.03	0.10	0.10
Highest Income:								
Shanghai	1.38	1.34	2.17	0.73	1.34	4.31	0.93	1.08
Beijing	1.12	1.06	1.91	0.59	1.01	3.80	0.72	0.80
Tianjin	1.44	1.45	1.27	1.10	1.04	2.50	2.16	1.14
High Income:								
Zhejiang	1.26	1.45	1.00	1.20	1.08	1.20	0.61	1.40
Guangdong	1.35	1.28	1.18	0.89	1.01	1.67	2.45	1.41
Jiangsu	1.32	1.36	0.96	1.34	1.06	1.76	1.54	1.32
Fujian	1.54	1.45	0.87	1.54	1.06	1.56	3.16	1.34
Liaoning	0.83	0.77	0.92	0.77	0.76	2.06	0.92	0.73
Upper-Middle:								
Shandong	1.22	1.31	0.69	1.66	1.19	1.20	1.09	1.04

Heilongjiang	0.97	1.00	0.79	1.32	1.29	2.29	0.46	0.50
Xinjiang	0.84	0.96	1.21	0.80	0.59	2.30	0.09	0.37
Lower-Middle:								
Hainan	0.87	0.47	1.47	0.25	0.68	1.84	2.63	0.97
Hebei	1.15	1.36	0.85	1.51	1.07	1.04	0.41	0.87
Hubei	1.17	1.38	0.98	1.51	0.99	1.14	0.40	0.87
Jilin	0.93	1.03	0.87	1.16	0.68	1.47	0.63	0.53
Lower Income:								
Inner Mongolia	0.88	1.06	0.85	1.13	0.82	1.19	0.13	0.47
Shanxi	0.92	1.09	0.74	1.35	0.98	0.76	0.18	0.55
Guangxi	1.12	1.28	0.67	1.59	0.86	0.78	0.90	0.80
Hunan	1.03	1.16	0.65	1.68	0.82	0.77	0.60	0.76
Henan	1.18	1.42	0.77	1.87	1.06	0.77	0.31	0.81
Anhui	1.18	1.36	0.66	1.89	0.97	0.80	0.56	1.01
Qinghai	0.66	0.82	1.19	0.72	0.17	0.63	0.01	0.26
Ningxia	0.80	0.97	1.13	0.79	0.79	0.70	0.03	0.47
Sichuan	0.88	0.98	0.74	1.34	0.76	0.62	0.27	0.87
Yunnan	0.95	1.05	0.90	1.14	2.34	0.46	0.11	0.36
Jiangxi	1.14	1.38	0.61	1.92	0.74	0.90	0.57	0.63
Shaanxi	0.83	0.95	0.86	1.06	0.58	0.58	0.55	0.49
Lowest Income:								
Tibet								
Gansu	0.87	1.03	0.88	1.28	0.79	0.68	0.22	0.41
Guizhou	0.81	0.97	0.77	1.25	0.75	0.40	0.14	0.42

* Econ.: Economic;Invest.: Investment;Effective.: Effectiveness;and Agri.: Agriculture.

Appendix 3.5 Evaluation on Provincial Social Development Level in China

Indicator	Overall Social Development	Urbani-sation Ratio	Medical & Healthy Supply	Education Condition	Population Quality	Social Security & Services
Year	**1995**	**1994**	**1995**	**1995**	**1994**	**1995**
Type of Value		**R**	**R**	**R**	**R**	**R**
Weight		0.3	0.2	0.2	0.2	0.1
Highest Income:						
Shanghai	2.12	1.68	2.09	1.87	2.97	2.25
Beijing	2.74	1.60	2.35	3.03	4.56	2.71
Tianjin	1.69	1.49	1.89	1.77	2.24	0.65

High Income:						
Zhejiang	0.98	1.37	0.95	0.90	0.47	1.02
Guangdong	1.00	1.47	0.89	0.87	0.36	1.35
Jiangsu	0.93	1.27	0.98	0.97	0.51	0.60
Fujian	0.77	0.99	0.85	0.99	0.38	0.32
Liaoning	1.35	1.56	1.78	1.28	1.10	0.54
Upper-Middle:						
Shandong	0.90	1.23	0.89	0.91	0.54	0.63
Heilongjiang	1.18	1.40	1.35	1.22	0.83	0.78
Xinjiang	1.20	0.84	1.64	1.43	1.09	1.13
Lower-Middle:						
Hainan	0.89	1.03	1.20	0.96	0.61	0.29
Hebei	0.78	0.78	0.93	0.93	0.52	0.68
Hubei	1.40	1.57	1.46	1.40	1.07	1.39
Jilin	0.99	1.29	1.10	1.12	0.46	0.70
Lower Income:						
Inner Mongolia	0.92	0.92	1.20	1.24	0.66	0.28
Shanxi	0.98	0.93	1.35	1.22	0.73	0.40
Guangxi	0.57	0.70	0.73	0.80	0.15	0.19
Hunan	1.08	0.80	0.88	0.96	0.38	3.96
Henan	0.71	0.69	0.81	0.85	0.38	0.96
Anhui	0.68	0.68	0.74	0.86	0.46	0.64
Qinghai	0.79	0.58	1.28	1.26	0.45	0.13
Ningxia	1.05	0.76	1.07	1.27	1.35	0.79
Sichuan	0.71	0.82	0.86	0.78	0.36	0.64
Yunnan	0.61	0.55	0.87	0.80	0.20	0.69
Jiangxi	0.83	0.81	0.88	1.01	0.36	1.34
Shaanxi	0.99	0.72	1.06	1.26	1.15	0.75
Lowest Income:						
Tibet	0.65	0.27	0.96	1.00	0.21	1.30
Gansu	0.79	0.68	0.93	0.97	0.55	0.95
Guizhou	0.59	0.64	0.69	0.78	0.15	0.70

Appendix 3.6 Evaluation on Provincial Living Quality in China

Indicator	Overall Living Quality	Consumption Level	Environment Quality	Urban Living Standard	Rural Living Standard
Type of Value	**R**	**R**	**R**	**R**	**R**
Weight		0.8	0.2		
Highest Income:					
Shanghai	1.80	2.21	0.15	1.47	2.34
Beijing	1.20	1.43	0.30	1.34	1.91
Tianjin	1.12	1.34	0.25	1.10	1.53
High Income:					
Zhejiang	1.36	1.45	1.02	1.32	1.72
Guangdong	1.38	1.49	0.96	1.52	1.53
Jiangsu	1.16	1.22	0.92	1.06	1.46
Fujian	1.16	1.18	1.10	0.94	1.16
Liaoning	0.96	1.12	0.31	0.89	1.07
Upper-Middle:					
Shandong	0.93	0.97	0.78	1.05	1.11
Heilongjiang	1.08	1.08	1.09	0.86	1.11
Xinjiang	0.99	0.95	1.19	1.04	0.87
Lower-Middle:					
Hainan	0.99	0.96	1.14	1.05	0.97
Hebei	0.90	0.93	0.79	0.98	1.16
Hubei	0.92	0.90	0.99	0.95	0.96
Jilin	0.96	0.95	1.02	0.81	0.98
Lower Income:					
Inner Mongolia	0.84	0.76	1.16	0.76	0.77
Shanxi	0.78	0.76	0.88	0.87	0.84
Guangxi	0.89	0.83	1.13	1.06	0.91
Hunan	0.87	0.82	1.09	1.07	0.86
Henan	0.76	0.74	0.84	0.85	0.88
Anhui	0.83	0.81	0.95	0.89	0.89
Qinghai	0.82	0.73	1.21	0.81	0.69
Ningxia	0.70	0.71	0.66	0.86	0.66
Sichuan	0.80	0.72	1.10	0.93	0.71

Yunnan	0.77	0.68	1.15	0.94	0.68
Jiangxi	0.84	0.80	0.99	0.84	0.97
Shaanxi	0.75	0.67	1.08	0.84	0.69
Lowest Income:					
Tibet	0.76	0.65	1.22	1.05	0.76
Gansu	0.66	0.54	1.13	0.80	0.54
Guizhou	0.73	0.63	1.12	0.92	0.68

Appendix 3.7 Evaluation on Provincial Infrastructure Condition in China

Indicator	Overall Infrastructure	Transportation	Water Supply	Communication Facilities	Electricity Supply
Year	**1995**	**1995**	**late 1980s**	**1995**	**1995**
Type of Value	**R**	**R**	**R**	**R**	**R**
Weight		0.30	0.2	0.3	0.2
Highest Income:					
Shanghai	4.23	8.07	1.41	4.23	1.28
Beijing	4.05	8.53	0.40	4.27	0.63
Tianjin	3.30	7.16	0.35	3.07	0.78
High Income:					
Zhejiang	1.53	1.98	1.25	1.68	0.90
Guangdong	1.89	1.87	1.50	2.67	1.12
Jiangsu	1.49	1.95	1.75	1.35	0.73
Fujian	1.31	1.32	1.05	1.50	1.29
Liaoning	1.79	2.98	0.40	2.08	0.96
Upper-Middle:					
Shandong	1.55	3.47	0.52	0.72	0.94
Heilongjiang	1.02	1.13	0.70	1.07	1.08
Xinjiang	0.90	1.05	0.66	0.83	1.03
Lower-Middle:					
Hainan	1.85	2.80	2.20	1.44	1.28
Hebei	1.54	3.13	0.50	0.78	1.31
Hubei	1.20	1.50	1.40	0.87	1.02
Jilin	1.21	1.64	0.50	1.23	1.24
Lower Income:					
Inner Mongolia	0.81	0.96	0.60	0.74	0.91
Shanxi	1.29	2.21	0.15	0.63	2.04

Guangxi	0.90	0.64	1.55	0.56	1.13
Hunan	1.06	1.29	1.38	0.61	1.07
Henan	1.29	2.84	0.40	0.42	1.16
Anhui	1.09	1.94	0.75	0.55	0.97
Qinghai	0.95	0.98	0.45	0.64	1.88
Ningxia	1.44	1.37	1.11	0.92	2.64
Sichuan	0.60	0.62	0.40	0.46	0.96
Yunnan	0.76	0.75	0.45	0.47	1.54
Jiangxi	1.06	1.19	1.55	0.67	0.97
Shaanxi	0.91	1.39	0.28	0.62	1.27
Lowest Income:					
Tibet	0.70	0.25	0.20	0.40	2.31
Gansu	0.87	1.08	0.35	0.60	1.47
Guizhou	0.80	0.65	0.43	0.26	2.19

Chapter 4

THE TAX DIVISION SCHEME: CENTRAL-LOCAL FINANCIAL RELATIONS IN THE 1990S

LIU Yi
Professor & Deputy Director, Institute of Geography,
Chinese Academy of Science
LIN Chusheng
Associate Professor, Department of Geography,
University of Hong Kong

Central-local relations refer to the vertical power relations between central and local governments, which are developed in the administration of state affairs and public facilities. Such relations exist widely among various nations whether their government structures are monarchical or federal. Central-local relations in China include public administration, personnel appointment and finance. Central-local financial relations concern taxation, fixed asset capital ownership, allocation of resources and materials, approval of large scale projects and capital investment. As central-local relations virtually concern the reconciliation and balance between the general well-being of the entire nation and local interests, particularly the regional allocation of economic benefits, Central-local financial relations have not only become one of the most important and sensitive issues but also served as a major operating force behind many regional development problems in contemporary China.

An independent state normally has a certain amount of fund for a certain period of time. Central-local financial relations within the country and their impacts on regional development are therefore related to two important respects. First, how should the central government collect tax from local governments and how much should be collected? What are the taxes that should be collected for the central government? What is the standard for such tax collection and where and from whom should the tax be collected? Secondly, how much fund should the central government allocate to regions after necessary financial disposal for the central state machinery? In what manner should central funds be redistributed, to whom, and by what standard? The first issue concerns tax collection whereas the second concerns the redistribution of central funds.

Important steps have been taken by the central government of China since the economic reforms to reorganize the central-local financial relations, ranging from the ownership of state property to the allocation of key industrial materials, granting of large development projects, investment decision making, and taxation. However, the most significant, efficient, and effective reform of the fiscal system actually took place in the first half of the 1990s (He, 1994; Ning, 1994; Xiao, 1997). Building on the previous fiscal responsibility system, a new tax division scheme was introduced to reorganize the central-local financial relations. Tax division represents the most difficult, complicated, and yet most significant change in the central-local financial relation in the history of the People's Republic (Wen and Wang, 1995; Hu, 1996).

This chapter reviews the changing central-local financial relation and discusses the tax division scheme introduced in the 1990s including its objective, contents, and effects as an important factor in regional development. We also analyze the existing problems after the fiscal reforms and offer our suggestions for policy making in redressing regional imbalance in development.

1. Evolution of the Central-Local Financial Relations

China's central-local relations have experienced constant changes involving the setting of standards and methods for tax collection and central fund allocation. Three major phases of important change can be identified.

1.1 Central Control (1950s and 1960s)

Financial arrangements during the initial period of post-1949 economic development were characterized by tight control of the central government. Several models of arrangements were practiced at different stages. There was the "unified revenue procurement and expenditure" model adopted during 1949-53 and by which the central government maintained tight control of all financial resources. This was followed by the First Five Year Plan when the "proportional division by categories" model was put in place to allow local governments of various levels to participate in financial management under the control of the central government. Later in 1958, a new model was introduced in which expenditure was determined by revenue and financial arrangement was fixed for five years. In the 1960s, the system was changed into one in which "revenue was proportionally redistributed between the central and local governments and rearrangements were made once a year. " Regardless of what model was adopted, China's national finance during this period remained under the tight control of the central government. Income of the central government took up more than 30 per cent of the total revenue for most of the years in the 1950s and 1960s except the difficult period of the Great Leap Forward and the three years of natural catastrophes (Table 4.1).

Table 4.1 Percentage of Central Revenue in Total Revenue, 1953-1998 (%)

Year	Central Revenue	Year	Central Revenue	Year	Central Revenue	Year	Central Revenue	Year	Central Revenue
1953	44.8	1961	21.5	1969	32.5	1981	26.5	1989	30.9
1954	45.4	1962	29.7	1970	27.6	1982	28.6	1990	33.8
1955	47.8	1963	23.1	1975	11.8	1983	35.8	1991	29.8
1956	46.0	1964	25.2	1976	12.7	1984	40.5	1992	28.1
1957	43.2	1965	33.0	1977	13.0	1985	38.4	1993	22.0
1958	20.8	1966	35.2	1978	15.5	1986	36.7	1994	55.7
1959	20.5	1967	31.6	1979	20.2	1987	33.5	1995	52.2
1960	22.5	1968	29.6	1980	24.5	1988	32.9	1996	49.5

Source: The State Statistical Bureau, China's Statistical Yearbook (1985-1997), China's Statistical Press.

1.2 Transitional Stage (1970s)

During the Cultural Revolution, central-local financial relations were maintained in various manners including the unified revenue procurement and expenditure in the late 1960s, local financial responsibility during 1971-73, and local proportional retention of revenue and expenditure responsibility during 1974-75. A new financial responsibility system was initiated in Jiangsu Province after 1976. The system allows local governments to take responsibility for financial management provided that they handed over a fixed portion of revenue to the central government for a number of years. After 1978, another system was introduced in ten provinces to allow local governments to retain part of the surplus revenue they generated and to link fund allocation with revenue generation. Despite these innovative arrangements, central revenue suffered from a substantial decline because of the detrimental effects of the Cultural Revolution. Central fund accounted for only 12 per cent of the total revenue of the entire nation in the year 1976 when the Cultural Revolution came to an end.

1.3 Decentralization and Financial Responsibility

The economic reforms instituted in 1978 have brought profound change to the central-local financial relations. The centre of change was the implementation of a financial responsibility system. Beginning in 1980, new financial arrangements were made according to the principle of "cooking in separated kitchens" (*fenzao chifan*). Local governments were allowed to look after their revenue and expenditure after financial needs of the government are met. The new system specified financial responsibility for governments of different levels. It linked government authorities directly with financial management and stimulated the enthusiasm of both the central and local governments for revenue generation. The practice of financial responsibility had created remarkable impacts at the beginning of its implementation. The proportion of central income in the total revenue of the whole country increased from 20 per cent in 1979 to 41 per cent in 1984.

To respond to the conversion of profits into tax, the State Council decided in March 1985 that the existing financial arrangements be further specified according to the principle of "tax division, revenue/expenditure specification, and responsibility to every level. " The total revenue was divided into three main categories: central fund, local fund, and fund to be shared by central and local governments. Expenditure was determined by two different administrative systems: enterprises and public organizations, but a new category of special expenses was added to the existing account. For those regions where local revenue is greater than expenditure, a portion of the surplus should be handed over to the central government. Those regions where revenue is smaller than expenditure could retain a portion of the revenue to be shared by central and local governments. For those regions where expenditure is greater to the sum of local revenue and the shared revenue, the central government would offer certain subsidyze. This model was applied to the entire nation except the two provinces of Guangdong and Fujian where financial arrangements were made separately.

The system of financial responsibility continued to be practiced in various manners after 1988. There were six main types of financial responsibility system:

i) *Responsibility according to increasing revenue*: the proportion of remitted fund and local retained fund are determined according to the actual revenue and expenditure of the 1987 budget and local fund generation capacity. A growth rate of revenue generation in the future is also estimated. This system was introduced in Beijing, Hebei, Liaoning, Shenyang, Harbin, Jiangsu, Zhejiang, Ningbo, Hunan, and Chongqing.
ii) *Division according to total revenue and expenditure*: The proportions of local retained revenue and central remittance are determined according to the percentage of local expenditure in local revenue. The actual financial situations of the 1986 and 1987 were

used as a baseline for determination. This system was introduced in Tianjin, Shanxi, and Anhui.

iii) *Division according to both total revenue and increased revenue*: Retention and remittance of revenue were determined not only by the percentage of local expenditure in the total revenue generated but also by the increased revenue generated every year. Division of the increased revenue between central and local governments is determined by an agreed ratio. This was practiced in Dalian, Qingdao, and Wuhan.

iv) *Increasing revenue remittance*: Using the situation of 1987 as a baseline, submitted fund is increased according to a ratio. This was implemented in Guangdong and Hunan.

v) *Fixed revenue remittance*: The amount of fund for submission is fixed and is taken from the surplus of local finance after local expenses have been deducted. This arrangement was made in Shanghai, Shandong, and Heilongjiang.

vi) *Fixed subsidize*: The central government provides fixed amount of subsidyze according to the deficit of local finance. This was applied to Jilin, Jiangxi, Gangsu, Shaanxi, Fujian, Inner Mongolia, Guangxi, Tibet, Ningsia, Xingjian, Guizhou, Yunnan, Qinghai, and Hainan.

2. Characteristics and Problems

Despite the various changes made with the central-local relations, China's financial system before the late 1970s was dominated by a highly centralized command economy in which local finance was reliant upon the central government. This situation was changed significantly after major reform measures were introduced in the early 1980s. The financial responsibility system was a product of economic reforms. It helped stimulate local incentive for revenue generation and promote regional economic development. However, the financial responsibility system only represents a temporary and transitional arrangement and does not meet the needs of further marketization of the economy. As economic reforms proceed, problems associated with the financial responsibility system have surfaced.

2.1 Weakening Capacity of the Central Government for Macro-Economic Control

Under the financial responsibility system, the portion of revenue to be handed over to the central government is usually fixed without necessary increment according to the increase of the total revenue. As a result, the percentage of central fund in the total revenue or the total gross domestic product has declined. As listed in Table 4.2, the ratio of central fund versus gross domestic product has dropped from 31 per cent in 1978 to only 13 per cent in 1993 (Fig. 4.1). This pattern deviates from the norm of both the developed and developing worlds where the ratio of central fund versus GDP has gradually increased since the 1980s. The current percentage of central fund in the total GDP is also significantly lower than the international standard according to statistics relieved by the IMF.

Table 4.2 Ratio Revenue to GNP 1978-1996 (%)

Year	Ratio	Year	Ratio	Year	Ratio	Year	Ratio
1978	31.2	1983	23.0	1988	15.8	1993	12.6
1979	28.4	1984	22.9	1989	15.8	1994	11.2
1980	25.7	1985	22.4	1990	15.8	1995	10.7
1981	24.2	1986	20.8	1991	14.6	1996	10.9
1982	22.9	1987	18.4	1992	13.1		

Source: The State Statistical Bureau, China's Statistical Yearbook (1985-1997), China's Statistical Press.

Fig 4.1 Ratio of Revenue to GDP (%)

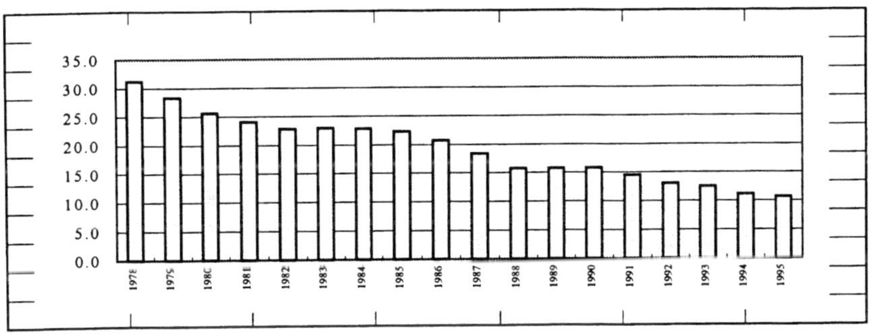

Moreover, the proportion of central fund in the total national revenue continued to drop from 40 per cent in 1984, 34 per cent in 1990, to only 22 per cent in 1993 (Fig. 4.2). Of the increased revenue, the bulk goes to local governments (96.7 per cent in 1988) and the central government only gets a tiny portion (3.3 per cent in 1988). Because of the shortage of income, the central government has suffered from deficit since 1979. The deficit increased from 6.9 billion yuan in 1980, 14.6 billion yuan in 1990, to 29.3 billion yuan in 1993. When debts were included, the central deficit was 98.5 billion yuan in 1992, which represented 20 per cent of the total national expenditure. During the years 1979-92, central deficit reached 422.0 billion yuan (Table 4.3, Fig. 4.3).

Fig 4.2 Percentage of Central Revenue in Total Revenue (%)

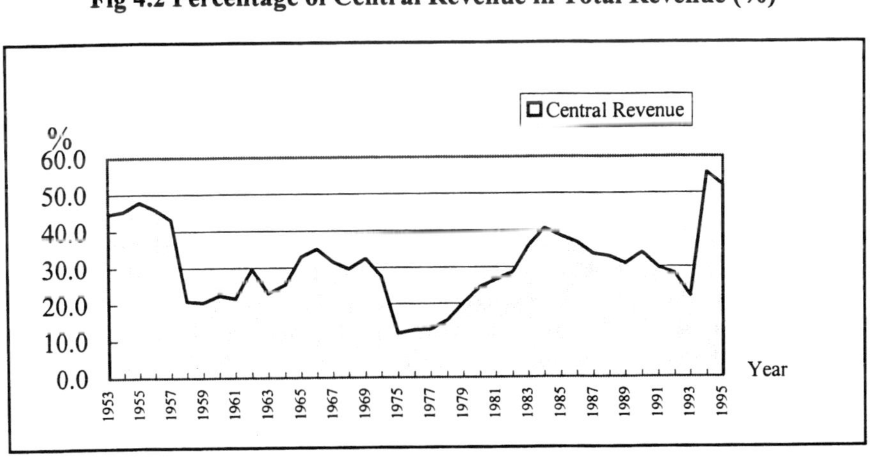

Table 4.3 China's Changing Deficit (100 Million Yuan)

Year	Deficit	Year	Deficit	Year	Deficit	Year	Deficit	Year	Deficit	Year	Deficit
1950	-2.89	1957	5.98	1964	60.52	1975	-5.27	1982	-17.65	1989	-158.88
1951	10.65	1958	-21.8	1965	6.99	1976	-29.62	1983	-42.57	1990	-146.49
1952	7.73	1959	-65.74	1966	17.15	1977	30.93	1984	-58.16	1991	-237.14
1953	2.74	1960	-81.85	1967	-22.49	1978	10.17	1985	0.57	1992	-258.83
1954	16.05	1961	0.04	1968	1.41	1979	-135.41	1986	-82.9	1993	-293.35
1955	-21.44	1962	8.3	1969	-99.1	1980	-68.9	1987	-62.83	1994	-574.52
1956	-18.31	1963	2.62	1970	13.49	1981	37.38	1988	-133.97	1995	-581.52

Source: The State Statistical Bureau, China's Statistical Yearbook (1985-1997), China's Statistical Press.

Fig 4.3 China's Changing Deficit (100 million Yuan)

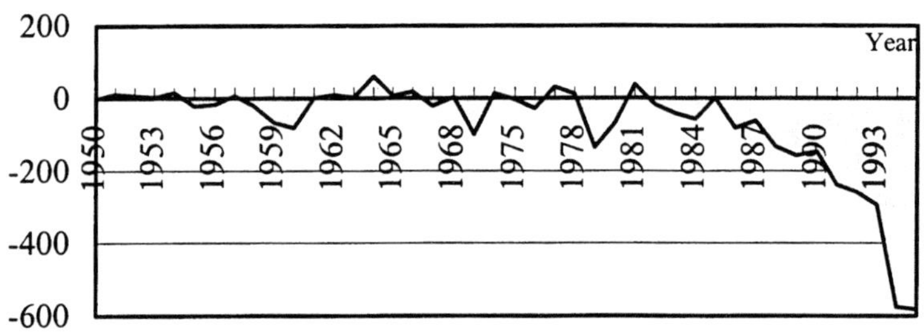

While the percentage of central fund in gross domestic product and in total national revenue continued to drop, the revenue handed over to the central by local governments has grown at a pace much lower than the growth of the total national income. The result has been a shrinking central fund unable to meet the demand of expenditure. However, the central government requires some indispensable fund to maintain necessary government functions. To balance up income and expenditure, the central government had to borrow from domestic and foreign sources. For instance, the central government borrowed 13.8 billion yuan through selling bonds and foreign loans in 1986. This, plus a bank loan of 9.5 billion yuan, accounted for 30 percent of total income of the central government. Government bonds and foreign loans reached 84.3 billion yuan which accounted for 86 percent of its total income in 1992. In view of the increasing debts of the central government, there was a great demand for changing the fiscal responsibility system so that the central government would not go into a financial crisis.

Tax revenue has been the main source of China's national revenue. The contribution of taxes to the national revenue has increased since the reforms, rising from 49 per cent in 1980 to 97 per cent in 1993 (Fig. 4.4, Table 4.4). However, the implementation of fiscal responsibility has distorted the financial relationships between the state and enterprises and reduced tax revenue. For instance, tax revenue collected from non-agricultural sectors has continued to drop since the mid-1980s. Tax revenue from non-agricultural sectors accounted for about 44 per cent of the national income in 1985, but it dropped to only 22 per cent in 1992. Moreover, under-reporting and cheating on tax returns have become a growing problem. It has been estimated that cheating on tax returns is practised by 50 per cent of the state-owned enterprises, 60 per cent of township and village enterprises, 80 per cent of private businesses, 90 per cent of joint ventures and co-operative enterprises, and 95 per cent of individual residents. The largest source of tax cheating is actually the state own sector. The central government has conducted investigations of tax reporting since 1985. About 100 billion yuan was collected during the nine years from 1985 to 1993 as a result of these investigations (Sheng, 1995).

Fig 4.4 Percentage of Tax Revenue in Total Revenue (%)

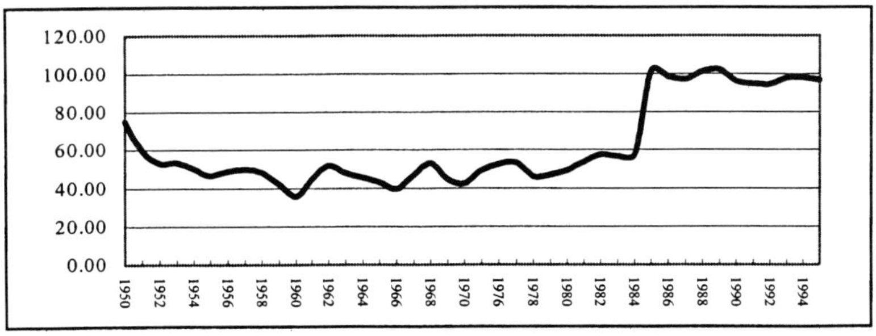

Table 4.4 Percentage of Tax Revenue in Total Revenue (%)

Year	Ratio	Year	Ratio	Year	Ratio	Year	Ratio	Year	Ratio	Year	Ratio
1950	75.13	1957	49.93	1964	45.55	1975	49.38	1982	57.74	1989	102.35
1951	59.69	1958	48.34	1965	43.16	1976	52.53	1983	56.74	1990	96.08
1952	53.00	1959	42.02	1966	39.73	1977	53.55	1984	57.66	1991	94.94
1953	53.52	1960	35.59	1967	46.89	1978	45.86	1985	101.79	1992	94.65
1954	50.25	1961	44.59	1968	53.03	1979	46.91	1986	98.53	1993	97.85
1955	46.79	1962	51.69	1969	44.70	1980	49.29	1987	97.32	1994	98.25
1956	49.01	1963	48.01	1970	42.42	1981	53.57	1988	101.41	1995	96.73

Source: The State Statistical Bureau, China's Statistical Yearbook (1985-1997), China's Statistical Press.

Because of its declining financial capacity, the power of the central government has been weakened. Its capability to maintain macro-level control of the national economy has been reduced and its ability to implement central policies has deteriorated. Local governments are now in a better position to bargain or negotiate with the central government to protect local interests. They have often asked for preferential treatments from the central government, or requested greater capital investment from state budgetary expenditure, or sought the permission of the central government for reduced tax revenue to be handed over to the central fund. On the other hand, local governments have tried to get around with or boycott central policies as a means to protect local interests. Many central policies were distorted or unable to be implemented once they have reached the local level. In light of these growing problems, there was a growing demand for changing the existing central-local relations to raise the proportion of central revenue in the total national income and strengthen the capability of the central government to control the national economy.

2.2 Growing Capability of Local Governments for Self-Determination

The fiscal responsibility system introduced in the 1980s has significantly reduced the financial capacity of the central government, substantially raised local revenue, and greatly strengthened the capability of local governments for self-determination. As the division of responsibility was centered around the separation of revenue generation and expenditure rather than the rational division of taxation and administration, implementation of the system has created profound impacts on regional economic development in the country.

The function of a government is mainly to control and allocate resources of the society. In the Maoist era of centralized decision-making, the central government was the main power house for the control and allocation of resources. After economic reforms, decentralization of decision-making has weakened the power of the central government and increased the capability of local governments in resource allocation. The result is a downward transfer of power from the central to local governments. In the late 1980s and early 1990s, local governments benefited significantly from the downward transfer of power. For instance, the Directives on the Reform of the Management System for Public-Owned Enterprises disseminated in 1992 granted individual enterprises autonomy in 14 respects. However, most of the granted autonomy was taken away by local governments. In addition to their rising financial capacity, local governments have greater capability to control resources because they have been granted the power of self mobilization and investment of capital. Many local officials now enjoy luxurious offices, transportation, modern telecommunications and other facilities superior than what is available to central government officials.

The fiscal responsibility system was organized according to different administrative relations of enterprises and agencies. Its implementation has given rise to a special manner of resource allocation which aims at the maximization of local interests. An important outcome is the separation and protection of local economies, which jeopardizes the development of a general national market. It also hinders necessary competition among enterprises. To maximize local interests, local governments and individual enterprises develop profitable enterprises disregard of local conditions. Because the central government was no longer able to coordinate local production decisions effectively, there was unnecessary duplication of production among regions. Another consequence of the weakening power of the central government has been growing regional economic disparity because of the implementation of differentiated policies. On the one hand, coastal provinces received preferential treatments and have not handed over to the central government sufficient revenue. On the other hand, backward regions in the interior received diminishing subsidy from the central government because of the shrinking central budget. Regional inequality has therefore become a persistent phenomenon for which the central government can do little to change before a new central-local financial relationship is established (Yang, 1992; Hu, 1995; Wei, 1995; Wang, 1996).

2.3 Central-Local Economic Interdependence and Irrational Contribution Among Regions

Implementation of the fiscal responsibility system has significantly changed the interdependence between the central and local governments. In addition to a shrinking central revenue, irrational arrangements are found concerning central budgetary allocation to local governments and revenue contribution from regions. Surprisingly, change in central-local relations that occurred during the years 1980-94 has significantly increased financial dependence of local governments on the central government. Among the 31 provincial administrative units of the country, 17 were financially self-sufficient in 1980. The remaining 14 in the inland and border regions were in deficit and reliant on financial support of the central government (Table 4.5; Figure 4.5). This situation changed five years later in 1985. The number of self-sufficient provinces and autonomous regions dropped from 17 to 10. In other words, about two-third of the country had to rely on financial support of the central government. The situation became even worse in 1990 when the number of self-sufficient administrative units dropped further to 6. Four-fifth of the country now has to depend on the support of the central government.

Fig 4.5 Financial Self-sufficiency of Provincial Regions (1995)

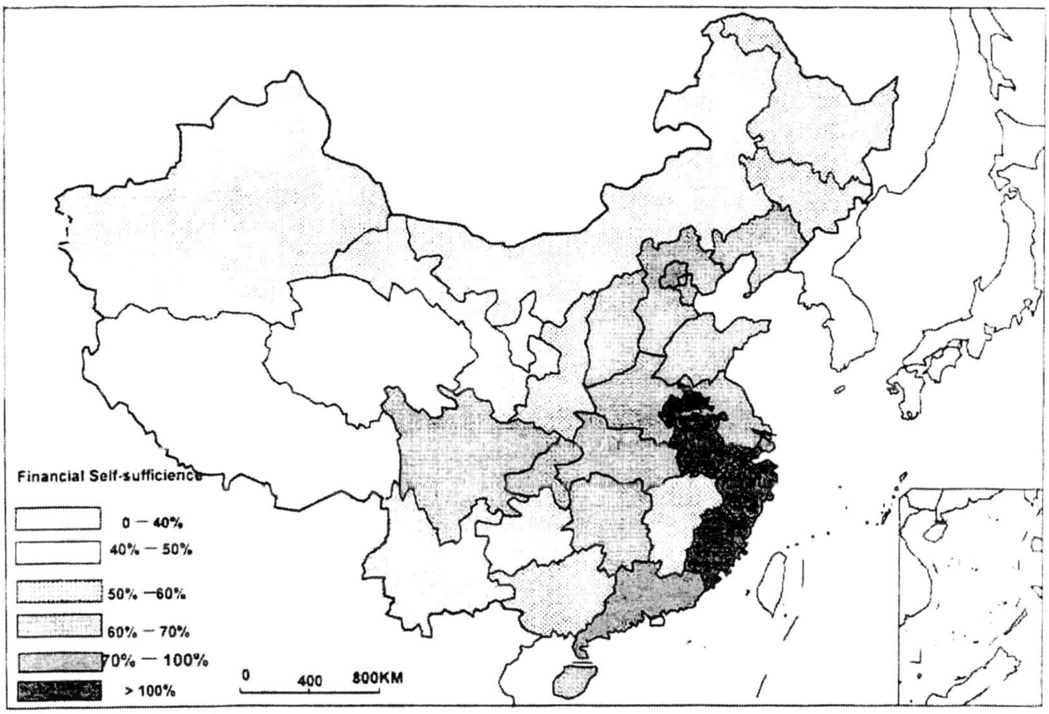

The central government functions as a chief agent for coordinating national development at a macro level. Local governments have an obligation to make financial contribution to maintain central functions. However, the fiscal responsibility system does not provide rational and scientific specifications on how local contribution should be made. The result is an unequal distribution of financial responsibility among regions based on 'rule of the thumb', or even personal subjective impulsive decisions.

Conventionally, advanced regions with an above-the-average level of economic development should make a correspondent financial contribution and backward regions with a below-the-average level of development should be the recipients of central subsidies. This has not been the case in China. An analysis of the correlation between per capita GNP and financial self-sufficiency among regions reveals that there is an unequal and unreasonable division of financial responsibility among these regions. Of the 17 provinces and municipalities that made financial contributions to the central government in 1980, 10 were found to have a level of per capita GNP lower than the national average (Table 4.5; Fig. 4.5). On the other hand, provinces like Heilongjiang and Qinghai where per capita GNP was higher than the national average had actually received financial support from the central government. The situation was improved slightly in 1990 when the six provinces that made financial contributions to the central government were all in the advance group with a per capita GNP above the national average. There were, however, still six provinces having a per capita GNP higher than the national average and receiving central financial subsidy, including Shandong, Guangdong, Fujian, Heilongjiang, Jilin, and Xingjiang.

Because of the lack of rational standards to divide regional responsibility, there was an unjustifiable determination of financial contribution and subsidy among regions in the ten year practice of the fiscal responsibility system. There were provinces such as Guangdong and Shandong which should have but have not made due contribution and therefore were able to accumulate more capital for further economic growth.

Moreover, there is no uniform standard for determining of central and local financial responsibilities. Different regions made different arrangements with the central government according to their special situations. This has given rise to much room for bargaining, negotiation, and complaints about unequal treatments. For instance, Shanghai handed over to the central government 25 billion yuan in the late 1980s, including 10 billion yuan produced by enterprises owned by the central government and 15 billion yuan generated by the local government of Shanghai. The bulk of revenue generated by Shanghai was handed over to the central government and Shanghai retained only 10 to 20 per cent of its revenue for local expenditure. About 16 billion yuan of revenue was generated by Shanghai in the year 1984. Only 2.77 billion yuan, or 17 per cent of the total revenue, was retained for local expenses. For the same year, Beijing generated 4.56 billion yuan and retained 2.722 yuan, or 59.7 per cent of the total amount, for local use. Although Shanghai was allowed to increase the proportion of revenue for local expenses from 10 per cent to 25 per cent after 1985, this proportion was still far below the one granted to Beijing. There is another case, i.e. Guangdong where great autonomy was given by the central government on the condition that the amount of revenue handed over to the central government should increase at 10 per cent a year. By 1989, Guangdong had already reached an industrial capacity similar to that of Shanghai. However, Guangdong only handed over to the central government a revenue of 1.5 billion which is far less than the contribution of Shanghai. At the beginning of economic reforms, Guangdong enjoyed a preferential treatment of the central government. However, Guangdong had not made greater contribution to the central government when it reached a stage of accelerated economic growth. The existence of unequal treatments among provinces by the central government has resulted in competitions, tensions, and separation among regions.

During the 1980s, the number of regions reliant on central subsidy had increased. However, because of the continued decline of the proportion of central revenue in both GNP and total national income, financial support from the central government had been much weakened. With the exception of Tibet and Qinghai, provinces that were financially deficient only received a financial aid of less than 10 yuan per capita in 1980. By 1990, most provinces on central subsidy only received an average financial aid of 10 yuan per capita except regions of minority ethnics which still received a subsidy of 12 to 60 yuan per capita (Table 4.6; Fig. 4.6; Fig. 4.7). As a result, most of the financial aid from the central government has actually been used for living expenses and has not contributed significantly to the promotion of genuine economic development.

Fig 4.6 Financial Surplus of Provincial Regions (1995)

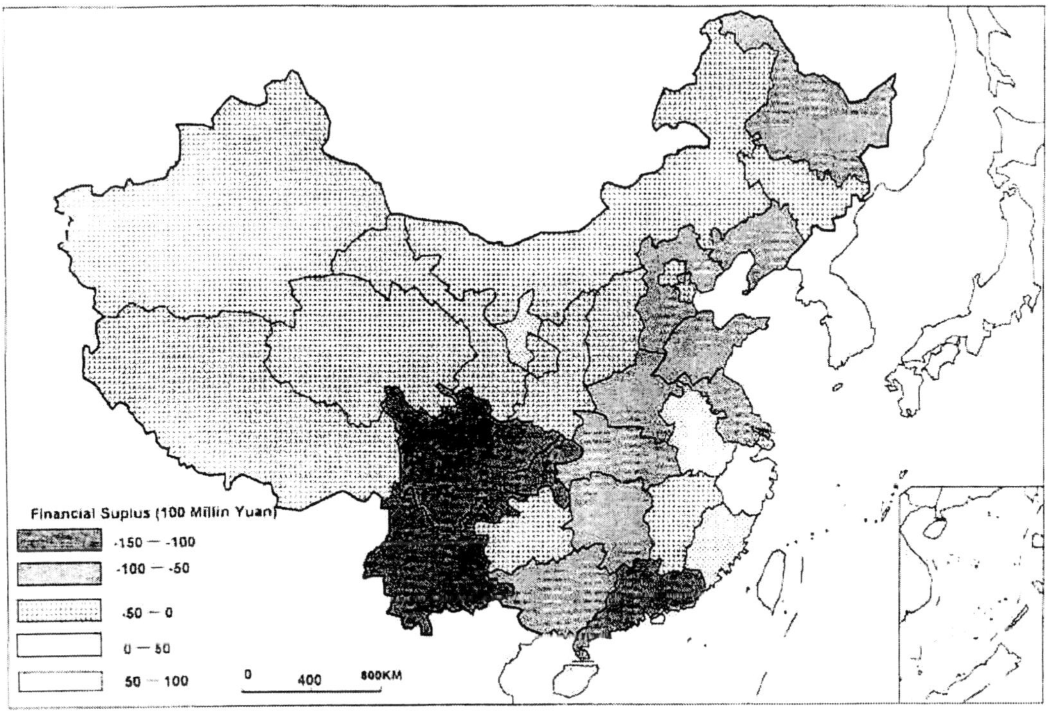

Fig 4.7 Per Capita Financial Surplus of Provincial Regions (1995)

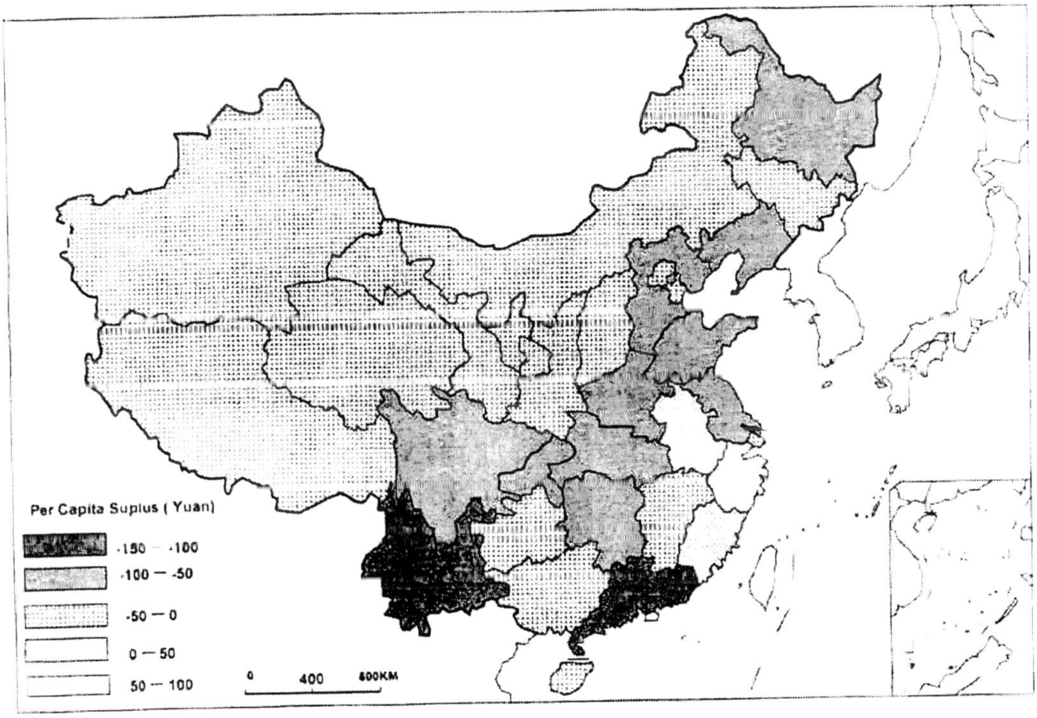

3. Tax Division Scheme and Its Implementation

3.1 Background

Although China's economic reforms had been successful, the reform of the financial system has never been satisfactory. The centre of financial reform since 1979 has been the separation of power between the central and local governments. The structure of decision-making has been changed from one that depends on vertical/departmental linkages into one dependent on horizontal/regional linkages. However, the essence of this economy remains to be reliant on administrative organization rather than free market forces. The result has been the growth of regional "feudal lords" to hinder the development of an integrated national market and weakened central capacity for macro level coordination. Thus, the country has run into the dilemma of relaxing control followed by chaos or tightening up control followed by the crashing of local enthusiasm.

In his visit of South China in 1992, Deng Xiaoping addressed the importance of developing a socialist market economy. This opened the door for the introduction of the tax division scheme. The Third Plenum of the Fourteen Central Committee of the CCP played a decisive role in the introduction of the tax separation system. The Resolution of the Central Committee of the CCP on A Number of Issues relating to the Building of a Socialist Market Economy, passed in the meeting, stipulated that: "...... Reforms of financial and taxation systems should be vigorously pursued. In the immediate future, attention should be directed first to the transformation of the existing financial responsibility system into a tax division scheme based upon a rational division of administration between the central and local governments so that the two systems of central taxation and local taxation could be developed. Central tax is defined as taxes required for the maintenance of central authority and performance of macro-level coordination. Central-local shared tax refers to taxes that are directly related to economic development. The amount of local taxes should be fine-tuned for the purpose of raising local tax revenue. The ratio of financial income in gross national income should be increased by quickening economic growth, improving profits, and expanding resources of fund generation. The proportion of central and local income should be determined rationally. A system of central refund and transfer payment should be developed......"

3.2 Resistance

Implementation of the tax division scheme did not occur without resistance. The system was designed in such a way that the central government will benefit from it in the short run and both the central government and backward regions will benefit from it in the long run. Resistance to the system has therefore came mainly from those regions that had benefited from previous financial arrangements.

Prior to the practice of the tax division scheme, local governments relied on tax exemption, retained financial surplus according to the responsibility system, and tax in exchange for the return of loan as the three main means for supporting local enterprises. These preferential treatments no longer exist in the tax division scheme. Furthermore, regions in the country were divided into areas of remittance and subsidy according to the amount of tax they handed in to the centre before. For those regions that used to pay a large amount of tax to the centre, they are classified as regions for remittance and are required to give a large amount of remittance despite the fact that their local financial retention was already very small. On the other hand, those rapidly growing economic regions whose previous tax base was small, they are classified as regions to receive subsidy although these regions have substantially increased their tax base and local retention as a result of rapid growth.

The implementation of the tax division scheme has significantly undermined financial capability of local governments. It has greatly raised the proportion of tax revenue in the total income of the central government to as high as 60 per cent. Those taxes that have great potential of expansion, such as circulation tax, are mostly included into the central revenue. On the other hand, taxes controlled by local governments are mostly unstable, piecemeal, or difficult to augment, such as agricultural tax and profit tax. The local extra-budgetary expenditure item no longer exists after the introduction of the new tax system. Many taxes that used to be entirely controlled by local governments are now shared by both the central and local governments. As a result of these changes, the financial control capability of local governments has been significantly reduced.

3.3 Principles and Procedures

The tax division scheme was introduced on 1 January 1994. The scheme rationally demarcates fiscal expenditure for governments of every level according to the division of authorities between the central and local governments. The principle was to combine administration with fiscal responsibility. All taxes were grouped into three categories, namely central tax, local tax, and tax shared by central and local governments. Two subsystems of taxation were established under the central and local administration to look after central and local taxation. The volume of local revenue and expenditure was scientifically determined to form the basis for taxation. A more standardized system of tax refund and transfer payment was also gradually developed. Moreover, a budgeting system was established and strengthened for governments at all levels.

3.4 Principle and Major Targets

The tax division scheme was designed according to a number of principles including:

a) Balanced centralization and decentralization, i.e. to ensure central financial capacity and allow local authorities for financial control.
b) Stimulating central and local incentives
c) Combined administration and financial responsibility
d) Revenue division between central and local governments and efficient tax collection. The division of tax control between central and local governments depends on whichever can collect the tax more efficiently.
e) Appropriate tax division according to the tax base. Taxes that have a larger base are collected by the central government whereas taxes whose bases are smaller are collected by local governments.
f) Fairness: division of taxes is to ensure a fair distribution of tax responsibilities
g) Maximized economic returns: tax division is determined in such a way that economic development is promoted and economic returns are maximized.

The objective of implementing the tax division scheme is to increase central financial capacity and raise the ratio of central revenue versus gross domestic product and the percentage of central revenue in total revenue. The major target is to increase the proportion of central revenue in total national revenue to 60 per cent. This target has been set in view of the historical trend which indicates that a 60 per cent central revenue is necessary to ensure national security and social stability.

3.5 Steps for Implementation

The tax division scheme was implemented in three steps: Initially, the new system is installed on the basis of the taxation year of 1993. Local financial capacity as of the year of 1993 remained unchanged. Then, the new and old systems were merged and fine-tuned to ensure a smooth transition. After a basic framework of deciding central-local relations was developed, vertical and horizontal linkages between the central and local governments were to be resolved. After the fiscal relationship between central and provincial governments is clarified, taxation reforms can then be carried out for governments below the provincial level. Finally, a mature tax division system was established and implemented after 1997.

3.6 Contents of the System

3.6.1 Division of Taxes Between Central and Local Governments According to their Administrations.

a. *Central fixed financial revenue* refers to revenues required to maintain national interests and practice macro control of the national economy, including custom duty, custom-collected consumption tax and value-added tax, consumption tax, central-government enterprise tax, income tax of local banks, foreign banks, and non-bank financial enterprises, revenue collected by railway departments, central branches of all banks, insurance companies (including operating tax, income tax, profits and urban construction maintenance tax), profits submitted by central-government enterprises etc. The refund of tax to foreign enterprises involved in export processing will be responsible by the central government with the exception of 1993 when local governments took up 20 per cent of the tax refund.
b. *Local fixed financial revenue* refers to taxes collected by local governments, including operating tax (excluding railway departments, central branches of banks, and insurance companies), income tax of local enterprises (excluding local banks, foreign banks, and non-banking financial corporations), profits handed in by local enterprises, personal income tax, urban land use tax, fixed asset capital investment direction tax, urban construction maintenance tax(excluding central branches of banks, railway departments, and insurance companies), estate tax, vehicle and shipping tax, animal slaughter tax, agricultural tax, cultivated land conversion tax, lease tax, capital gains tax on properties, and income from using state-owned land.
c. *Central-local shared revenue* refers to taxes directly related to economic development, including value-added tax, natural resource tax, and stock exchange tax. Value added tax is shared by central (75 per cent) and local governments (25 per cent). Natural resource taxes are divided between central and local governments according to the types of resources. Most resources are taxed by local governments with the exception of ocean petroleum which is taxed by the central state. Stock exchange tax is equally divided between the central and local governments (Yang, 1995).

3.6.2 Tax refund is Determined by Profits

To facilitate the gradual and progressive implementation of the tax reform, the central government introduced a special system of tax refund in 1994 as a transitional arrangement to maintain the original interest of local governments. The system of tax refund was based on the tax amount of 1993. The net local-to-central remittance was calculated according to the actual local revenue, profits and central-local tax division (i.e. consumption tax+75 per cent value-added tax-central budget allocation). The calculated net local-to-central remittance was returned to local governments to maintain the existing local financial capacity. This returned amount was used as the base for future tax refund. After 1994, tax refund to local governments has been gradually increased in 1:0.3 based on the increase of national value-added and consumption taxes. In the case when local remittance was below the 1993 base amount, central tax refund would be reduced. This arrangement suggests that, although the framework of tax division was brought in place in 1994, the essence of the previous tax responsibility system remained unchanged except the fact that the central government can now collect certain portion of the increased revenue from local governments to satisfy the financial need of macro-level economic coordination.

3.6.3 Division of Tax Collection is Made According to Administration

The collection of taxes was assigned to governments of different levels in accordance with the division of administration. Central taxes are collected by the State Council and its departments and central-local shared taxes are controlled according to the sources of revenue generation. Local taxes are collected by provincial governments and their departments. With the exception of Special Economic Zones and regions of ethnic minorities, all local governments are required to collect local taxes. Autonomous and minority regions can be granted special tax control authority (e.g. adjustment of taxation rate for national taxes and control of special taxes) provided that it does not affect central revenue collection and macro-level coordination and that it is approved by both the Standing Committee of the National People's Congress and the State Council. Special Economic Zones can also enjoy special authority in the collection and control of taxes.

3.7 Effects of Implementation

The tax division scheme represents a major breakthrough in the central-local relations since the founding of the People's Republic. The effect of implementation is well summarized by Liu Jibin, Deputy Minister of Finance, "Over the past 5 years we have achieved major breakthrough in fiscal and taxation reforms. A new circulating taxation system supported mainly by value-added tax and supplemented by consumption and business taxes has been developed. A tax refund system has been introduced and the central-local relations have been fine-tuned" (Xiao 1997). Some significant effects can be identified from the implementation of the tax division scheme in the first three years since 1994.

The percentage of fiscal revenue in GDP has stopped to decline. The two main goals of implementing the tax division scheme are to raise both the percentage of fiscal revenue in GDP and the ratio of central revenue to total national revenue so as to maintain a steady growth of national and central revenue (Zhou, 1997). The record of the first three years of implementation of the new tax scheme since 1994 has shown that there is an increase in the percentage of fiscal revenue in GDP. Total fiscal revenue reached 736.7 billion yuan in 1996, which accounted for 10.9 per cent of GDP for the same year. This was compared with 10.7 per cent of 1995.

The ratio of central revenue to total national revenue has increased first and then dropped. This ratio was 22 per cent in 1993 and was raised to 55.7 per cent in 1994 due to the new tax division scheme. However, it did not last long as the ratio dropped to 52.2 per cent in 1995 and 49.5 per cent in 1996 (Table 4.1). The reduction of the ratio of central revenue to total national revenue was compounded by a growing expenditure and the resulted deficit has to be balanced by bonds issued. The central government has therefore found it difficult to increase the transfer of payment to underdeveloped regions in the western interior.

Local governments have become increasingly reliant upon the support of the central government. In 1995, two year after the implementation of the new tax division scheme, only three provinces were able to become financially self-sufficient. The remaining 28 provinces and special municipalities had to rely on the central government for financial support. This transition is especially evident in the coastal region where many advanced provinces have turned from being relied by the central government into reliant upon the central government. This transition suggests that the central government is no longer dependent on advanced regions. On the contrary, it is now the advanced regions that have to ask for support from the central government. The capacity of the central government for macro-level control has therefore been greatly strengthened (Wang, 1995).

3.8 Problems to be Solved

Using the 1993 revenue as the base for determining tax refund has been far from being fair. It proliferates the previous unfair taxation situation and does not meet the requirements of the new tax division scheme. There are two major problems that need to be tackled. First, there is unfair allocation of central subsidy among the advanced and backward regions. Many advanced regions have got more subsidy than the backward regions. Second, per capita financial assistance offered by the central government has been poorly correlated with per capita GDP at the provincial level. Financial assistance offered in the means of transfer payment has not been standardized. One of the main objectives of the new tax division scheme is to establish an effective system of fiscal adjustment. More specifically, the central government coordinates the financial capacity of local governments through transfer payment and ensures funds necessary for the operation of local governments are available. However, the new tax division scheme thus far has shown that the system of transfer payment has not been well developed beyond the refund of taxes from central to local governments. As there is only a limited amount of fund to be returned from central to local governments, the function of the central government to effectively regulate regional development has therefore been very constrained. Moreover, the selection of the 1993 revenue as the base for tax refund has had some negative effects. The central government has given up a large portion of revenue to safeguard the existing financial capacity of some local governments. This arrangement was abused by some local governments that filed in false report on the base revenue. The consequence has been a widening of differences between economically advanced and backward regions.

The principle for implementing the tax division scheme has been to maintain existing local interests so as to avoid possible resistance to the new scheme. Indeed, to protect local interest has not only been a top priority of the central government but also an important task for which the central government has made significant compromise. However, the fiscal reform should be a rearrangement of central-local interests. It would be difficult to achieve the goals of the taxation reform without making fundamental rearrangement of economic interests between the central and local governments.

3.9 Causes and Shortcomings

There are several reasons explaining why the tax division scheme has not satisfactorily achieved its goals. First, major tax revenue collected by the central government was unable to increase in parallel to the growth of the national economy. Both the value-added tax and consumption tax were growing at a rate lower than that of economic growth. More importantly, tax revenue from state-owned enterprises, the second largest source for taxation, has expanded very slowly. Secondly, there was a growing burden for the central government for tax refund to support export activities. At the moment, tax refund for export has been the sole responsibility of the central government. The amount of tax refund for 1996 has increased by 30 billion yuan compared to that of 1995. This, and the rapid increase of tax refund for export have been the most important factors responsible for the reduction of central revenue. Thirdly, local taxes have been growing at a pace faster than taxes collected by the central government. Local taxes such as business operation tax, personal income tax, and non-state-owned enterprises tax have great potential of growth after the adoption of a new management system.

There are several shortcomings associated with the new tax division scheme. First, in the respect of the division of taxes, those collected by the central government are behind taxes collected by local governments in terms of elasticity or potential for growth. For instance, the centrally collected consumption tax was unable to expand in parallel to economic growth because it only applied to a limited number of commodities that are under the macro control of the state. By comparison, personal income tax and non-state-owned enterprise tax (e.g. township and village enterprises, privately owned enterprises, joint venture and export processing enterprises, etc.) all have great potential for tax generation after proper management has been adopted. Second, in the respect of tax collection and management, local taxes are usually collected more efficiently and sufficiently than central taxes because of the enforcement of local governments. On the other hand, many enterprise taxes are collected by local governments first and then passed onto the central government and they were not collected efficiently and effectively. Finally, regarding the distribution of tax revenue, the central government has been heavily burdened by tax refund for export for which local governments take no part.

4. CONCLUSION AND SUGGESTIONS

The fiscal relations between the central and local governments represent a sensitive area of central-local relations. It touches on the economic interests of both the central and local governments and has been one of the driving forces for regional development. Central-local relations have gone through a number of changes since the founding of the People's Republic, but the tax division scheme introduced in the early 1990s has been the most profound change. The above analysis can lead to the following concluding remarks:

a) Central-local relations were dominated by a highly centralized planned economy before the 1990s. There has been significant historical variation in central-local fiscal relations, but three major phases can be identified during the period from 1949 to 1990. Despite the historical variation, the central-local relations before the 1990s, whether the unified procurement and expenditure before the reforms or the financial responsibility system after the reforms, were characterized by a highly centralized administrative and economic system without rational and scientific design for revenue collection and expenditure.

b) The financial responsibility system is not a good system to last for long because of the many problems it has caused. The financial responsibility system was a product of economic reforms. It contributed to regional economic development during the first decade of reforms. The system consolidated local interest and motivated local enthusiasm for economic development. However, the system was a temporary and transitional arrangement and does not meet the requirement of rational resource arrangement according to free market forces. It is a system that shifted the emphasis from vertical to horizontal linkages without changing the nature of the central-local relations. The system resulted in continued reduction of the percentage of central revenue in the total national revenue, weakening of the central capacity for macro economic control, and produced a financially weak central government. On the other hand, the decision-making power of local governments was greatly augmented and local interests were strengthened. Local protectionism was exacerbated and development of a uniform national market was hindered. Regional development has thus suffered from unnecessary duplication of economic structures among regions. There was also an uneven distribution among local governments in terms of their financial contribution to the central government and a subsequent widening of regional differences.

c) The implementation of the tax division scheme has shown that the ratio of fiscal revenue to gross domestic product has stopped to decline, the percentage of central revenue in total national revenue has raised, and the central function for macro-economic control has been strengthened. Moreover, the tax division scheme has increased the dependence of local governments on the central government. Twenty-eight out of the thirty-one provinces and special municipalities in the country have become deficient in local revenue. Even economically advanced provinces in the coastal zone had to rely upon financial support of the central government. The consequence has been weakening of local interests, which should be beneficial to the development of a standardized system of transfer payment to reduce regional inequality.

d) The mechanism for tax refund under the tax division scheme requires further improvements. The tax division scheme has been a major breakthrough in dealing with central-local relations. Although significant progress has been made, the new system still faces many problems such as tax classification, tax collection and management, and redistribution of tax revenue. In other words, the tax division scheme has resolved the issues of how and from whom taxes should be collected, but it has shortcoming in tax refund and redistribution of tax collected. The base amount for tax refund is not standardized. There is an unfair allocation of central financial subsidy between the economically advanced and backward regions. Per capita central subsidy to provinces is poorly correlated with per capita gross domestic product. The transfer payment system needs further improvements (Wang, 1996; Liu, 1996).

China is currently in a transitional stage moving from a planned economy to a market economy. The tax division scheme represents a major step to reorganize central-local relations according to the requirements of a free market economy. However, several important issues need to be tackled.

4.1 Tax Classification should be Adjusted According to Administration

More need to be done to further increase the percentage of central revenue in the total national revenue. Collection of enterprise tax should not be determined simply by administration. As the market mechanism develops, there is a growing demand for inter-regional and inter-enterprise horizontal exchange. As a result, to collect enterprise tax according to vertical administrative relation will become increasingly problematic. Enterprise tax should therefore be changed from a local tax to a central-local shared tax so that the relationship between local government and enterprises be weakened, cross-enterprise exchange be facilitated, and a unified national market be eventually formed.

Land use tax should be changed from a locally collected tax into a central-local shared tax. Land is constitutionally owned by the state. To reclassify land use tax as a central-local shared tax will facilitate central control over the massive loss of cultivated land. The percentage of central and local tax division could be determined in a way that local interests will be properly maintained.

Personal income tax should be changed into a central-local shared tax. In most of the developed countries, personal income tax is collected either solely by the central government or jointly be both the central and local governments. Personal income tax has great growth potential and will be an important source for raising fiscal revenue by the central government. The tax collection and remittance function of local governments should be strengthened by effective measures such as linking tax refund with tax remittance.

4.2 Establish a Scientific and Standardized System for Regional Transfer Payment

The existing temporary arrangement of tax refund based on the 1993 revenue figure should be abandoned. When determining the base amount for taxation, the standard should be set on the basis of objective calculation of regional capacity for revenue generation and expenditure rather than on the basis of the revenue amount of a particular year. All provinces, autonomous regions, and special municipalities should be ranked in the order of their gross domestic product and they should be treated equally in terms of tax collection and refund. Tax division should be determined after a clear definition of function and administration for governments of various levels so that the central function for macro-level economic control be strengthened and differences in per capita income be reduced.

4.3 The Power for Tax Legislation should be Delineated According to Rational Division of Administration Between the Central and Local Governments

With a vast territory, China is characterized by an uneven regional development and diversity in economic resources. To have one single set of tax legislation will not suit various specific local situations. As the implementation of the tax division scheme proceeds, there is a need for decentralizing certain power for tax legislation to local governments so that some special taxes be collected to enlarge the local tax revenue. This will not only facilitate the utilization of local comparative advantages but also speed up development of the local taxation system. The Chinese taxation system can also be brought up to a level compatible with international standards. The definition of tax legislation power should be determined by China's supreme power organs, The National People's Congress and its Standing Committee. Specifically, the division of tax legislation power between the central and local governments could be done in the following manner: (1) The National People's Congress should be responsible for the legislation of taxes that apply to the entire nation. (2) People's congresses at the provincial level could have legislation power for local taxes outside nation-wide taxes on the condition that such legislation does not contradict with national legislation, that it does not affect remittance of the central revenue, and that it does not hinder the formation of a unified national market. Local tax legislation should be filed with the National People's Congress before implementation. (3) Local tax legislation should be controlled by provincial governments and should not be further decentralized to any lower level governments. Such tax legislation could be applied to the entire province or to certain regions within the province. Thus the present disparity of regions and their vitality for economic growth awaits major amendments and improvements in the taxation system as one important dynamic for positive change.

References

1. He Zhengyi. 1994. Prospect of the effect of three major actions in tax reforms. *Reforms,* 2 (Chinese text).
2. Hu Angang. 1995. *China Regional Difference Report.* Liaoning Press (Chinese text).
3. Hu Angang. 1996. The tax division scheme: assessment and suggestions. *Strategy and Management,* 5 (Chinese text).
4. Liu Yongchang. 1996. A perspective of reconstructing China's inter-governmental transfer payment system. *The Management World,* 4 (Chinese text).
5. Ning Xueping. 1994. An assessment of the tax reform proposal. *Reforms*, 2 (Chinese text).
6. Sheng Sianyang. 1995. *Feudal lords in a great nation: China's central-local relations.* Beijing: China Society Press (Chinese text).
7. Wang Chingyun. 1996. A study of national regional policy. *Economic Issues,* 4 (Chinese text).
8. Wang Hongyang. 1996. The tax division scheme: more needs to be done on tax refund. *Economic Reference Daily.* August 22 (Chinese text).
9. Wang Jingsong. 1995. *Government and Politics of the People's Republic of China.* Beijing: CCP Central Committee University Press (Chinese text).
10. Wei Wei. 1995. Regional differences and regional coordination in China's economic development. Anhui People's Press (Chinese text).

11. Wen Wu and Wang Yongzhu. (eds). 1995. *Major issues on the transition of China's taxation system in the 1990s*. Beijing: China Statistical Press (Chinese text).
12. Xiao Yi. 1997. Major progress in macro-level control in China. *Guangming Daily*, September 1 (Chinese text).
13. Yang Shiuqing. (ed.). 1995. *National Taxation*. Beijing: China People University Press (Chinese text).
14. Yang Weiming. 1992. An empirical study of changes in regional income disparity. *Economic Study*, 1 (Chinese text).
15. Zhou Fuxiang. 1997. A dynamic analysis of two ratios associated with fiscal revenue. *Macro Economic Management*, 6, p. 20 (Chinese text).

Chapter 5

REGIONAL ECONOMIC AND TECHNOLOGICAL COOPERATION

PANG Xiaomin

Associate Professor, Institute of Geography,
Chinese Academy of Science

Becky P.Y.LOO

Assistant Professor, Department of Geography,
University of Hong Kong

1. BACKGROUND

In the 1980s, rapid economic growth of the country was accompanied by widening regional disparity. Problems of uneven regional endowment have become increasingly apparent with the deepening of reforms. In particular, these problems are manifested in the following aspects:

1) Different geographical advantages, especially in terms of relative location and economic linkages.
2) Widening disparity in the demand and supply of regional resources, technological equipment and human resources.
3) Lack of capital funds for promoting economic development at the local level.
4) Dramatic increase in the demand for grain, food and industrial crops in the coastal region where industrialization and urbanization have been taking place rapidly.
5) Reliance of inland provinces on the "windows" of the coastal region to participate in the international economy.
6) Pressure for industrial relocation from the coastal to inland region. In particular, the enormous pressure that large industrial and urban conglomerates imposed on the environment has aroused much concern.

On the one hand, these problems can be seen as conflicts of interest among the localities. On the other hand, they may be considered as golden opportunities for regional economic and technological cooperation. With limited regional cooperation, local protectionism had become one of the outcomes in the 1980s. From 1980 to 1982, many local governments promulgated discriminatory measures to protect local "infant" industries, particularly processing industries. From 1985 to 1988, the rapid expansion of processing industries in certain areas had led to keen competition for raw materials, like tobacco, wool and silk. Beginning from the latter half of 1989, the slackening of national economic growth had led many local governments to introduce a series of measures to protect the local market for local producers (Wei, 1995).

Towards the 1990s, it became increasingly clear that reliance on governments at upper levels to give policy and financial supports was no longer sufficient to ensure satisfactory economic performance. The search for capital, market, raw materials, equipment, human resources and information has to rely increasingly on regional cooperation. Thus, both the central and local governments were trying to formulate and implement a system for fostering regional co-operation and establishing regional economic integration policies and measures to suit local circumstances.

However, what are the best ways of achieving regional co-operation in China? There is still no consensus. Besides, the fundamental rationale and policy objectives of regional economic and technological cooperation have been challenged on many grounds. The key issue hinges on the "impossibility" of regional economic and technological co-operation to achieve the tasks of increasing market efficiency and simultaneously promoting balanced regional development. According to western theories, market forces will lead to widening regional disparity by increasing the concentration of labour, capital and goods in the more-developed areas, at least at the early stage of development (Williamson, 1965). These negative impacts, or backwash effects, of the more-developed regions would outweigh the positive impacts, or spread effects, on the less-developed regions until some complementarity has developed between the two areas, usually only at a late stage of development. During the early stage of development, polarization seems to be the rule as the initial advantage of the former becomes more firmly established. In view of such trend of regional divergence, Tan (1996) insists on the need for central government intervention to implement regional co-ordination so as to achieve more balanced regional development, in particular, central fiscal transfer should be used to redress regional disparity. However, such a recommendation is difficult to implement, at least in the near future, as the financial ability of China's central government has been declining under the Open Policy (Wu, 1996). Quite differently, Lu (1996) has attributed the problems of regional protectionism to the fundamental conflict of interests among the regions in the transition to market socialism. According to this view, the solution lies in the setting up of an administrative system compatible with the market economy.

Against this background, this Chapter analyzes the major policies of regional economic and technological co-operation and their outcomes since the Eighth Five-Year Plan (FYP). The 2 basic aims are to provide a basis for assessing the rationale and effectiveness of regional policies and measures and to gain deeper insights into regional differences by examining regional relationships. In comparing the targets, goals and development conditions at different levels of government, we can obtain a clearer idea on the balance of power between the central and local governments. Then, we would proceed to examine the rationale, feasibility and effectiveness of government policies to promote regional economic and technological cooperation.

2. National Policy Framework

In the 1980s, regional economic and technological cooperation has targeted at resolving problems of departmentalism and promoting regional development. Nonetheless, the successfulness of regional policies was doubtful. For instance, Li (1996) has pointed out that the entrenched problems of departmentalism and the tendency for localities to build "large and complete, small but complete" (self-contained) production systems have persisted in China despite policies to encourage regional co-operation. Since the Eighth FYP, the central government has adjusted its regional policy in two major aspects. Firstly, it has gradually shifted the major foci of its preferential policies from the coastal areas to the Central and the West. These policies were similar to those implemented by the United States Federal Government to redress regional disparity in the 1960s. Secondly, the central government has strengthened guidance for closer regional relationships, and has formulated policies to encourage regional co-operation to solve local problems. In the "Outline of the Ninth FYP and Future Outlook in 2010" promulgated on 17 March 1996, a regional co-ordinated development strategy was, for the first time, put forward as an official direction. Specifically, it demanded attention to 1) the relationship between overall national economic performance and regional economic development; 2) the formation of regional (inter-provincial/regional/municipal) economies; and 3) the relationships among different regions.

In retrospect, the aim of the central government in promoting regional economic and technological cooperation has been changed in the 1990s. Regional cooperation has since been seen as a way of achieving national economic development goals and contributing towards national political and social stability. In the process of deepening reforms, the aims of breaking away from regional and departmental "fragmentation", creating a unified market for national production factors and making use of market mechanisms in resource allocation, building a national economic system of high efficiency and international competitiveness were put forward (IMF 1994). Regional co-operation can be seen as a foundation for establishing a national market and to promote the free movement of commodities, materials, technology, information, labour and capital. In the Ninth FYP, there were official calls to eliminate artificial barriers for the formation of a national unified market.

However, economic efficiency is still the top priority. The resources which the central government can put in the West without taking money from and dampening the economic vitality in the East are very limited. Thus, the central government has to rely on the initiatives and resources of local governments to promote regional cooperation and to maximize the triggering-off effects of the more- to the less-developed areas. The moves of the central government to achieve "common prosperity" were still mainly restricted to the following 4 major aspects, which did not require much financial commitments.

2.1 Major Directions and Principles of Opening Internally

In 1994, the "Regulation of the Central Committee of the Chinese Communist Party (CCP) on Economic System Reforms" has emphasized the need for opening not only to the outside world but also internally among different regions. Chairman Jiang Jimen has expounded on this personally by stressing the need of devising more positive policies and measures to allure capital to invest in the Central and the West. Furthermore, the principle of "relying on the market for directions, enterprises as the main agents" was established. In other words, this new round of economic cooperation since the mid-1990s would follow the socialist market economic principles and modern enterprise system. In addition, the major emphasis on regional cooperation has shifted from the areas of economic production, trade and technological cooperation to the new areas of structural adjustments and market expansion. The "healthy" enterprises in the East were also encouraged to take over enterprises in the Central and the West. In the "Report of the Fourteenth National Congress of the CCP", the regional cooperation principle of "mutual benefits, complementarity and common development" was proposed.

2.2 Planning and Project Plans

Regional restructuring and the establishment of regional markets were embodied into the national planning framework. In the "Outline of the Ten-Year Plan and the Eighth FYP", there were proposals to develop different types of regional cooperation in different areas and at different levels. The need for speeding up the formation and development of a unified market was highlighted. In the Ninth FYP, it was proposed that the process of devising a strategy for regional industrial relocation should be speeded up so that some of the raw material oriented processing and labour-intensive industries in the East would gradually relocate to the West.

The "Cotton Industry Relocation Project" is representative of this industrial relocation policy. Since the Eighth FYP, the government has already put forward plans to adjust the spatial distribution of traditional industries, such as the cotton industry. In 1991, the State Council has decided to make structural adjustment to the cotton and weaving industry first by gradually shifting the production capacities to the West, where most of the important raw materials and labour come from. In the major cotton producing areas, the strategy is to depend on the advantage of low-cost cotton to establish a new base for the cotton and weaving industry. The goal of relocating at least 5 million weaving machines to the West was set. In early 1994, a national decision to start the relocation of 500 weaving machines from Shanghai to Xinjiang in 1995 was made. In March 1995, the State Council has further given directions to shift the cotton processing industry from coastal cities to cotton producing areas and confirmed that the number of weaving machines in major cities of the coastal areas should be reduced by another 3.4 million from 1997 to 1998.

In early 1994, the Agricultural Ministry also tried to seek approval of the State Council to begin "exemplary" projects for East-West cooperation among rural enterprises. The major aim was to encourage successful rural enterprises in the Eastern coastal region to invest and enter into joint projects with enterprises in the Central and the West.

Since the Sixth FYP (1981-1985), the "coastal-inland" and 'East-Central-West" divisions have been used. In the Spring of 1992, the State Council has decided to begin the planning of another official regional division -- the 7 macro-regions. Up till 1996, the State Council has already called upon meetings in all but one (the central economic region) of these 7 macro-regions. The basic aim of the new division was to understand and firmly establish the advantages and development potentials of different regions and to propose plans and measures for regional economic integration and to undertake major inter-provincial projects on energy, transport, communications, finance, water supply and information services.

2.3 Assignment of Administrative Responsibilities

With the ultimate aim of eliminating poverty in China by the end of the Century, the central government has strengthened works in providing financial assistance and mobilized the economically more developed regions to help the less-developed ones. Two major programmes were launched. The first one aims at providing "matching support" to resettle residents affected by the Three Gorges Projects. The second one aims at giving "matching support" for poverty relief. Under these programmes, the State Council has assigned most of the more-developed provinces/municipalities to help specific less-developed provinces/municipalities. At the same time, there was emphasis on encouraging domestic migration to open up new areas and a shift of policy direction to help the poor by providing them with financial subsidies to assisting them to create their own wealth.

2.4 Special Government Subsidies and Services

During the Eighth FYP, the central government has established a limited co-operation fund of approximately 100 million yuan to support major projects of regional cooperation. However, it was stipulated that the funds would cease to exist after the Plan. The central government has also provided administrative and managerial support to regional cooperation through the State Planning Commission, State Economic Commission and other offices of the State Council. For instance, the State Economic and Trade Commission and Yunnan Government have held a working conference on the preparation of the "First Meeting on Cooperation among Enterprises in the East, Central and West". In April 1997, the State Council Special Economic Zone Office and the Governments of Jiangsu, Shanghai, Tianjin and Shaanxi had organized the "97 Conference on Cooperation, Investment and Trade between the East and the West".

3. Cooperation Strategy and Policy Measures Among Local Governments

Since the 1980s, the central government has gradually devolved its decision-making power to local governments. Provincial governments have become an increasingly independent economic agent of investment and production. Hence, local government policies and economic system can directly affect the development directions and models of regional economic development.

3.1 Main Aims and Motives

Generally, the attitudes of local government have become increasingly positive and less divergent about regional cooperation. In practice, there is a tendency for local government to maximize local interests and guard against potential regional competition. They have to strike a balance between local interest and the collective interest of cooperation by developing a regional cooperation strategy. The different strategies that have been proposed by local governments can be divided into the following major categories:

3.1.1 Attract Business and Investment

Moving into the late 1990s, the competition for domestic capital and market has become increasingly keen among regions in China. Since 1993, government loans have been relatively tightly controlled. The ability of different localities to meet the demand for capital by local enterprises has differed considerably. Thus, there is an urgent need for some local governments to overcome shortfalls in local capital. Under such circumstances, some local governments are having high expectation on the possibility of attracting and utilizing outside capital through regional cooperation. This aim was usually pursued together with an attempt to open the domestic market and enlarge local market share. Such strategy has been widely adopted by provinces/municipalities in the Central and the West.

3.1.2 Capitalize on Local Comparative Advantage

In many provinces and municipalities, regional cooperation was seen as a strategy to capitalize on the local comparative advantage. The major emphasis that each region placed on its own comparative advantage is different and has changed under the dynamic situation of the Open Policy. The emphases and changes are summarized in Table 5.1.

Table 5.1 A Typology of Local Government Goals for Developing Regional Comparative Advantages

Type	Local Government Goals
Locational Advantages	*Coastal Locations* • Well-developed information network • Close linkages with the international market. *Central Locations* Include Anhui and Hunan • "Middleman" between the East and the West *Border Locations* Include Heilongjiang and Yunnan • Development of border trade *Others* • Proximity to major developed regions/provinces, such as the circum-Bohai rim, Shanghai-Jiangsu-Zhejiang region and Guangdong
Policy Advantages	*East* • Earlier reforms in trade system, financial system and property rights *Central and West* • Recent government policy of strengthening East-West cooperation
Natural Resource Advantages	*Resource-Rich Areas* Include Shanxi, Inner Mongolia, Xinjiang, Gansu, Ningxia, Tibet and Guangxi • Inter-provincial cooperation among Heilongjiang, Anhui, Hebei and Guangxi to exploit natural resources together
Other Production Factor Advantages	*Complementarity in Capital, Technology and Human Resource* • Shanghai's recent slogan of "using capital to exchange for market"

3.1.3 Exploit the Benefits of Large-Scale Regional Corporations

Large-scale regional corporations can be a major source of benefits for local governments. The major advantages for cross-region corporations can be generalized into three major aspects. Firstly, cross-regional corporations can fully utilize the various types of sectoral and locational preferential policies of the central government towards different areas. A case in point is the proposal put forward in the "Economic Co-ordination Meeting of Central Cities along Changjiang" to make use of the preferential policies for the exploitation of Pudong Open Region and to receive "matching support" for resettlement out of the Three Gorges catchment area. Similarly, regional cooperation among Zhejiang, Guangdong and Jiangxi was made on the premise of fully utilizing different preferential policies offered to different areas ranging from the Special Economic Zones to the interior areas. Secondly, large-scale regional corporations can help to create a regional image and extend regional influence. The aim is to increase the recognition and social influence of the region so as to encourage more emphasis and support by the central government to improve the regional investment environment. Lastly, regional corporations allow tasks which cannot be done by one municipality or province alone to be achieved. This type of cooperation can help to work towards complementarity and spatial division of labour by partnership. Within this broad framework, developing an international economy by exploiting regional comparative advantages and strengthening the co-ordination of ports and hinterlands are 2 major targets. For instance, the urban conglomerate in central Liaoning have made use of its locational advantages of facing northeast Asia, the West-Fujian-South-Jiangxi-Southeast-Guangdong Economic Region has developed its locational advantage of facing Taiwan and proximity to Hong Kong and Macau, and the Huaihai Economic Region has emphasized the "New Asian and Europe Land Bridge" and the role of Lianyungang Port as a "window".

3.1.4 Follow Central Policy to Support the Less-Developed Regions and Achieve Regional Balanced Development

Since the 1990s, many local governments have put much emphasis on supporting the Three Gorges Projects and for "matching support" for poverty relief programmes. So far, 20 provinces (municipalities), 10 large- and medium-sized cities and more than 50 central ministries have participated in the resettlement works related to the "Three Gorges Projects". The basic pattern that one province (municipality) and one large- to medium-city would support one county (city/region) in the catchment area of the Three Gorges is formed. Under the national policy of East-West co-operation and the strategy of "matching support" for poverty relief, the Central and the West have become more open and cooperative towards the East.

Besides, there is also increasing awareness of the widening gap within the coastal provinces. For instance, northern Jiangsu, western Fujian and western Guangdong are increasingly lagged behind when compared with their more developed counterparts in the same province. In 1993, Xiamen, Zhangzhou and Quanzhou were assigned to help the Longyan area and Sanming municipality. Starting from 1995, 5 municipalities and regions in southwest Fujian have jointly proposed the "cooperation between the hilly and coastal areas, economic diffusion, complementary advantages and common development".

3.2 Major Policy Measures

Starting from the 1990s, the concept of using enterprise as the major agent for regional economic and technological cooperation has been widely accepted. Individual provinces, municipalities and regions have reoriented towards stronger macro-regional policy and guidance, improvement of policy environment and creation of conditions and services to foster enterprise cooperation, as well as the establishment of a foundation for regularizing and formalizing regional cooperation. The major aspects of the policy measures are as follows:

3.2.1 Planning and Management

When compared to the Eighth FYP, individual provinces and municipalities have put more efforts on regional economic cooperation in their development strategies. Tianjin, Hebei, Anhui and other provinces and municipalities have studied special planning topics on economic cooperation. Some regional organizations had gradually implemented regional economic planning through the setting up of information exchange and enquiry systems. Some of the examples included the Nanjing Regional Economic Co-ordination Meeting, Wuhan Economic Cooperation Region, Chongqing Economic Region, Huaihai Economic Region, Ecumenical China Economic Region, Southeast-Fujian Economic Region and Southwest-Fujian-North-Jiangxi-East-Guangdong Economic Region.

In recent years, the relatively more developed provinces and some Central provinces have reached a series of multi-cooperation agreements on economic and technological cooperation, e.g. between Tianjin and Hebei and between Shanghai and Anhui. Hebei province alone signed more than 30 mutual cooperation agreements. In early 1997, the State Council has stipulated that all provinces which were assigned to participate in the "matching support" for poverty relief programmes should enter into written agreements to establish the leading concepts and principles of poverty relief by matching means and economic cooperation.

3.2.2 Formulate Relevant Preferential Policies and Implement Supportive Measures

Apart from preferential tax terms to encourage horizontal cooperation, there were also policies to help to handle the household registration of non-local people and the education of their children. In the East, many of the preferential policies for regional co-operation or help were related to the "Three Gorges Project". Since 1995, many provincial governments in the Central and the West have introduced new preferential policies, including simplified procedures, tax reduction, opening of local market and other preferential terms for the use of their land and facilities, to the East.

Besides, the rules and regulations for regional cooperation or support have also shifted from broad sectoral guidance to relatively specific projects and enterprise co-operation proposals. Many regional cooperation organizations have formulated and implemented regional preferential policies and measures to promote the free movement of production factors across space. In the Central and the West, the aim of "attracting business and investment" has been obvious. The Fujian-Zhejiang-Jiangxi-Sichuan Nine Parties Economic Region is one of such organizations. The preferential policies introduced by this organization from 1986 to 1994 included the followings:

- Co-ordination of tax, commodity prices and loans among the nine municipalities with their neighbouring provinces
- Formulation of "Certain Opinions on Trade among the Nine Municipalities"
- Priority in allocating stalls in regional trade and exhibition events
- Formulation of flexible price-setting policies by relevant price bureaux to give due considerations to prices in neighbouring provinces
- Formulation and implementation of tax reduction methods by relevant tax bureaux
- Setting up of an industrial and commercial system to implement the policies of "centralizing capital, allocation by branches, offering interest rate preferences and allowing easy transfer" among members so as to ensure higher economic efficiency and larger impacts.

3.2.3 Set up Exclusive Support Funds

Some provincial governments have put aside limited funds for regional cooperation. In 1995, Yunnan province has designated 30 million yuan for supporting joint projects across provinces. Some other funds set up in the East and more developed provinces were used to subsidize the less developed areas. Up to 1996, the provincial government and 13 municipal governments in Guangdong have allocated 90 million yuan out of the local budget for such purposes. In Jiangsu, Shanghai, Shandong and Zhejiang, the amounts have also reached 30 to 40 million yuan a year.

Apart from the above measures, some provinces have further moved towards the new areas of official training, recruitment, building of schools and mitigation works against natural disasters. The measures introduced by Shenzhen to northern Guangdong in this respect are notable. These measures include:

- The setting up of economic cooperation funds.
- Starting from August 1990, Shenzhen municipality has chosen the hilly areas in North Guangdong to begin its works in the "matching support" for poverty relief programme. By the end of 1995, the accumulated amount of funds used in "matching support" has reached more than 500 million yuan. The total number of subsidized projects in Meizhou, Heyuan, Shaoguan and Qingyuan in Guangdong, Aba (Ngawa) Zangzu Autonomous State in Sichuan, Jinggangshan in Jiangxi, Yenan in Shaanxi and Wan Xian, Wushan and Badong in the catchment areas of the Three Gorges has reached over 40.

- Starting from 1993, it has organized 18 inter-regional training classes, and more than 2,000 officials from the interior have been trained.
- It was assigned by Guangdong to help Taipu, Wuhua, Fushun and Zijin counties and by the national government to help Guizhou in the "matching support" for poverty relief programme.
- It recruited more than 1.5 million labour from the Central and West. Among them, 0.8 million came from poor areas.
- It helped to build more than 142 primary schools in 11 regions and helped more than 0.64 million children to continue education.
- It also helped to fight natural disasters in these areas.
- Lastly, it was to act as a "window" of technology, culture and information for the poorer provinces.

3.3 Major Goals and Forms

In terms of organizations, there are 3 major forms of regional economic and technological cooperation. They are inter-provincial organizations, self-initiated cooperation organizations within a region and those set up under the "matching support" for the poverty relief programme.

3.3.1 Preferences for Regional Cooperation Partners

The preferences for partners in regional cooperation have been mentioned by many provincial governments during the Eighth and Ninth FYPs. Table 5.2 summarizes the major emphasis of individual provinces on regional cooperation.

Table 5.2 Strategies for Regional Cooperation in Selected Provinces and Municipalities

	Major Economic Cooperation Strategies Put Forward
Tianjin	*1994* • Put more emphasis on Beijing-Tianjin-Hebei and circum-Bohai economic cooperation • Strategy of "importing from the West and exporting to the East"
Hebei	*1996* • Put emphasis on the circum-Beijing-Tianjin and Bohai area and opening internally to all provinces *1997* • Began to put emphasis on strengthening relationships with the more developed areas, like Shanghai
Liaoning	• Put emphasis on strengthening regional relationship with the Central and the West, such as Inner Mongolia, Xinjiang and Shaanxi • Fully utilise the national policies for the Central and the West *1997* • Shenyang Municipality put forward the policies of strengthening cooperation with the southeastern coastal areas
Shandong	• Put emphasis on "Going North" (i.e. Northeast) and "Entering the West" (i.e. Southwest and Northwest)

Shanghai	*1995* • Develop horizontal economic co-ordination with the Changjiang Delta and of the provinces and municipalities along Changjiang • Cooperate with other important economic central cities and regions • Cooperate with the Central and the West, especially provinces and/or municipalities which the State Council has assigned links under the "matching support" programmes
Fujian	*1996* • Put forward the "along the coast, river, border and transport line" development strategy
Guangdong	*1995* • Strengthen cooperation in resource exploitation and development in the Northeast and Northwest • Strengthen market and material linkages with the East and Central-South *1997* • Emphasis on resource-rich provinces of Hebei, Jilin, Heilongjiang, Guangxi, Sichuan and Gansu • Open the entire national market
Heilongjiang	*1995* • Further tightening cooperation and co-ordination with southeastern coastal provinces and municipalities • Fully utilise the capital, technology, labour and product advantage • Relocation of rural and private enterprises • Invite more-developed provinces and municipalities of the southeastern coast to join the "Harbin Conference Meeting"
Shanxi	*1995* • "Attract from the East, form alliance with the West, face the entire country" • Actively promote horizontal economic cooperation and co-ordination with the coastal region
Hunan	*1992* • Emphasis on the coastal region and fully develop inter-provincial cooperation and co-ordination • "Attract from the East, form alliance with the West, export to the South and import from the North" • Emphasis on cooperation with provinces and municipalities along Changjiang River and in the coastal region • Develop linkages with Yunnan, Guizhou and Sichuan • Exploit the market in the Southwest
Anhui	*1992* • Emphasis on co-ordination with Shanghai and the coastal provinces and municipalities • "Attract from the East, form alliance with the West, encourage north-south interflows" *1997* • Provide "matching support" in the Three Gorges Projects

	• Develop from the lower to the upper and middle courses of Changjiang River, along the sea, the river and co-operate with the Central and the West • Set up economic co-ordination office in Beijing, Shanghai, Shenzhen, Zhuhai, Hehe and Manzhouli • Organize 319 2-way interprovincial visits, develop and establish 324 pairs of friendly cities (regions)
Jiangxi	*1995* • Put forward the proposal of relying on Beijing-Jiujiang-Kowloon Railways • Emphasis on attracting investment from the more developed coastal region
Sichuan	*1996* • Coastal developed provinces as major foci for opening internally • Foster East-West cooperation by the formation of Sichuan-Shanghai alliance and Changjiang production belt, the provision of "matching support" to the catchment areas of the "Three Gorges" by the eastern provinces and to "selected local areas" by the southeastern provinces • Launch East-West, Sino-foreign and West-West cooperation
Guizhou	*1996* • Emphasize the construction of external transport links • Fortify entrepreneurial links with Shanghai, Jiangsu, Fujian, Guangdong and Shandong • Establish more than 200 production, sales, technology and information counters in more than 20 coastal provinces and municipalities
Yunnan	*1995* • Emphasize coastal region as a focus of domestic cooperation • Put forward Shanghai, Guangdong and Shandong as the foci *1996* • Put forward the continual and strong development of the south-western provinces for economic co-ordination and cooperation • Try to expand the co-ordination and cooperation with Guizhou and Shanghai • Establish offices and enterprises in areas like Shanghai, Beijing and Guangdong
Ningxia	*1996* • Put forward the need for strengthening interchange and co-ordination with Shanghai, Zhejiang, Jiangsu, Fujian and Qingdao • Set up offices in places like Shenzhen, Shanghai, Hangzhou, Beijing, Xian and Xiamen

Based on the information, we may generalize a number of directions. (1) Coastal-inland cooperation seems to be the most important direction for regional links while the East-West cooperation strategy put forward by the central government has also been considered as the key policy direction. (2) Apart from East-West cooperation based on the official East-Central-West divisions, another emerging direction of cooperation is between the North and the South. Regional differences between the two have been particularly important in the East. Besides, the new trend for the Central to make use of its locational characteristics to foster cooperation both with the East, mainly for absorbing capital to exploit local resources and industrialization, and with the West, mainly for opening up market. For instance, Anhui has adopted the strategy of "competing with the East for market in the West and competing with the West for capital in the East". (3) Geographical proximity was not a major consideration for regional cooperation at this stage. The 7 macro-regions proposed by the government are not yet having important impact on the choice of partners in regional cooperation, except among the coastal provinces, such as Shanghai, Hebei, Beijing and Tianjin. And, finally, (5) the national investment strategy did have an impact on the choice of partners by local governments. The most important example was the emphasis on the development axis along the Beijing-Jiujiang-Kowloon Railway.

3.3.2 Efforts to "Attract Business and Investment"

The major ways of speeding up capital inflows into a province/municipality are either to attract capital from other provinces or to utilize foreign capital jointly with other provinces. Since the Open Policy, the latter has become more popular. Since 1992, many inland provinces, especially those along Changjiang, are also allowed to implement preferential policies to woo foreign investors and export-oriented enterprises. There is a common hope that regional cooperation can raise the area's competitiveness in attracting foreign capital.

At present, the utilisation of "outside" capital in the Central and West has mainly remained at the stage of utilizing capital domestically, especially those from the East. In fact, many northern coastal provinces, including Liaoning, Hebei and Tianjin, have also emphasized the benefit from the relocation of capital and industries from southeastern provinces. This is the so-called "Relocation of southern capital to the north" and "economic invasion of the north". Cooperation with the coastal southeastern more-developed provinces has been a major means for inland provinces to raise capital.

On the means to attract foreign investment by regional cooperation, the experience has still been relatively confined to the coastal provinces, especially Guangdong, Fujian and Shanghai. The proposal of Shanghai to strengthen financial cooperation with Hong Kong to develop large-scale enterprises or groups in the other areas of the country has been the most notable.

3.3.3 Cooperation in Natural Resource Exploitation and Processing

Since the Eighth FYP, many provinces and regions have stressed the importance of natural resource exploitation and further processing in regional economic and technological cooperation. In the first round of spatial cooperation in the 1990s, many natural resources deficit provinces developed on-site processing industries. For example, Shanghai, Beijing and Tianjin have tried to extend their cooperation with the natural resource rich regions. Such cooperation is important in promoting local economic development. In the 1990s, the strategy has entered a new phase. More provinces and municipalities have emphasized such complementarity of production factors.

Table 5.3 Cooperation Directions for Resource Exploitation in Selected Areas

	Strategies for resource exploitation
Guangdong	• Get in necessary resources and materials through inter-provincial enterprise cooperation
Shanghai	• Develop resource and processing projects in resource-rich areas to promote local economic development • Establish raw materials and processing bases for Shanghai's industrial development • Explore new sources of exports throughout the country • Support and utilize resources of the catchment area of the Three Gorges
Zhejiang	• Explicitly recognize the complementarity between the East and the interior as motives for the participation of Zhejiang enterprises in East-West cooperation
Fujian	• Establish resource bases, especially for green food products, commercial grain, chemical industry and construction materials • Obtain the much-needed energy and raw materials supply through cooperation • Obtain technology and advanced management experience and resources
Hebei	• Develop as an agricultural produce base for Beijing and Tianjin
Heilongjiang	• Transform the advantages in raw materials to production and try to build an export base jointly with other provinces • Welcome Guangzhou municipality to Heilongjiang to exploit raw materials and develop processing industries so as to enlarge exports of the local economy
Shanxi	• Fully utilise its advantage in raw materials • Support economic development of coastal provinces and municipalities with local advantages in energy, raw materials, coal and electrical materials • Work towards the support and cooperation in developing light industry, such as weaving, wearing and apparel, food and pharmacy, with coastal provinces and develop brand-name products
Anhui	• Develop poultry, water products, vegetables, fruits and other agricultural produce bases along the river • Jointly develop tourist resources with other provinces
Inner Mongolia	• Use local energy and raw materials as the base to establish agricultural produce and basic industrial base jointly with other provinces
Shanxi	• Use raw materials to exchange for technology
Gansu	• Rely on local resources to open up and develop
Ningxia	• Rely on the advantage of raw materials to obtain capital and technology

Thus, the cooperation strategies in the less-developed provinces are still basically in the relatively primitive forms of production factor exchange and/or establishment of joint production bases. Nonetheless, the strategies of the more-developed provinces which are lacking in resources have already shifted to the higher forms of cooperation by establishing cross-region enterprises for resource exploitation. The setting up of bases has mainly centered around agricultural produce, raw materials for processing industries and product sourcing for foreign trade.

3.3.4 Local Sectoral Adjustment

In the Eighth FYP, there was some cross-region production cooperation, though there had not been much sectoral bias. The major area for enterprise cooperation, especially in the "matching support" programme, was agricultural exploitation. Also, there were 2 major recent trends. Firstly, cross-region cooperation has shifted towards industrial and commercial cooperation. Secondly, there was a trend to emphasize cross-region investment and joint operation to achieve local restructuring. The most notable example is for Shanghai to relocate its cotton and weaving and mechanical (including general mechanics and automobile) industries outside the municipality. Conversely, Anhui, Hunan, Jiangxi, Sichuan, Guizhou and Xinjiang have given explicit priority for industrial relocation from the coastal regions. Besides, bilateral cooperation between Hubei and Anhui has also started with the hope of achieving production restructuring in the two provinces.

3.3.5 Joint Market Exploitation

Since the Eighth FYP, inter-regional competition for exploiting the domestic market has become increasingly intense. No wonder, opening up of the market has become a major goal in regional cooperation. Activities for opening the domestic market includes mainly organization of commodity trade and exhibition in other provinces, information disseminating meeting on purchase and product information, arrangement of local enterprise to participate in important commodity exchange and commercial activities, holding important national and regional trade and exhibition events and sending delegates to organize large-scale trade activities to outside provinces. These events mainly aim at "cultivating" the domestic market.

Besides, the establishment of offices in other provinces has also created "windows" or agents for local products and their sales. Among the pioneer provinces which have actively opened the domestic market are Guangdong, Zhejiang, Fujian and Shanghai and some other southeastern provinces. Anhui, Hubei and Hunnan in the coastal region have followed closely. The establishment of trade and business network by Shanghai government in the 1990s has been exemplary. In 1996, the Shanghai Industrial Production Sales Company, the first of its kind in China, was founded. It has since set up 6 sales centers in areas like Wuhai, Wanxian and Chengdu. Through the systems of membership and agents, the sales locations have shifted from the source of product supply to the market. There are now more than 350 similar projects and the total investment have amounted to over 1,300 million yuan.

3.4 Major Forms of Regional Cooperation Organization

When compared to the 1980s, the types of regional cooperation have basically remained unchanged. The four major types of spatial economic cooperation included the inter-provincial/regional, neighbouring province, intra-provincial and inter-municipality types (Zhang 1987).

3.4.1 Formation of Regional Cooperation Organizations

The major forms of regional cooperation include the arrangement of experience-sharing, visits by high level officials, the establishment of joint meeting by provincial (municipal/regional) heads and their delegates, the setting up of organizations to establish a system of communications and matching agents, to encourage enterprise cooperation and to strengthen publicity. Some organizations, like the Huaihai Economic Region (1993) and the Economic Region of Interior Municipalities in Liaoning (1997) have set up premanent economic region headquarters.

3.4.2 Major Activities and Construction Projects

As mentioned earlier, the forms of regional cooperation have already surpassed the limited forms of adjusting regional deficits and surpluses and simple exchange of production factors. The general tendency has been the formation of a regional production systems based on coordinated regional basic infrastructure, capital market and information network. Some of the major regional cooperation organizations and their activities are listed in Table 5.4.

3.5 General Development and Progress

According to incomplete statistics, the number of inter-provincial/municipal economic and technological cooperation projects has already reached 44,000 and the capital involved has reached 5.32 billion yuan. Among these, inter-provincial economic and technological cooperation has accounted for 32.2% of the projects and 40.7% of the invested capital. These regional economic and technological cooperation have, in turn, actively promoted the development of regional economies. In general, the following characteristics are identified:

3.5.1 Convergence of Production Factors in the Coastal Region

Since the 1990s, there were large inflows of capital, labour and energy resources to the coastal region, especially the Special Economic Zones (SEZs) and Open Cities (OCs). Among them, Shanghai and Shenzhen were the major areas of concentration. This phenomenon has been commonly termed as the "migration of the peacock to the southeast". From 1992 to 1994, the number of economic and technological cooperation projects that have attracted inflows to the East has increased dramatically. In contrast, the projects for "diffusion" to the Central and the West have declined. By 1994, the former was nearly 4 times that of the latter. The outflows of technological and human resource from the interior to the coastal areas have been taking place relatively quickly. It is estimated that about 80% of the higher level managerial personnel and technicians of the interior have already moved into Guangdong. Gao et al (1996) estimated that about 40% of the central government capital injection into the West since 1994 has flowed back to the East.

During the Eighth FYP, investments from other provinces and municipalities into Shanghai has been so large that it has become one of the major contributing factors for the municipality's rapid economic development. According to a 1995 survey with more than 7,000 non-local domestic enterprises in Shanghai. The total tax revenue generated by these enterprises was as high as 6.7 billion yuan, which was nearly 30% of Shanghai's fiscal revenue. Moreover, these enterprises employed nearly 0.25 million Shanghai labour. Besides, profit remittance from non-local joint enterprises in Shanghai was over 3 billion during the Eighth FYP. In the same year, the estimated amount of investment from other provinces/municipalities to Shanghai was as high as 60 billion yuan.

Table 5.4 Major Cooperation Directions for Selected Regional Cooperation Organizations

	Major Goals Stated
Southwest Five Province Seven Municipality Economic Co-ordination Meeting (12th Meeting in 1996)	• Utilize preferential policies offered by the country to speed up development in the West • Join together in fighting for and matching national policies • Join together in bidding for major regional projects • Speed up the formation of a system of regional markets • Establish a Regional Economic Development Bank • Jointly build up a comprehensive Foreign Trade Centre • Join together to increase the degree of opening to the outside world
Southwest Five Provinces Economic Co-ordination Meeting (1995)	• Join together to construct an international transport link • Join together in the exploitation and utilization of resources
Huanghe Economic Cooperation Region (7th Meeting in 1995)	• Join together to form an integrated "Huanghe market" • Deepen enterprise cooperation
Nanjing Region Economic Co-ordination Meeting (8th Meeting in 1996)	• Jointly develop regional basic infrastructural construction • Develop regional brand-name products • Expand the commercial and trade markets • Develop tourism
West Fujian-South Jiangxi-Southeast Guangdong Economic Co-ordination and Development Meeting (1995)	• Jointly fight for central government support to the three provinces • Institute regional development planning • Speed up transport network construction • Begin economic project co-ordination • Establish regional information network
Central Liaoning Economic Region (1996)	• Join together in establishing outside links • Formation of a regional market system based on the free flow of transport and communications • Regional restructuring
Southwest-Fujian Regional Cooperation (1996)	• Join together in exploiting resources • Join in building and enjoying the benefits of transport facilities • Outward-oriented economic cooperation • Join in "cultivating" a regional market • Centre around "pillar industries" to develop spatial division of labour and matching supply • Join to form large- and medium-scale industrial projects and groups

Circum-Changjiang Economic Co-ordination Meeting (8th Meeting in 1996)	• Jointly develop the Changjiang production belt • Join together in building major infrastructural facilities along the River • Put emphasis on establishing markets for technology, information and foreign trade • Promote regional cooperation • Fortify and develop cross-region and cross-industry enterprise groups

By the end of 1995, the number of non-local domestic enterprises in Shanghai has reached 14,000, which was roughly the same as the number of foreign enterprises there. Of this number, at least half of them are from the Central and the West. Sichuan, Hubei, Hunan, Jiangxi, Anhui, Jiangsu and Zhejiang have already accounted for 16.43% (2,300 enterprises). A closer look at the destinations of the investment reveals that nearly three-fourths (73%) of the non-local investment was concentrated in Pudong. Apart from the large number of enterprises in Pudong, which has risen to 4,200 in June 1996, the construction of office building was a major form of inter-provincial investment there. For instance, Anhui has invested 1.3 billion to construct the Yuean Building. Other provinces' participation in the Pudong Waigaoqiu Bounded Zone was also notable. Of the 2,224 projects in the Zone, 900 were directly under the other 28 provinces and municipalities or the central government. The SEZs and other coastal OCs have also attracted the inflow of capital from other coastal provinces. According to incomplete statistics, the number of enterprises established by the 30 provinces/municipalities or the central government in Shenzhen is more than 7,000 and the amount of investment involved is more than 5 billion yuan. In Zhuhai, the number of these enterprises has also reached 3,498 in April 1995. There were more than 700 offices of other provinces/municipalities in the city.

Imbalance development of the stock market in the coastal region has reinforced the tendency for the concentration of capital in the coastal region. In 1996, Shanghai has accounted for 40% of the total business turnover of overseas financial organizations in China. Table 5.5 shows the regional distribution of listed companies in China by the end of 1995. Nearly half of them were Shanghai or Shenzhen companies. The situation has not changed much. In May 1997, about 72.8% of the listed companies were still found in the East.

The socio-economic influence of the large central cities has been reinforced under the Open Policy. Many of these central cities have gradually become centers for inter-provincial/municipal offices and information exchanges. Regional connections, especially at the provincial level, have been fostered by such a development. Table 5.5 summarizes the number of inter-provincial/municipal and State Council offices established in various important cities.

Table 5.5 Spatial Distribution of Inter-Provincial/Municipal and State Council Offices

Beijing	530
Fuzhou	211
Guangzhou	1155
Shanghai	700
Tianjin	1200
Wuhan	600
Xiamen	341

3.5.2 Strengthening of North-South Economic Linkages

The strengthening of North-South economic linkages within the coastal region can be seen most clearly from the number of joint projects among provinces/municipalities and the amount of capital involved. Judging from the number of key joint projects Tianjin has with other provinces/municipalities from 1992 to 1994, Guangdong, Shandong, Fujian and Zhejiang in the South were found to have the closest economic links with the municipality. Conversely, 67% (6.5 billion yuan) of the capital inflows into these southern provinces came from Beijing and Tianjin in the North. For example, Tianjin has invested more than 3.94 million yuan and established 16 major export production bases in Hebei.

3.5.3 Tendency for the Central to Co-operate Both with the East and the West

As mentioned earlier, the Central has become increasingly aware of its locational advantage as a neighbour of the East, as well as the West. While the emphasis of the Central region is still on opening up the market in the East, it has also tried to open up the market in the West. This comparative advantage of the Central has manifested in the "1997 East-West Cooperation, Investment and Trade Conference". Take Anhui as an example. 84 enterprises there have co-operated with 24 groups in the East mainly in light industry, chemical industry and commerce. Most of the contracts with the Northwest have concentrated on agricultural projects and the transformation of technology or technological cooperation.

3.5.4 Beginning of Inflow of Capital and Technology to the West

In the 1990s, the West has begun to attract investment funds. Generally, these funds came from three major sources: foreign capital, domestic capital outside the province/municipality and local capital. In Sichuan, the shares of these 3 sources have roughly been the same in 1996. In Guizhou, domestic capital outside the municipality has been the major source of investment and foreign capital is still very scarce. Apart from the 4 municipalities, i.e. Shenzhen, Qingdao, Ningbao and Dalian, assigned by the central government to help Guizhou in the "matching support" for poverty relief programme, Shanghai, Nanjing, Fujian, Guangdong and Shandong have also invested in the municipality. From 1992 to 1996, Guizhou has entered into 2000 joint projects with these provinces/municipalities; and the capital involved was more than 3 billion yuan.

Apart from investment, remittance from labour working in coastal and more developed provinces has also been significant. The amount of remittance from people working in Guangdong and Shanghai to the Central and the West has reached more than 20 billion and 10 billion yuan respectively.

Another important source of capital inflow to the West was closely related to the "matching supply" to the "Three Gorges" and for poverty relief programmes advocated by the central government. According to incomplete statistics, the number of support projects from the central government and other related municipalities has reached 820 in 1996. The actual capital utilised was 1.59 billion yuan. Up till the end of 1996, the investment of Shanghai to the catchment areas of the Three Gorges and Tibet has reached 160 and 130 million yuan respectively.

Technological cooperation was a major aim of East-West cooperation. Nonetheless, it seems that advancement in this area has been slow and relatively limited. However, there were some breakthroughs in the 1990s with Ningxia entering into a series of technological cooperation projects with Zhejiang, Tianjin and Liaoning. In addition, Inner Mongolia has also fortified economic and technological links with the Northeast, Hebei, Beijing and Tianjin.

During the Eighth FYP, regional cooperation of the West with other regions was of a relatively low level. It is mainly concentrated in agricultural produce processing, land and resource exploitation, technology transfer and labour-intensive light industries. In the 1990s, regional cooperation has mainly followed two directions. Firstly, capital from the East has concentrated in the exploitation of natural resources, especially in projects in agriculture and mining. Secondly, joint development and technological transfer were common in the relocation of factories of brand-name products to poor counties as "matching support". Examples of the former included investment of 86.97 million yuan from Jiangsu, Anhui, Shandong, Sichuan, Hubei and Shenzhen to develop a cotton production base in Xinjiang in 1994. This amount accounted for 13% of all investments from other provinces or municipalities to the province. Other examples included the exploitation of iron and other heavy metals by joint ventures of Guizhou with other provinces/municipalities, like Shanghai, Jiangsu, Zhejiang and Guangdong and the setting of tobacco producing base in Guizhou jointly with Shanghai. Examples of the second direction included the relocation of Zhejiang Drinks and Shanghai White Cat Detergent Factories into the West.

3.6 Major Problems

There were many problems of regional cooperation at the local level.

3.6.1 Natural Resource Exploitation and Utilization

Regional co-operation can be advantageous both to the resource-rich and deficit regions. Nonetheless, the present moves, mainly by the encouragement and participation of the government alone were not very successful. Joint natural resource exploitation/development in the less-developed regions has, in some cases, not benefited the resource-deficit regions as the products were not sold to them. Conversely, there was also a situation when the outside partners refused to buy the products because of their low-quality. Of the 1.5 billion yuan Shanghai joint investment with the other provinces/municipalities, a large part was used for building natural resource bases. Yet, only about one-fifth has reached relatively high efficiency. In the joint exploitation of water resources by the Zhujiang Delta area, the Southwest and the Honghe region of Guangxi, there have been complains of unequal share of benefits. The exploitation by the coastal Eastern regions of the natural resources in the West still faced many such problems.

3.6.2 Utilization of Better International Trade Conditions

Regional cooperation in speeding up the formation of an outward-oriented economy and the development of international trade in the interior has only been developing very slowly. The participation of the interior in export processing industries and the development of bounded zones are still limited.

3.6.3 Industrial Relocation

Industrial relocation from the East to the West mainly involves second-hand technology and limited production scales. Its contribution to the development of medium and small-scale enterprises in some villages has been substantial. However, this is usually not welcomed in the large cities. In addition, most of these relocations are limited to villages/counties which the central government has assigned to the more developed coastal regions.

3.6.4 Formation of Regional Corporations

Though the number of regional groups has mushroomed after the Open Policy, many of them are not very successful and faced hardship. The major obstacles included property rights and the under-development of financial market. Basically, the property market is still limited, and the chance for interregional loan was very slim. Non-local enterprises do not have the right to raise fund locally. Due to these constraints, the costs of cooperation have been relatively high for these regional corporations.

3.6.5 Commercial and Trade Cooperation

With increasing competition for the domestic market, different provinces/municipalities have tried to increase their market shares and become national and regional commodity trade centres. Hence, regional commercial and trade cooperation has become increasingly difficult. For instance,

Shanghai's regional commercial business has faced increased difficulties. Shanghai Number 1 Department Store once has opened more than 20 chain stores in the country. By now, only 3, i.e. the ones in Hefei, Wuhan and Xian, have remained. The limited success was attributable to high costs of operation because of the harsh conditions to operate in "alien" provinces.

4. Future Development and Difficulties for Regional Cooperation Organizations

The stage of regional cooperation in China has lagged much behind the national goals of building a regional market system and regional economic integration. Of the different types of regional cooperation organizations, the difficulties for developing inter-provincial/municipal regional cooperation organizations are the most prominent.

4.1 Major Progress Made

4.1.1 Improve Regional Identity and Increase Regional Influence

Since the Eighth FYP, there were more than 100 regional cooperation organizations in China. Only about one-third were relatively active. Most of the others were formed for information exchange, friendship and publicity. By far, the greatest success of spatial cooperation was achieved among cities and counties of the same province in the coastal region. Examples included the Southeast-Fujian Economic Region and the Southwest-Fujian Economic Region. Both intra- and inter-provincial corporations are relatively underdeveloped in the less-developed provinces. In recent years, the formation of such corporations and their activities in cities/provinces along Changjiang are also notable. Table 5.6 briefly categorized the regional cooperation organizations into 4 types.

Table 5.6 Some Relatively Active Regional Economic Cooperation Organizations since the Eighth FYP

Type	Examples
Inter-Provincial (regional)	• Southwest-Five-Provinces-Seven-Municipalities Economic Co-ordination Meeting • Southwest Five Province Region Economic Co-ordination Meeting • Huanghe Economic Cooperation Region • Central-South Economic Cooperation Region
Adjoining Provinces/areas	• Nanjing Economic Cooperation Region • Huaihai Economic Cooperation Region • Central Economic Cooperation Region • Penghai Zanxi belt municipality (state) head meeting • Southwest-Fujian-East-Jiangxi-East-Guangdong Economic Development Cooperation and Development Meeting • Hunan-Hubei-Sichuan-Guizhou Border Region • Fujian-Zhejiang-Jiangxi-Anhui-Nine-Municipalities

	Economic Region
Adjoining Municipalities within the Same Province	• Central Liaoning Economic Region • Northeast-Fujian-Five-Municipalities Economic Cooperation Region • Southwest Fujian Regional Cooperation • Central-Yunnan-Eight-States Region
City Economic Cooperation Network	• Jiangsu Suzhou-Wuxi-Changzhou-Tangzhou Economic Cooperation Region • Hunan Changchutan Economic Cooperation Region • Circum-Changjiang Central Cities Economic Co-ordination Meeting • Changjiang Delta 14 Cities Heads Economic Co-ordination Committee Meeting • Major Towns along Silk Road Economic Connection Meeting

4.1.2 Wider Scope of Regional Cooperation

Table 5.7 summarizes the progress made by the major regional cooperation organizations.

Table 5.7 Major Progress Made by Selected Regional Cooperation Organizations

	Major Progress
Five-Provinces-Seven-Municipalities Economic Co-ordination Meeting	*1984-1994* • More than 160 projects in energy, transport, communications and agriculture • 6478 cooperation projects • Extending and linking up 49 dead-end road at regional boundary, totalling nearly 1,000 km. • Fund raised: More than 5 billion yuan • Commercial, trade and material co-ordination: More than 40 billion yuan • Labour exchange: More than 10,000 • New production value: 6.25 billion yuan • Tax contribution: 1.5 billion yuan
Huanghe Economic Cooperation Region	• Established friendship and cooperative relationship with 10 provinces/districts and 70 cities • Held joint meetings • Over 10,000 cooperation projects • Raised several 100 million yuan • Material cooperation flow: 20 billion yuan • Strengthening of joint resource exploitation

Nanjing Region Economic Co-ordination Meeting	• Established 67 industry networks and 86 groups *1986-1994* • 4220 economic cooperation projects • New production value of 8.95 billion yuan • Tax contribution: 1.26 billion yuan • Material flow: 2.65 billion yuan • Fund raised: 2.07 billion yuan • Attracted 5.1 billion yuan
Wuhan Economic Cooperation Region	*1987-1995* • 1833 Economic Cooperation projects • Newly generated production value: 1.2 billion yuan • Tax contribution: 200 million yuan • 8165 labour exchange program worth 830 million yuan • Establish 10 major commercial groups and set up more than 200 chain shops. Sales revenue: 32.4 billion yuan. • Use financial network of central cities along Changjiang to raise more than 230 billion yuan and increase foreign exchange of more than 4 billion yuan.
Chongqing Economic Cooperation Region	*1988-Early 1996* • 2789 cooperation projects (newly produced value: 3.81 billion yuan, tax contribution: 84 million yuan, material, commerce and trade value: 8.18 billion yuan, raise fund: 1.96 billion yuan, number of joint enterprises formed: 532).
Hunan-Hubei-Sichuan-Guizhou Border Region	• Newly established specialized markets and wholesale markets: more than 80 • Material cooperation flows: Approximately 8 billion yuan
Fujian-Zhejiang-Jiangxi-Anhui Nine Party Economic Region	*1996* • Total capital raising ability of the banking system: more than 2 billion yuan • Material cooperation flows: 81.4 billion yuan • Extending and linking up 10 dead-end road at regional boundary, totally 693 km.
Southwest Fujian Economic Development (1995)	• A Regional Economic Cooperation Office • Formulated a Regional Development Plan and a Regional Development Strategy • Development of national and provincial roads by investing nearly 5 billion yuan • Completed the construction and transformation of 1,800 km of national and provincial roads • Foster "Mountain-Sea Cooperation" and enterprise cooperation in 5 municipalities

Circum-Changjiang Economic Co-ordination Meeting	• Development of a Changjiang banking network. In 1994 alone, loans extended amounted to 308.5 billion yuan and the foreign exchange earned amounted to 22.5 billion U.S. dollars

Thus, areas of regional cooperation have gradually extended to material commercial and trade co-ordination and financial system development. There is also major development in the area of transport infrastructural construction. In particular, the extension and linking up of dead-end roads at regional boundaries to form a road network has been achieved both by the Southwest-Five-Provinces-Seven-Municipalities Economic Co-ordination Meeting and the Fujian-Zhejiang-Jiangxi-Anhui-Nine-Municipalities Economic Region. Another progress in basic infrastructural development and co-ordination achieved is the development of the navigation channel of Huanghe and the construction of Sanmenxie and Fengning Highway bridges by the Shanxi-Shaanxi-Henan-Huanghe Economic Cooperation Region. Similarly, control of environmental pollution among the municipalities in the upper and lower courses of Huanghe has been co-ordinated by the Huaihe Economic Cooperation Region. Besides, friendly negotiation on the supply and protection of water in Hebei, Tianjin and Beijing within the circum-Bohai region has begun.

Another major achievement of regional cooperation was the promotion of free capital flow across regions. At present, Shanghai Pudong Development Bank and the Changjiang financial network seems to be the most successful. The Shanghai Pudong Development Bank was formally established as a regional share-holding commercial bank in January 1993. The major aim is to make full use of the different forms of credits and increase the flow of domestic and foreign capital to assist the development and opening of Pudong and the Changjiang river basin. Up to June 1996, its capital assets have reached over 40 billion yuan and it has set up more than 22 branches along the Changjiang River.

Based on the spirit of the Joint Meeting of the Municipality heads of Shanghai, Nanjing, Wuhan and Chongqing in October 1986, the People's Bank in the 4 municipalities and the Headquarters of Communications Bank have called for the 28 cities along Changjiang for a horizontal financial network meeting in Wuhan to establish a Changjiang financial network (Changjiang Financial Network Office). The aim was to promote the movement of capital and a regional capital market along Changjiang. The major achievement includes the development of loan-breaking activities for participating banks. According to incomplete statistics from October 1986 to June 1995, the total amount of capital materialized through the network has reached 1008.7 billion yuan. Its ability in stabilising the foreign exchange market before the floating of foreign exchange rate is also very significant.

4.2 Major Difficulties Encountered

4.2.1 Large-Scale Regional Basic Infrastructural Projects

The need for regional cooperation in large-scale infrastructural projects spanning different administrative units is more than apparent. The problems of sources of funds, benefits allocation and land resumption and compensations are still prominent in preventing full-scale regional cooperation. These problems are particularly serious in the construction of new railway lines, large-scale airports and co-ordinated development of ports.

Due to the existence of multiple systems, inter-linked through transport services along Changjiang has still remained at an early negotiation and preparatory stage. The proposal to jointly develop deep-water ports and the achievement of modern and high-speed transport linking Shanghai and other cities along Changjiang have stagnated because of disagreement among Shanghai, Jiangsu and Zhejiang on the sharing of benefits. Similar problems have occurred in the joint efforts of Fujian, Guangdong and Jiangxi to speed up the construction of inter-provincial railways and highways in the region.

4.2.2 Regional Financial Cooperation System

The system of finance in China was still very much based on administrative divisions. Hence, local control and quotas on loans and the rules of forbidding inter-regional loans still existed. Moreover, the development of inter-regional bond market is still heavily restricted by the administration and the management boundary for capital movement. In view of recent macro-adjustment measures of the central government, rise of local protectionism and the low level of computerization in the banking system of China, future advance in the area seems difficult.

4.2.3 Regional Industrial Cooperation

The national system of project assessment and approval and restrictions on land use for cross-region projects have equally deterred regional cooperation. Multi-lateral investment projects in China are thus still limited and, hence, the progress made towards regional economic unification is slow. Based on the statistics we have discussed above, the progress made in regional cooperation seems very remarkable. Nonetheless, these figures released by the regional cooperation organizations have actually concealed much difficulties in the cooperation. For instance, among the more than 20,000 economic cooperation projects reached by Shanghai with other municipalities along Changjiang, only 169 involved actual investment. Up to June 1996, Shantou has set up more than 72 administrative units within the Southwest-Fujian-East-Jiangxi-Southeast-East-Guangdong Economic Cooperation Region; yet, the number of projects involving production was only around ten.

Of the many regional cooperation organizations, the Changjiang Economic Cooperation Region has been one of the earliest and most promising regional cooperation organizations. In the early 1990s, the relevant ministries have set up special research themes on the ways to co-ordinate industrial development in the entire delta region. Nonetheless, most cities in the region are still developing similar industries. For instance, fourteen cities of the Delta area have designated automobile, petroleum, electronics and new construction material industries, which utilise high technology and of high value, as their pillar industries. The hope of co-ordinated development or division of labour still seems to be relatively dim.

5. CONCLUDING COMMENTS: CONFLICTS BETWEEN LOCAL AND CENTRAL GOVERNMENTS

In future, the success of regional economic and technological cooperation seems to rest on the degree of match of goals and direction of the central and local governments on regional cooperation. No doubt, the policies and measures of the central government in encouraging regional economic and technological cooperation are important. However, the intentions of local governments to set up regional strategies and in establishing and developing inter-regional links are as important in achieving regional economic and technological cooperation.

As the central government gradually devolved its power to local governments, the ability of the central government in investing or subsidizing local governments has also become increasingly weak. Thus, the need for raising funds locally and developing a sectoral strategy in accordance with regional comparative advantage becomes apparent. Increased regional cooperation and exchanges are obviously important in these two aspects.

The outflow of capital and production factors to the East seems inevitable under the current market situation despite the aim of the central government to achieve regional balance and to reduce regional disparity by hoping the West to "attract business and investment". This process seems to be an integral part of the internationalization and marketization of the national economy. Both the locational comparative advantage and the accumulated development potentials of the coastal region dictate that it has been and will continue to be China's engine of economic growth. At present, it has been estimated that the return for per 100 yuan investment in fixed assets in the East was 2-4 times the return in the West. There has also been certain beneficial triggering-off effects from these more-developed areas under the "poverty relief" programme of the central government. In the near future, regional cooperation will still be mainly happening in the major coastal (port) cities and their hinterlands, and then between the coastal cities and the Central and the West.

Lastly, the age-old problem of departmentalism, mainly manifested in the separation of ministrial direction and local/regional interests in China, is still affecting the country and can easily stiffen the early development of regional cooperation and lead to re-emergence of localism or cellular economy in China. The constraints of the banking system on the formation of a system of regional cooperation has been very significant. Besides, the problem of the tax system, which stipulates that all values created by non-local enterprises have to pay tax to the local government and included in the local National Income only, has been another major obstacle to the achievement of higher levels of regional cooperation.

REFERENCES

1. ---. 1993. *Collection of Documents of the Conference of the Eighth Municipal Heads in Huaihai Economic Region,* October 1993 (Chinese text).
2. ---. *Outline of the Planning of Economic and Technological Cooperation in Anhui in the Ninth Five-Year Plan* (Chinese text).
3. Chen Yaobang. 1996. Promote regional economic co-ordinated development. In *Future and Development* 5 (Chinese text).
4. Heibei Economic and Technological Cooperation Office. 1996. *Report on the Progress of Hebei Province in Opening Internally* (Chinese text).
5. Heilongjiang Economic and Technological Cooperation Office. 1997. *Strengthen Cooperation with the South* (Chinese text).
6. Hu Angan. 1996. Tax Division: Evaluations and Suggestions, in *China's Social Sciences* 8 (Chinese text).
7. Jin Feng. *et al.* 1992. *Analysis of Regional Economy and Policies*. N.p.: China's Environmental Science Press (Chinese text).
8. Land Resources Office of the State Economic Commission. 1997. *China's Regional Development Report* N.p. (Chinese text).
9. Land Resources Office of the State Economic Commission. N.d. *Report on Regional Economic Co-ordinated Development and Regional Economic Cooperation*. N.p. (Chinese text).
10. Li Peng. 1996. *Li Peng's Speech on the Three Gorges Projects, October* (Chinese text).
11. Li Siming. 1996. Changing regional disparities in China. In Li Siming, Tang Wing Shing, Qiang Lanhong and Zhou Suqing (eds) *Different Aspects of China's Regional Development.* Taipei and Hong Kong: University of Taiwan, Centre of Human Resources and Hong Kong Baptist University, David C. Lam Institute for East-West Studies, 19-44 (Chinese text).
12. Li Youwei. 1996. Prepare for the Return of Hong Kong, Strengthen Cooperation, Work for Common Prosperity, in *Circum-Bohai Rim Development* 5 (Chinese text).
13. Liu Shucheng. *et al.* 1994. Measuring regional income disparity and policy suggestions. *Research on China's Regional Economic Developmen.* (Chinese text).
14. Liu Tiemin. 1996. System innovations and regional economic co-ordinated development. In *Future and Development* 6 (Chinese text).
15. Lu Dadao. 1997. Regional cooperation and regional economic co-ordinated development. In *Regional Studies and Exploitation* 1 (Chinese text).
16. Lu Xianxiang. 1997. Four major problems in the China's industrial development. In *China's Industrial Economy* 3 (Chinese text).
17. Malmberg, A. and Solvell, O. N.d. Spatial Clustering, Accumulation of Knowledge and Firm Competitiveness. Conference Paper.

18. Market and Price Department of the State Economic Commission. *Outline on the Development of Markets and Planning in the Ninth Five-Year Plan*. N.p. (Chinese text).
19. National Currency and Capital Committee. 1996. *China at the Doorstep of Market Economy*. N.p.: National Financial Press (Chinese text).
20. Research Institute of the State Economic Commission on Land Resources and Regional Economic Cooperation. 1996. *Strengthen Horizontal Economic Cooperation and Promote Regional Economic Development*. N.p. (Chinese text).
21. State Council. *Notice of the State Council on the "Matching Support" to Migrants From the Three Gorges Catchment Areas*, Document No. [94]58. (Chinese text).
22. State Council. *Notice of the State Council on the Organization of the More-Developed Regions to Help the Less-Developed Regions*, Document No. [96]26. (Chinese text).
23. Tan Chenglin. 1996. Ph.D. Thesis of People's University. (Chinese text).
24. Wang Yang. 1995. *Historical Breakthroughs in the Exploitation of Wanjiang*. N.p.: Anhui's People's Press. (Chinese text).
25. Wei houkai. 1995. *New Pattern of Regional Economic Development*. N.p.: Yunnan People's Press (Chinese text).
26. Wei houkai. 1997. Regional Economic Growth and Decline in China, in *China's Industrial Economy* 3 (Chinese text).
27. Williamson, Jeffrey G. 1965. Regional inequality and the process of national development: a description of the patterns. In *Economic Development and Cultural Change* 13(4, part 2): 44.
28. Yang Kaizhong. 1994. Analysis of changing regional disparities. In *Economic Research* 12 (Chinese text).
29. Zhang Wanqing. *et al.* 1987. *Network of Regional Cooperation and Economy*. N.p.: Economic Science Press (Chinese text).

Chapter 6

REGIONAL APPRAISAL OF THE TRANSPORTATION AND COMMUNICATION INFRASTRUCTURE

JIN Fengjun
Associate Professor, Institute of Geography,
Chinese Academy of Science
Jixian WANG
Associate Professor, Department of Geography,
University of Hong Kong

Transportation and communication (T&C from here onwards) are not only the infrastructure and conditions for regional economic growth, they themselves are also one major area of regional economic development, particularly because they provide basic environment for living and for all substantive investment. To assess the T&C infrastructure development in China during the Eighth Five-Year-Plan (FYP) (1991-1995), several government organizations and some research institutes have reached similar conclusions. Among them, the comment from China Transport and Traffic Association (1996) is representative: "In general, it was the period of fastest and greatest changes since the founding of PRC in T&C. The major achievements can be summarized into five aspects: (1) a steady growth in transport volume, and a fast growth in telecommunications; (2) transport infrastructure keep improving, (3) vehicles and facilities for passengers and freights movement and information transmission have been enhanced; (4) a general upgrading of technologies in use; and (5) the reforms in the T&C sector have gone farther. " Similar comments were made by the State Statistics Bureau and Institute of Territorial and Regional Development under the State Planning Commission. These assessments however, did not make any systematic analysis on the spatial variations of T&C provision. Since the existing literature only provides relevant analysis for transport infrastructure development up to the 1980s, we devote this chapter to an analysis of China's T&C infrastructure in the first half of the 1990s, with special focus on its regional characteristics and impacts.

Centering on the relationship between infrastructure construction and regional development, this chapter will be divided into five sections. The first section introduces the general characteristics in the T&C infrastructure in China and its changing trends since 1990. It is followed by a discussion on the interrelation between regional infrastructure provision and economic growth at different spatial scales. The third section classifies China's provinces into four categories based on their existing T&C infrastructure. With the recognition of these regional deviations and the linear nature of T&C development, the forth section suggests 5 major T&C infrastructure zones or corridors that are more advanced than other regions. The last section concludes the analysis by pinpointing current development mechanisms in the sector, future trends, and suggests major aspects for improvement.

1. Development Trends of T&C Infrastructure in the Eighth FYP (1991-95)

1.1 General Characteristics

- While the inputs for construction increased enormously, the regional investment scale also varied enormously.
- Multimodal transport systems were in place for moving both goods and people; cooperation among transportation modes and between transport and communication sectors began to establish, providing much better services, although the coastal being much more advanced than inland regions.
- In terms of service quality, product variety and timing service were two areas that have improved most.
- A nationwide hierarchy of T&C networks was forming, which resulted in the establishment of a few large corridors with comprehensive infrastructure capable for intensive investment and development.
- All the above trends enlarged the regional gap in T&C infrastructure.
- T&C facilities and equipment put in place in this period were have upgraded China's T&C systems in general and narrowed the technology gap between China and the advanced countries.
- The pressure from social demand has largely been released and the backward development of T&C sector in the last few decades was changed.

1.2 Development Trends

During 1991-95, China's total investment in T&C infrastructure was 680 billion RMB yuan by the state and state-owned enterprises, which was 3.6 times that of the previous FYP period. This trend in fact has been sustained in 1996 when 267 Billion yuan was put in place as fixed capital investment for these two sectors (see Table 6.1). With this amount of investment, the construction of T&C infrastructure with better equipment and newer technology has speeded up, which can be seen in the following seven aspects.

Table 6.1 Investment in Transport & Communication Infrastructure 1986-1996

	1986-1990 (the 7th FYP)	1991-1995 (the 8th FYP)	1996
Basic Construction	95.6	460.6	183.8
Upgrading	33.6	191.4	83.3
Others	7.1	28.0	-
Total Investment:	**146.3**	**680.0**	**267.1**

Note: The above statistics include only the capital investment from the government and state-owned enterprises.
Sources: Complied from *China Transportation Statistics Yearbook*, various years, Beijing: China Transportation Statistics Yearbook Press;and *China Communication Statistics Yearbook*, various years, Beijing: China Communication Statistics Yearbook Press, the Ministry of Communication.

1.2.1 Effective Operation of Railways on New and Upgraded National Railway Trunk Lines[1]

There were 5,900 km of newly built railways. Of these, 3,719 km had gone into operatation, 3,885 km were double-tracked, and 2,762 km were electrified in the five years. They include the double-track lines linking Hongzhou in Zhejian with Zhuzhou in Hunan, Datong in Shanxi with Baotou in Inner Mongolia, and Jiaozuo in Henan with Zhichen in Hubei. The electrified sections were previous bottlenecks at areas of topological barriers, such as Xiamen-Yingtan, Chengdu-Guiyang, Guiyang-Kunming, Changsha-Guiyang, Lanzhou-Wuwei, and Beijing- Zhongzhou. More importantly, for long distance rail transport, the national trunk lines such as Beijing-Guangzhou, Beijing-Shanghai, Longhai-Lanxin, and Hongzhou-Zhuzhou have had been upgraded to an annual through capacity of more than 40 million tons. These have greatly released demand pressure in general. The special demands in individual areas have also been handled well. For examples, the first unit-heavy-duty train service has started along the rail from Datong in Shanxi to the port city of Qinhuandao in Hebei for coal export from Shanxi; special high-speed passenger rail are running between Guangzhou and Shenzhen; and 5,000-ton heavy-duty trains for breakbulk goods are running on Beijing-Shanghai and Beijing-Guangzhou.

1.2.2 New Stage of Highway Development in China: High Standard for High Speed[2]

Since the founding of PRC, the highway system has experienced four stages of development: (1) fast expansion of road networks in the 1950s; (2) continual expansion of the network while upgrading the existing roads in the 1960s and 70s; (3) mainly upgrading the existing highways in the 1980s; and (4) construction of expressways as the National Trunk Roads System (NTRS) in the 1990s. As the backbone of national highway system, the NTHS was designed to consist of 12 trunk highways linking with all cities of million population, and most cities of population between 0.5 to 1 million. Five of them are called as 'vertical' (north-south) and the other seven 'horizontal' (east-west) corridors for development.

The five vertical trunk highways include Tongjiang (in Helongjiang) - Sanya (in Hainan), Beijing -Fuzhou, Beijing -Zhuhai (in Guangdong), Erlianhaote (in Inner Mongolia) Hekou (in Yunnan), Chongqing Zhanjiang (in Guangdong). The seven horizontal trunk lines are: Suifenhe (in Helongjiang) - Manzhouli (in Inner Mongolia), Dandong (in Laoning) - Lasa (in Tibet), Qingdao (in Shangdong) - Yingchuan, Lianyungang (in Jiangsu) - Huergusi highway (in Xinjiang), Shanghai-Chengdu (in Sichuan), Shanghai - Reili highway (in Yunnan), and Henyang (in Hunan) - Kunming (Yunnan).

The NTRS has been planned as the major task for China's highway construction in the 1990s and the early 21st century. Some 5,000 km were completed during the Eighth FYP. Most existing expressways and class I highways built up so far are part of the NTRS. By the end of 1995, China had 2,141 km express highways, and 94,500 km Grades I and II highways (i.e. all-season paved roads). The expressways and highways of Grades I and II are together called "high-standard highways". The 1990s has witnessed the most rapid growth of these high-standard highways. From 1993 to 1995, about 500 km expressways were built annually, and in 1996 the number jumped to 1,117 km, which was more than half of the existing expressways at the time.

[1] Data quoted in this section are from *1996 China Transportation Statistics Yearbook*, Beijing: China Transportation Statistics Yearbook Press.

[2] Data quoted in this section is from Zhang, Baisheng (1994) "National Trunk Road Planning", in *1994 China Transportation Statistics Yearbook*, Beijing: China Transportation Statistics Yearbook Press, page (?) and *1996 China Transportation Statistics Yearbook*, Beijing: China Transportation Statistics Yearbook Press.

Figure 6.1: National Trunk Road System in China

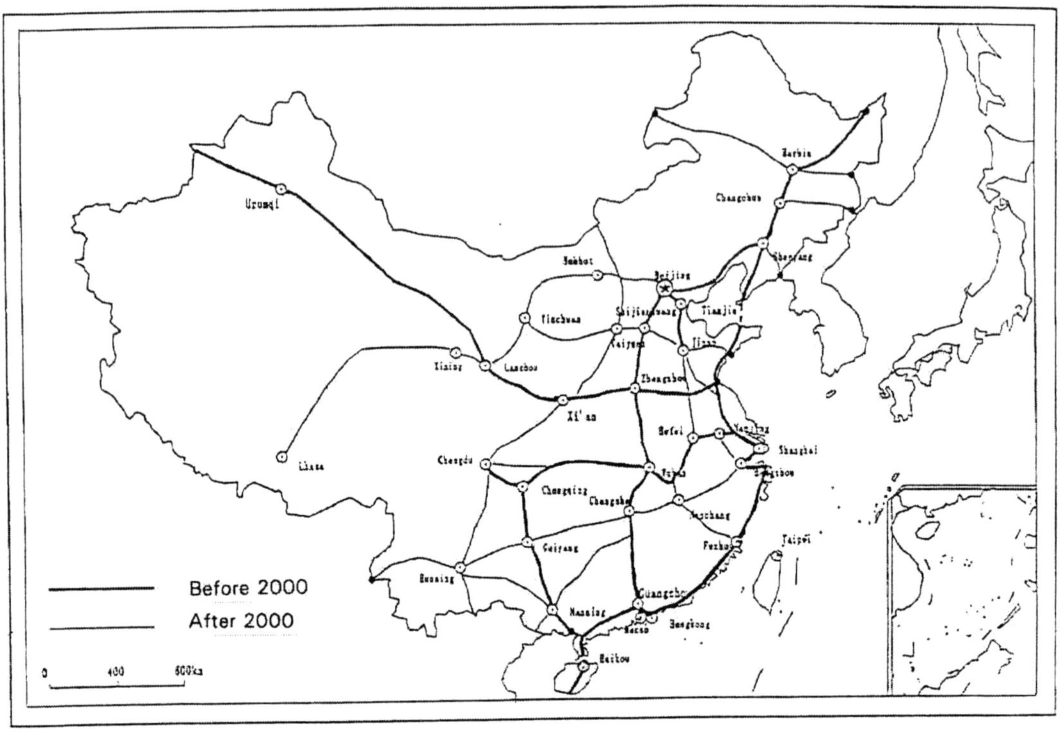

1.2.3 New and Existing Coastal Ports Equipped with Specialized Hi-tech Terminals

The first half of the 1990s witnessed also the fastest development in port construction in Chinese history. There were 100 deep-water berths for vessels of 10,000 dwt in size, or one third of the deep-water berths even built before. By 1995, plus the existing 196 berths of various sizes, the total annual throughput capacity reached 120 million tons. Among the newly built deep-water berths, 13 are for containers, 11 for coal, and 12 for bulk goods, all being operated with the state-of-art equipment. The specialized terminals for coal transport are mainly for regional development and local needs. By 1995, 44 deep-water berths for coal were in use, with a handling capacity for coal loading/unloading and transshipment of 100 million tons per year (Zhang, J, 1996).

1.2.4 A Port-centered Container Transportation System in the Making

In the mainland, container ports were not constructed until the end of the 1970s. The formation of container port system was accelerated in the 1990s. To catch up with the development in containerized shipping in the world, China had, for the first time, invited foreign investors and operators to renovate and update its port facilities and operations. By the end of 1995, there have been 39 berths running, and four major ports for international trade, i.e. Dalian, Qingdao, Ningbo, and Shenzhen are capable of handling third generation container vessels (with size of 4,000 TEU, or Twenty-foot Equivalent Unit). Shanghai as China's largest seaport is limited by its water depth, but it has 10 container berths and has doubled its throughput in the five-year period.

It needs to point out that Hong Kong, the world busiest container hub, now becomes part of China's port system. This is not just due to the return of sovereignty, but also because of the large volume it handles for the mainland as well as the integration of the port with the nearby port Yantian, forming a twin-port load center for South China region (Wang and Slack 2000).

Land connections from these coastal ports also started to take shape. For example, the highways container collecting points increased from 170 in 1990 to 250 in 1995, and the numbers of railway stations capable of handling some business for containerized transportation increased to 476 in 1995 from 61 in 1990 (Zhang, J. 1996).

1.2.5 A Boom of Airports and Air Transport

In the first five years of 1990s, there were 360 new international and domestic flight routes started to operate, 19 new airports constructed, and 15 expanded or upgraded. In 1995, the total volume of passengers flown hit a record high of 50 million, and jumped from the eighteenth in the world in 1990 to the eleventh in terms of such total turnovers.

As building and upgrading of new airports are no longer tightly controlled by the central government, local governments want to stimulate their economy by attracting more business activities and tourists by constructing a new airport, newly established regional airlines also want to have a better "home" as headquarters. The trend for larger and better airports, too, continued in the last two years.

1.2.6 Limited Improvement in River Transport

In comparison to the other modes, improvement of river transport had made less progress. Some 3,000 km of river waterways have been re-dredged, and 116 new piers or decks built during 1991-96.

1.2.7 Rapid Expansion of Long-distance Telecommunication Networks

There has been important development in several aspects of long-distance telecommunication. First, all provincial capital cities except for Tibet have been connected by fabric-optic cables through the establishment of Telecommunication Trunkline Network (TTLN). The TTLN is the backbone of China's telecommunication network, consisting of inter-provincial fabric-optic cables, digital microwave cables, concentric cables, and satellite ground stations. The 22 inter-provincial fabric-optic cables and 19 satellite ground stations installed during the Eighght FYP have fundamentally changed the backward situation of China in long-distance telecommunication. The 22 fabric-optics (FO) are: Shanghai-Nanjing, Shanghai-Fuzhou, Fuzhou-Guangzhou ("Southern Coastal FO"), Beijing-Jinan-Nanjing, Beijing-Wuhan-Guangzhou (elevated), Zhengzhou-Xi'an, Xi'an-Chengdu, Chengdu-Kunming, Guangzhou-Haikou, Guangzhou-Nanning, Beijing-Shenyang-Harbin, Xi'an-Lanzhou-Wulumuqi, Wulumuqi-Yining, Nanning-Kunming, Zhengzhou-Xuzhou, Changsha-Nanning, Wuhan-Chongqing, Beijing-Wuhan- Guangzhou (underground), Beijing-Taiyuan-Xi'an, Beijing-Huhehoute-Yinchuan-Lanzhou, Hangzhou/Fuzhou-Guiyan. There are a few lines still under construction such as Beijing-Jiujiang-Guangzhou, and Dalian-Shanghai ("Northern Coastal FO") (Liu, S. 1992).

Second, most public telecommunication networks have been upgraded from manual exchanges to automatic exchanges and from analog mode to digital mode. Third, 100,000 km fabric-optics and 786,000 km microwave lines have been newly installed. 315 cities and 1920 counties are now connected with international long-distance services, double the number of connected cities of 1990.

Figure 10-2: Telecommunication Truck-line Network in China

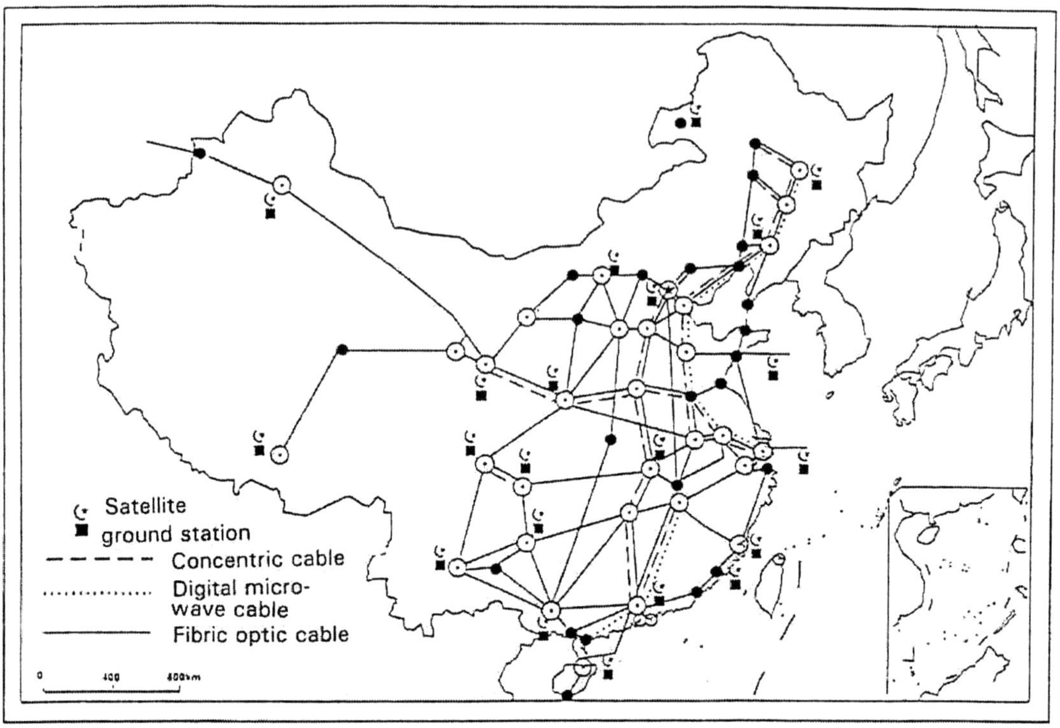

2. Increased Regional Disparities

2.1 The Investment Gap

T&C infrastructure, to a large extent, determines the infrastructure foundations of a region and hence its development potential and environment to attract more investment. During the 8th FYP, about 80% of state investment for basic construction of fixed capital goods were put into T&C infrastructure. However, as shown in Figure 6.3 and Table 6.2, regional gaps of investment in this sector are obvious and which have enlarged regional disparities of development potential in general.

Guangdong had the largest (70 billion Yuan in 1991-95) and the biggest proportion (24.9%) of its capital investment in T&C infrastructure. Meanwhile, Qinghai, one of the poorest interior provinces, only had about 1 billion Yuan for the same purpose.

Regional disparities are somewhat different when considering the population size of each province by using the index of investment per capita during 1991-95:

- Above 1,000 Yuan: Guangdong (1078), Hainan (1953)
- 500~1,000 Yuan: Beijing (915), Tianjin (600), Shanghai (675), Fujian (528), Tibet (988)
- 200~500 Yuan: Shanxi (312), Inner Mongolia (295), Liaoning (273), Heilongjiang (253), Jiangsu (259), Zhejiang (366), Shangdong (217), Hubei (238), Guangxi (214), Yunnan (237), Qinghai (257), Ningxia (414), Xinjiang (313)
- Below 200 Yuan: Hebei (157), Jilin (133), Anhei (119), Jiangxi (192), Henan (127), Hunan (151), Sichan (136), Guizhou (61), Shaanxi (141), Gansu (63).

Table 6.2 Provincial Variation of Transport & Communication Basic Construction Investment, 1991-95

Share* / Scale	≥20%	15%~20%	10%~15%	<10%
≥15,000 million Yuan	Fujian Guangdong	Jiangsu Zhejiang Shandong		
10,000~15,000 mil.			Henan Hubei Sichuan	Liaoning
5,000~10,000 mil.	Jiangxi	Shanxi Hunan Guangxi Hainan Yunnan	Beijing Tianjin Hebei Inner Mongolia Heilongjiang Anhui	Shanghai
<5,000 mil.	Tibet	Ningxia	Shaanxi	Jilin Guizhou Gansu Qinghai Xinjiang

- The share means the proportion of T & C infrastructure investment in the total basic construction investment.

Source: *1996 China Transportation Statistics Yearbook*, Beijing: China Transportation Statistics Yearbook Press.

The numbers show that among the most intensively invested provincial units, Tibet received a special treatment from the state, as it is listed along with three southern coastal provinces (Guangdong, Hainan, and Fujian) and the three metropolises (Beijing, Shanghai, and Tianjin). Among the lowest that had less than 200 yuan per catipa, Shaanxi is more or less a surprise. If we look at the Table 6.2, this province is also located in the bottom right corner, showing little investment into the T&C infrastructure. It seems to reveal the slow pace of development and low degree of openness of the economy in the entire northwest China where Shaanxi is the leading province.

Figure 6.3: China's Investment in Transport & Communication Infrastructure

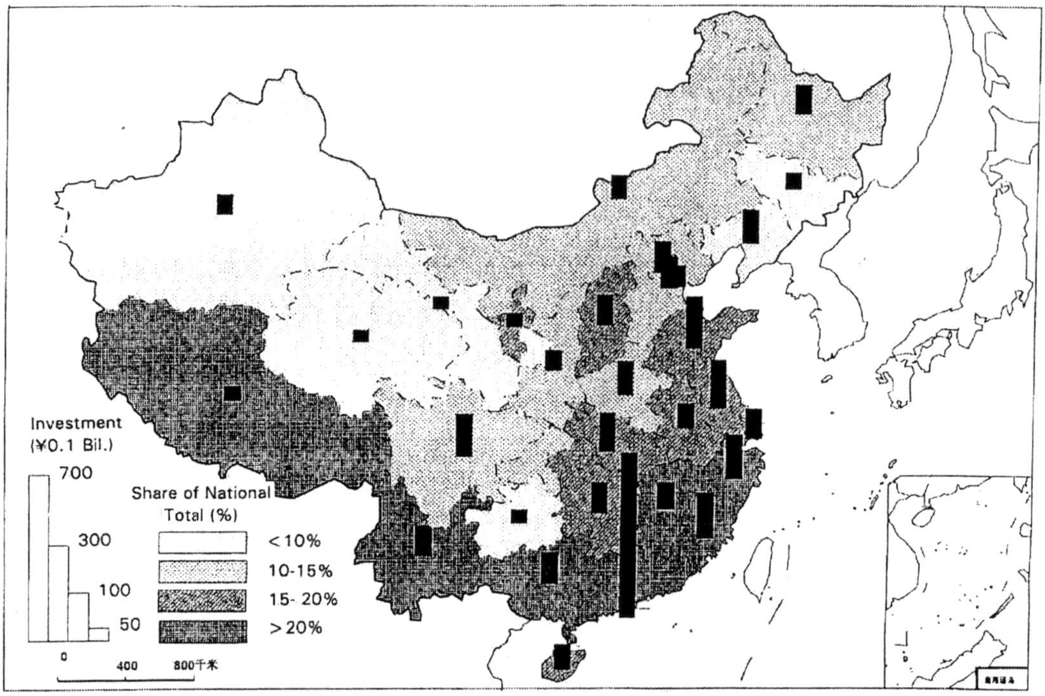

In general, the scale of capital investment in T&C infrastructure in the coastal region was significantly larger than in the Central and West regions during 1991-95, as indicated by Table 6.3. Roughly, more than half of the T&C infrastructure investment was gone to the coastal region, although it makes up of less than 15% of the territory of China.

Table 6.3 Regional Variation of T & C Share in Total Basic Construction Investment, 1991-1995

Regional Share	1991	1992	1993	1994	1995	1991-95
East (Coastal)	36.8	53.1	37.7	41.3	41.3	42.4
Central	13.2	15.0	13.1	15.5	15.5	14.8
West	10.8	13.5	10.5	9.5	9.5	10.4
Indivisible	39.2	18.4	18.4	33.7	33.7	32.4
Total	100.0	100.0	100.0	100.0	100.0	100.0

Source: Compiled from *Statistics Yearbook of China*, various years, Beijing: China National Statistics Bureau.

Reinforcement of transport and communication infrastructure in the coastal provinces seems to continue when examining the list of "key infrastructure projects of 1995-96" considered by the central government given in Table 6.4. Among the 44 transport or communication related projects, 20 projects were located in individual coastal provinces, and 11 out of the 13 "cross-provincial" projects were related to or within the coastal region. The Central and West regions together got only 13 or less than 1/3 of the projects listed in this central government plan.

The new investment obviously will enlarge the gap between the coastal and the other two regions that already existed prior to 1990. This is not only due to the differences in economic strength among the regions and provinces, but also attributed to the special investment policies.

Table 6.4 Numbers of Major Infrastructure Projects in China, 1995-96, by Region

	Railways	Waterways	Highways	Airports	Post and Telecom-munication	Dams and hydrographic projects	Total
East	3	12	3	2	1	1	22
Central	2	1	1			3	7
West	5		1			2	8
Cross-region	7		1		5		13
Total	17	13	6	2	6	6	50

Source: Compiled from *China Statistics Yearbook*, various years, Beijing: China National Statistics Bureau.

2.2 The Growth Gap

The investments may be materialized into different forms of infrastructure to improve different transportation or communication systems. In this section, we describe how the projects constructed in the T&C sector vary spatially, which will help us to see more clearly the general trend of "the good becoming the better" in China's T&C infrastructure development.

The growth gap is confirmed by figures in Table 6.5 that the regional split of the net growth in transport and communication sector among three large economic zones was similar to that of GDP, while was in reverse to the population growth. For example, while the coastal region accounted for 38.8% of total net population growth of China during 1991-95, it generated 63.5% of the net national GDP, and accounted for 60% increase in roads, 41% in railroads, 65% in telephone subscribers, 62% in urban and 77% in rural telephone exchange stations in the same period.

Table 6.5 Regional Contribution to National Net Growth in Transport and Communication (%)

Region	Highway	Railway	Long-distance Phone Lines	Telephone Exchange Stations			Telephone Subscribers	GDP	Population
				Total	Urban	Rural			
East	59.7	41.3	65.2	62.0	57.6	77.3	65.4	63.5	38.8
Central	19.8	25.2	21.3	25.0	27.3	17.0	23.3	23.7	33.3
West	20.5	33.5	13.5	13.0	15.1	5.7	12.8	12.8	27.9

Source: Compiled from *China Statistics Yearbook*, various years, Beijing: China National Statistics Bureau.

Regional disparities in installed facilities are found mainly in the following five aspects that all favored the coastal region:

- Construction of high-standard highways
- Construction of trunk railway lines
- Hub ports development
- Airports development
- Diffusion of Telecommunication

2.2.1 Rapid Construction of High-standard Highways in the Coastal Region

The first half of the 1990s witnessed the most rapid growth of high-standard highways in China's history. The so-called high-standard highways include the expressways and the highways that meet the national standards as Grade I highway. In the five-year period, some 50,000 km of high-standard highways were built, that doubled both the total length and density of the highways of high quality, the latter reached to 1.01 km/100km^2 in 1995. By province, most rapid growth is found in Guangdong, Shandong, and Liaoning, each with a contribution of 14.7%, 13.7%, and 7.3% respectively to the national net growth in high-standard highways. The disparity becomes more obvious if we measure the share of net high-standard highway growth by the three geographical zones: 60% of newly built highways and 55% of new high-standard highways are concentrated in the coastal region that has only 15% of the national territory. The density of high-standard highways there increased from 1.81 km/100 km^2 in 1990 to 3.95 km/100 km^2. In contrast, the West region of 56.6% of the national territory has only 20% of newly built highways and 14% of high-standard highway growth.

2.2.2 Upgrading and Construction of Trunk Railway Lines Mainly in the Coastal and Central Regions

Five years are not long for railway construction. However, through the partial improvement and upgrading and trunk lines under construction, one still confirms the regional bias trend of "the winner takes it all". The Beijing-Hongkong railway connecting ten provinces and the Nanning-Kunming railway linking Guanxi with Yunnan are among the key projects favoring the coastal and central regions. The two major new lines were put into operation in this period are the Datong-Qinhuangdao and Houma-Yueshan lines for improving coal export from Shanxi. Other new lines include the Guangzhou-Mazhou-Shantao and Sanshui-Maoming in Guangdong, the Jining-Tongliao line in Inner Mongolia, and the Hefei-Jiujiang line in Anhui.

Major upgrading projects are also concentrated in the East and Central regions. Except for the double-tracking program for the Lanzhou-Xinjian railway to improve the basic railway accessibility of the West region, all other projects such as the upgrading of the Beijing-Guangzhou railway, double-tracking of the Hangzhou-Zhuzhou, Xuzhou-Lianyungang, and Qingdao-Jinan lines, and electrifying of theYingtan-Xiamen lines all occur or are related to the coastal region. Statistics show that 41% of the newly completed railways are located in the coastal, 25% in the Central, and 34% in the West regions.

2.2.3 Coastal Hub Ports as A New Emphasis of Development

It has been a natural outcome of the country's open-door policy to speed up the development of coastal ports, especially major national and international hubs, to facilitate ever- increasing trade. During 1991-95, 100 deep-water berths were built in 56 coastal ports, 57 of which are located in the seven major ports of Dalian, Tianjin, Qingdao, Shanghai, Ningbo, Shenzhen, and Guangzhou. The new container terminals in these ports have a significant impact on the spatial pattern of China's containerized transport. Shanghai, Shenzhen, and Tianjin have by now established direct calls from major international shipping alliances. As a result, the role of Hong Kong as China's only international transshipment hub port for containers has ended. At the same time, the provinces along the coast began to compete for regional hub role in North, East, and

South China. An inland multi-modal transport system based on these existing or potential ports also has started to take shape.

2.2.4 Concentration of New Airports in the Coastal Provinces

The trend of "the winner takes it all" is easily observed in airport concentration. Among 19 newly built or relocated airports during the Eighth FYP, 11 are located in the coastal region. Since most of these new or upgraded airports are large in size, they have led to a much larger handling capacity of the coastal cities than the interior cities. For example, in 1995, 15 out of 24 airports that handled more than one million passengers were in coastal cities.

2.2.5 Diffusion of Telecommunication in Coastal Rural Areas

Telecommunication is among the sectors that have gained nationwide rapid growth since the economic reforms began in 1978. Besides, its actual and target user groups have become different spatially, indicating a clear diffusion process of the technology in geography. Two indicators – the growth of telephone exchange stations and the growth of telephone sets used in rural areas tell the story. Table 6.6 shows that during 1991-1995, four coastal provinces (Guangdong, Jiangsu, Zhejiang, and Shandong) together took 67% of the national net growth of telephone exchange stations and 82% of the net growth of total telephone subscribers in the rural areas of China. This implies that while the Central and West still had their telecommunication limited in their cities, the same technology was wide spread in the rural areas in the East.

Table 6.6 Top Four Provinces for Net Growth and Share of Rural Telephones in China, 1995

	Rural Exchange Stations		Rural Telephone Subscribers	
	Net Growth	**Share (%)**	**Net Growth**	**Share (%)**
Guangdong	3,655,000	26.9	2,301,000	28.6
Jiangsu	1,807,000	13.9	1,314,000	16.3
Zhejiang	1,606,000	11.8	1,190,000	14.8
Shandong	978,000	7.2	5,852,000	7.3
National Total	*13,602,000*	100.0	*8,056,000*	100.0

Source: Compiled from 1996 *China Statistics Yearbook*, Beijing: China National Statistics Bureau.

3. Classification of Provinces by T&C Infrastructure

3.1 Classification Methods

It is necessary but difficult to assess comprehensively the provision of T&C infrastructure at the provincial level. We try to classify the provinces both qualitatively and quantitatively to measure their "hardware" that existed prior to 1991 and their improvement in 1991-95. The provinces are categorized according to five aspects of the "hardware" of T&C for further investment and development. Then we weigh each aspect to establish a ranking that indicate relative positions of the provinces as a reference. The following are the five aspects for consideration:

- Railway Infrastructure: four categories are considered as follows, regarding to the pre-existed foundation and improvement:

A. of above-average foundation, or having national trunk line(s) upgraded or new trunk line(s) constructed, significant improvement of accessibility;
B. of above-average foundation, or having some lines upgraded and some improvement of accessibility to reach that level of foundation;
C. of average foundation after some improvement;
D. of below-average foundation and no obvious improvement.

- Highway: Provinces are classified by four indicators, i.e. overall density, net growth in overall density, high-standard highway density, and its net growth, as shown in Table 6.7:

Table 6.7 Classification of Provincial Highway Condition

	Density	Net growth of density in 1990-95	High-standard highway density	Net growth of high-standard highway density in 1991-95
A	25km/100km^2 (=2 time of national average) or above	4km/100km^2 (=3 time of national average) or above	3km/100km^2 or above	1.5km/100km^2 (=3 time of national average) or above
B	18~25km/100km^2	3~4km/100km^2	2~3km/100km^2	1.0~1.5km/100km^2
C	10~18km/100km^2	2~3km/100km^2	1~2km/100km^2	0.5~1.0km/100km^2
D	<10km/100km^2	<2km/100km^2	<1km/100km^2	<1.0km/100km^2

- Telecommunication: Three indicators are used for classification: (1) number of sockets for long-distance calls, (2) number of exchange stations, (3) number of telephone subscribers. The four categories are:

A. of good foundation and keeping pace with or above the national growth rate;
B. of above-average foundation and keeping pace with the national growth rate;
C. of average foundation but keeping pace during 1991-95 with the national growth rate;
D. of lower-than-average foundation but catching up with the national growth rate.

- Ports (both coastal and river): The four categories are based on the size and function of the ports:

A. having national or international hub port(s);
B. having important regional port(s);
C. having important local port(s); and
D. having ports.

- Airport: Similar to the port categories, four classes are derived based on having or not having the hardware facilities and services for operating the systems as (A) international airport, (B) national airport, (C) regional airport, or (D) local airport only.

3.2 Development Categories

Putting the above indicators together as shown in Table 6.8, the 30 provincial units of China can be divided into four categories according to their T&C infrastructure by the end of 1995. These categories are described as below, and are also mapped in Figure 6.4:

Figure 6.4: Regions of China Classified by their Situation in Transport & Communication Infrastructure Development

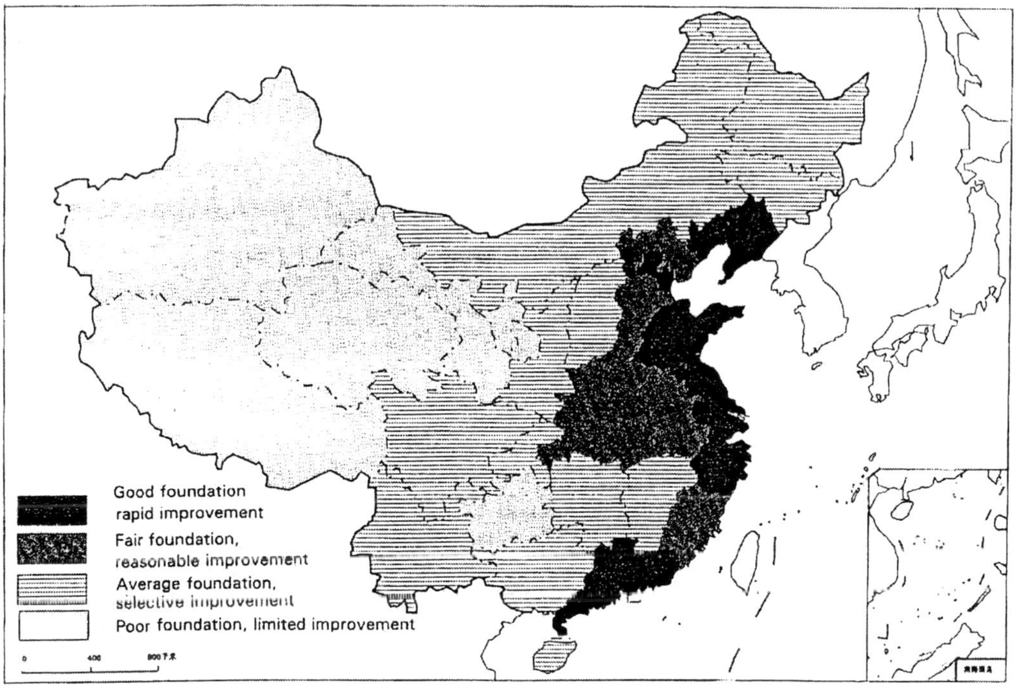

I. those of good foundations, developed very fast with significant improvement: Beijing, Liaoning, Shanghai, Tianjin, Jiangsu, Zhejiang, Shandong, and Guangdong;

II. those of relatively good foundations, and developed fast with important improvement: Hebei, Anhui, Fujian, Henan, and Hubei;

III. those of average foundations, and with significant improvement in some areas: Shanxi, Inner Mongolia, Jilin, Heilongjiang, Jiangxi, Hunan, Hainan, Guangxi, Sichuan, Yunnan, and Shaanxi;

IV. those of poor foundations and slow in development: Tibet, Gansu, Ningxia, Qinghai, Xinjiang, and Guizhou.

Provinces in Category I are all located in the coastal region and had good foundations prior to the 1990s while still had gained great improvement. It is worth attention however that some provinces in the Central region, such as Anhui, Henan, and Hubei in Category II have had significant changes in their infrastructure and hence have great development potentials. If there is any single province that stands out in the period, it is Guangdong. Since in this five-year period she turned herself into the most advanced province in the country in terms of T&C infrastructure.

We may call this as "Guangdong phenomenon". With a great leap forward in the T&C infrastructure, Guangdong made a remarkable achievement during the 8th FYP. Her contribution to the national net growth in every major aspects for 1991-95 tell the same story: 23% of newly built highways (including 15% of high-standard highways and 21% of expressways); 30% of newly built coastal deep-water berths; 17% of sockets for long distance telephone calls; 14% of telephone exchange stations (including 10% of that in urban and 27% in rural areas); and 15% of total telephone subscribers (including 29% of those in the rural areas). In terms of rankings or absolute numbers, the progress made by the province is also phenomenal. By the end of 1995, Guangdong ranks number one in China in highway density (47 km/100 km^2), number two in highway length (84,600 km), number three in high-standard highway length (5.3 km/100 km^2) only after Shandong and Jiangsu. The telephone ownership reached to 10% of population, which is 2.3 times of the national average (4.66%). The major reason for this "Guangdong phenomenon" is the adoption of diversified financing and operation forms. For attracting necessary capital to finance the T&C projects, exchanging user-ship of land for highway or port construction, selling the old (infrastructure) for the new, have been a few most popular practices, among others. BOT (build, operate, and transfer) has been one of the most popular operational mechanisms.

During 1991-95, the total investment put into the T&C sector for infrastructure in the province was about 70 billion Yuan or 8 billion US dollars, which is equal to the total investment for the entire Central region for the same purpose in the same period. However, a lack of planning and coordination has been found not in a few cases. Over-development in the T&C infrastructure due to overestimating the demand was also seen in places, when a simple belief that "build roads if you want to get rich" has been overwhelmingly regarded as a politically and economically correct slogan.

Table 6.8 Qualitative Assessment of Transport and Communication Infrastructureby Provincial Unit

	Highway	Railway	Port	Airport	River Waterway	Telecommunication
Beijing	A	A	B	A		A
Tianjin	A	A	A	B		C
Liaoning	A	A	A	A		A
Shanghai	A	B	A	A	A	A
Jiangsu	A	A	B	B	A	A
Zhejiang	A	B	A	A	B	A
Shandong	A	A	A	A		A
Guangdong	A	A	A	A	A	A
Hebei	A	A	B	B		B
Anhui	A	A	C	B	B	C
Fujian	B	B	B	B	C	B
Henan	A	A		B		B
Hubei	A	A	C	B	B	B
Shanxi	B	A		C		C
Inner Mongolia	D	B		C		C
Jilin	D	B	D	C	D	C
Heilongjiang	D	B	D	C	C	C
Jiangxi	C	A	C	C	B	C
Hunan	A	B	C	C	B	C
Hainan	B	C	B	C		C
Guangxi	C	B	B	C	D	C
Sichuan	C	C	C	B	B	C

Yunnan	C	C	D	B	D	C
Shaanxi	C	C		A		C
Guizhou	D	C		C	D	D
Tibet	D			D		D
Gansu	D	C		C		C
Ningxia	C	C		C		D
Qinghai	D	D		D		D
Xinjiang	D	D		C		D

3.3 Quantitative Ranking of Provinces

Different modes or ways of transportation and communication do not have the same influence on investment decisions and on economic development in general. For this reason, we give weights here on the five aspects discussed earlier, based on our experience and understanding of the importance of each aspect for forming the environment for further investment and economic growth in the case of China. The baseline is set by the D category of highways. The categories of each aspects of T&C infrastructure are weighed as below:

Highways:	A=3,	B=2.5,	C=2,	D=1
Railways:	A=4,	B=3.5,	C=2.5,	D=2
Ports (coastal):	A=3.5,	B=2.5,	C=1.5,	D=0.5
Airports:	A=3,	B=2,	C=1,	D=0.5
Ports (river):	A=2,	B=0.5,	C=0.2,	D=0.1
Telecommunication:	A=2,	B=1.5,	C=1,	D=0.5

Applying this weighting system to the provinces, we have the following ranking as a reference to indicate the development status of T&C infrastructure as investment environment for further development:

Guangdong (17.5),
Shanghai (17.0),
Liaoning (15.5),
Jiangsu (15.5),
Shandong (15.5),
Zhejiang (15.5),
Beijing (13.5),
Tianjin (13.5),
Hebei (13.0),
Hubei (12.5),
Fujian (12.2),
Anhui (12.0),
Henan (10.5),
Guangxi (10.1),
Hainan (10.0),
Hunan (10.0),
Jiangxi (10.0),
Sichuan (10.0),
Shaanxi (9.0),
Yunnan (8.6),
Shanxi (8.5),
Heilongjiang (7.2),
Jilin (7.1),
Inner-Mongo.(6.5),
Ningxia (6.5),
Gansu (6.0),
Guizhou (5.6),
Xinjiang (5.0),
Qinghai (4.5),
Tibet (2.0).

4. Node-Corridor Development

Having evaluated T&C infrastructure by province, we must point out that in essence, transportation and communication tend to connect places or nodes in linear forms. When an economy develops rapidly, it is seen often that the variations of T&C provision appear to form some geographical corridors along which accessibility between nodes is much higher than that outside the corridors. In this section, we are going to portray these corridors formed or further developed in 1991-95. We have thus identified five linear concentrations of T&C construction by the end of 1995: along the Changjiang (Yangtse) River, along three major railway corridors and along the coast. Such node-corridor linear development with relatively advanced T&C infrastructure is significant in China's regional economy, and is also important for us to understand better the spatial dynamics of the country.

4.1 The Changjiang (Yangtse) River Corridor

Changjiang (Yangtse) River could be the transportation corridor of the longest history in China in linking the prosperous coastal areas with its most populated inland province of Sichuan. However, for one reason or another, the fundamental improvement and upgrading of this transport corridor had not been made until the 8th FYP. It was decided in that plan to: (1) revitalize river trade, (2) develop Shanghai as the largest trade center and port city, and (3) start to construct high-standard highways, including one parallel to Changjiang to connect Shanghai with Chengdou, Sichuan, and (4) construct some railways along the River. These decisions are important, knowing that previously the central government had insisted on a principle to relay mainly on the river transport along this corridor and which failed due to a variety of practical reasons.

In the new age of multi-modal containerization and with Shanghai at the river mouth as the busiest container port on the mainland, the container river trade has been developing very fast. It involves some 30 river ports and 40 cities, including Nanjing, Wuhan, and Chongqing. By 1995, Shanghai already jumped into number 19 in the world container ports in terms of TEU throughput.

Linking the river port cities are several expressways completed in the 1990s. They include the Nanjing-Shanghai, Shanghai-Hangzhou, Chengdu-Chongqing, Hefei-Nanjing, Nanchang- Jiujiang lines, plus the passing ones such as the Beijing-Zhuhai line that passes Wuhan. Fabric-optical telecommunication lines were also constructed along the River from Nanjing to Wuhan and then to Chongqing.

We notice however, after having discussed to developments achieved, that the most uncertain and probably most under utilized transport capacity could still be the Changjiang River itself. There are two major uncertainties: (1) how and to what extent the river transport could be better organized to meet the demand and new challenge for containerized transportation; and (2) how and to what extent the construction of the Three Gorges Dam would affect the river transport.

4.2 An Improving Corridor along Longhai-Lanxin Railways

The Longhai-Lanxin railway refers to a connection of two railway lines of Longhai and Lanxin, which form the longest east-west trunk railway in China, i.e., from Lianyungang in Jiangsu, via Zhengzhou, Xi'an, Lanzhou, to Ulumoqi in Xinjiang. The importance of this line is that it is part of the Euro-Asia Railway Bridge, the land "short-cut" linking East Asia to the European continent. The line has been kept improving since it opened in 1990, particularly after the double-tracking project for Lanzhou-Ulumoqi section has completed. This longest east-west railway has operated 35 container collection stations for the Euro-Asia trade, and interchanges with 11 north-south railway lines at various cities. As all these railways are connected in the middle section of the Longhai line, the most accessible and hence economically active areas along the line are the cities between Xuzhou in Jiangsu to Lanzhou in Gansu, mostly in Shaanxi, Hehan, and Anhui provinces.

The longest east-west high-standard national trunk highway from Lianyungang to Hoerguosi parallel to Longhai-Lanxin railway is now under construction. A few key sections of it, such as the expressways of Xi'an-Baoji, Xi'an-Lintong, and Kaifeng-Luoyang, have been completed and started to operate in the 8th FYP. Other sections in Henan province are about to be completed. These new expressways connecting the capitals and major cities of the interior provinces with the coastal region just starts to pick up its traffic. It has large capacity surplus waiting for potential development in future. Similar capacity surplus is also found in telecommunication facilities such as the newly completed fabric-optics corridor from Jiangsu to Xinjiang along Xuzhou-Zhengzhou-Xi'an-Lanzhou-Wulumoqi-Yining.

As this corridor has been upgraded for more trade along the Euro-Asia land bridge, we have reservation about the future demand for the bridge. The uncertainties involve many problems in more than 10 countries that it passes, particularly those of the former Soviet Union. In the near future, the new T&C infrastructure along this corridor will serve mainly domestic development. Further utilization and growth of Lanyungang port for example, will be largely depended on whether it can provide better connection for the inland provinces along the corridor.

4.3 Upgraded Harbin-Dalian Corridor in Northeast China

The Northeast has been the oldest industrial region with the highest railway connections and density in China. Although the region has suffered seriously for the downturn of its heavy industries and the problems accumulated in its state-owned enterprises since 1990, several important development programs in transport and telecommunication have upgraded the infrastructure there, particularly along the Harbin-Dalian corridor. First of all, the expressway connecting Shenyang with Dalian built at the end of 1980s is China's longest trunk expressway. Since 1990, the construction of high-standard highway systems around this expressway has continued and has made the region achieve the highest expressway density in the whole country.

Secondly, Dalian port has expanded with three new deep-water container terminals during the 8th FYP, and has kept itself within the top four ports in the country and become the load center for the entire Northeast China. For its unique geographical location in the region, the port city has been well connected by all the major means of T&C, including expressways and double-tracked rails, airlines, pipelines, and fabric-optic cables to all the three northeast provinces.

The undergoing programs for T&C infrastructure upgrading there include the electrification of the Harbin-Dalian railway started in the early 1990s, the completion of which will further the connecting capacity within the Northeast, and reinforce the leading position of Dalian in an opened market economy with more trade by sea. Indeed, the better connection a city or area is established with these coastal cities favored by the current economic structure and state policies, the more likely for its economy to follow the general trend in China toward a market-oriented development.

4.4 The Most Connected North-South Corridor:Beijing-Guangzhou

The Beijing-Guangzhou T&C infrastructure corridor is the most important among all the corridors discussed here. Being aware of its importance, the central government and local authorities keep investing to strengthen the T&C infrastructure there. In 1991-95, the Ministry of Railways upgraded the Beijing-Guangzhou railway and it can now handle 70 million tons of freight per year and 50 pairs of passenger trains per day (Ni 1997).

The most important national trunk express highway from Beijing to Zhuhai is partially completed and put into operation, such as the sections of Beijing-Shijiazhuang in Hebei, Guangzhou-Zhuhai in Guangdong, Zhengzhou-Xuchang in Henan, and Changsha-Xiangtan in

Hunan. In terms of telecommunication, the Beijing-Guangzhou elevated fabric-optics and Beijing-Wuhan-Guangzhou underground fabric-optics connections were built and put in use in the first half of the 1990s.

Recognizing that the Beijing-Guangzhou railway has been operated for more than a half century connecting many large cities, we expect that the completion of the parallel express highway will stimulate significant economic development along the corridor. The role of the railway there will be affected by the highway system. The impact of the new expressway on the rail is largely dependent on whether the rail can upgrade itself to effectively and efficiently handle multi-modal container transportation.

4.5 Advanced Coastal Port Cities with Strong Connections to Nearby Provinces

Port cities along the coast do not form a transport corridor by themselves. However, the cities have gained location advantage through the growing national and international trade and have together formed an interface between China and the rest of the world due to their port status. The special policies such as setting up Special Economic Zones (SEZ) and Coastal Open Cities have attracted not only foreign capital but also inland provinces to invest for export-led development. It may be debatable that the motives of most investors were for establishing export-oriented production base or for China's market. However, an undebatable fact is that both the port facilities for export and the land, air and river transport for linking these cities with other part of the country have been markedly improved.

The most advanced coastal port cities on the Chinese mainland include Shanghai, Dalian, Tianjin, Qingdao (Tsing Dao), Ningbo, Shenzhen (including Yiantian), and Guangzhou, all of them playing more important role as China's economy goes global. Among them, Dalian, Ningbo and Yiantian in Shenzhen were recognized and planned by the central government as China's major hub ports for international container shipping. This has made Shenzhen as a port city for the first time, challenging Hong Kong's position as the load center for the entire South China.

The land transportation for these seven port cities, plus two smaller port cities specialized for energy export (Qinghuandao in Hebei for raw oil and coal, Rizhao in Shangdong for coal), has all been upgraded. By the end of 1996, the above seven cities all have double-tracked high capacity railways and express highways linking with nearby major cities and provinces. Their land transports, newly built or under construction, is also networking the second-tier port cities such as Xiamen in Fujian, Zhanjiang, Shantou, and Zhuhai in Guangdong, and Lianyungang in Jiangsu.

In the early 1990s, high-quality telecommunication with fabric-optical cables started to be formally installed and operated in China. The lines have been and will continue in the near future to be highly concentrated along the coast for connecting the cities there with each other. These lines include Shanghai-Hangzhou, Jinan-Qingdao, Beijing-Shenyang-Harbin, Beijing-Wuhan-Guangzhou (Canton), South Coast, and Hangzhou-Fuzhou.

In addition to the linear concentrations and node-corridor development discussed above, we have also identified some provincial infrastructure corridors, such as the one from Jinan to Qingdao in Shandong province, and some regional transport links that may have potential to form corridor development, such as the Beijing-Jiujiang-Kowloon railway which was completed in 1996.

5. Conclusion: A Summative Assessment

5.1 The Characteristics of T&C Infrastructure Development in the 1990s

5.1.1 Supply for Urgent and Short-run Needs

For a few decades from the 1950s to 70s, China had a planned economy and a general policy to minimize the total transportation as a societal goal, since transport was regarded as one of the 'non-production' sectors. The cumulated demand from that time and the rapid growth in the 1980s led to a status of constant shortage of supply in terms of T&C infrastructure. Investment and construction since then had largely been for urgent needs. As a result, the long-term plans tended to be made at low standard and then failed when the demands increased rapidly. For example, in the 1992 plan by the Ministry of Transportation for the national trunk roads system, expressways were not considered as an essential part, which was turned out to be a major mistake.

5.1.2 An Economic-growth Driven Model for Development

T&C construction in China has always been driven by economic growth, and it was not exception in the period of the 8th FYP. Although for a developing country doing that does not surprise anyone, it does create problems. For example, the equity and welfare issues in developing highways to facilitate the use of cars in cities have never been raised. The discussion on the large amount of cultivated land being consumed by roads has also limited to a small academic circle. Indeed, T&C construction in China has been treated as the infrastructure for economic growth only, not for social development.

5.1.3 Market-led Regional and Modal Bias

The above two characteristics are reinforced by regional and modal biases under market mechanisms in the 1990s. As discussed previously in detail that "the good becoming the better" model means that coastal provinces that are economically more active have captured much more investment in T&C infrastructure and the regional gap has been enlarged. Accelerated growth in highway faster than other modes also reflects the same development bias, as shown in Table 6.9. The major reason for this, as we are going to discuss below, is the diversified resources and market-oriented institutional setting for financing T&C infrastructure construction.

Table 6.9 Sectoral Structure of Investment in Transportation and Communication Construction by Planning Period

Unit: Percentage (%)

<table>
<tr><th></th><th>Railway</th><th>Highway</th><th>Waterway (River and Coastal)</th><th>Airway</th><th>Pipe Lines</th><th>Post and Telecom-munication</th></tr>
<tr><td>1st FYP (1952-57)</td><td>65.6</td><td colspan="2">27.9</td><td>1.1</td><td></td><td>5.3</td></tr>
<tr><td>2nd FYP (1958-62)</td><td>63.8</td><td colspan="2">30.3</td><td>1.2</td><td></td><td>4.7</td></tr>
<tr><td>Adjustment (1963-65)</td><td>63.2</td><td>23.1</td><td>6.1</td><td>3.6</td><td></td><td>4.0</td></tr>
<tr><td>3rd FYP (1966-70)</td><td>63.6</td><td colspan="2">30.4</td><td>1.4</td><td>0.2</td><td>4.6</td></tr>
<tr><td>4th FYP (1971-75)</td><td>54.5</td><td colspan="2">37.7</td><td>2.9</td><td>0.1</td><td>4.8</td></tr>
</table>

5th FYP (1976-80)	46.4	14.9	24.7	3.1	4.9	6.0
6th FYP (1981-85)	48.1	13.0	24.3	6.2	1.1	7.9
7th FYP (1986-90)	40.3	21.5	20.3	6.6	0.9	10.4
8th FYP (1991-95)	36.6	26.5	8.4	10.0	0.3	18.2

Source: Compiled from *China Statistics Yearbook*, various years, Beijing: China National Statistics Bureau.

5.2 The Major Causes for the Rapid Growth of T&C Infrastructure

It has been discussed in detail that large amount of capital were captured in the T&C sector during 1991-95, which has resulted in the greatest changes in a short period time that has ever happened in China. This trend of rapid improvement of T&C infrastructure has continued since. We believe that the following are the reasons for this rapid growth in the 1990s:

5.2.1 Imperative Demands for Higher Level of Domestic and International Trade, Travel and Exchange of Information Derived from the Economic Takeoff in the 1980s

In the 1980s, the economic reforms towards a market-oriented system undergone were not following the same pace for every sector and for every region. The T&C sector, although known for long as the bottleneck of country's economy, was behind many other sectors for adopting the means of market-oriented systems, such as privatization, deregulation and decentralization, and hence had been slow in responding to the rapid changes in other sectors. Consequently, the T&C bottlenecks became even more severe after the opening and reform, and demands for a fundamental upgrading of the T&C sector became imperative in the late 1980s.

5.2.2 Much Diversified Resources and Flexible Institutional Settings to Finance T&C Construction

The huge and urgent demand for more T&C provision in fact means an area of high potential return. When the finance systems reforms in the early 1990s opened this highly profitable sector in China, a great amount of investment came from diversified resources to finance T&C projects at both national and regional levels. The newly adopted mechanisms include establishing local collective funds, issuing stocks or bonds, loaning from large international financial institutions and foreign governments, and obtaining foreign direct investment (FDI) through BOT and other methods.

5.2.3 The Preferable Policies of Central and Local Governments for T&C Development

In the 8th FYP and the Ten-Year Development Strategic Planning (1991-2000), T&C and other infrastructure sectors were emphasized as preferred sectors for development. During 1991-95, there were also specific policies from transport-related ministries and local governments to stimulate T&C construction. Among them, the most important was to allow foreign investors to invest in highway, port, and airline industries.

5.2.4 The Cumulative Effects of Investment by Networking and Integration

Economy of scale in T&C infrastructure appears in forms of cumulative effects of networking and integration. For example, only when highways are all connected with each other, can they more effectively be used than when the lines are not linked with each other. In the 1990s, these cumulative effects of investment by networking nodes and integrating modes set in all T&C submarkets, particularly in highway and telecommunication systems. The improvement in port-highway connections for multi-modal container transport has also been significant.

5.3 Trends of Growth in Future

5.3.1 Rapid Growth of Supply Will Continue

In the Ninth FYP period or 1996-2000, the provision of T&C is expected to grow in at least by the same rate as the previous five years for three reasons. First, the demand induced by the economic growth will continue. Second, many of newly built T&C facilities such as toll highways have shown attractive and steady returns within a relative short term, which will lead to more interested investor to join the market. Third, the reforms of state-owned enterprises and the impact of financial crisis in the 1997/98 in many Pacific Asian countries have resulted in an important decision of China's government to invest more than ever in T&C construction, in order to absorb labor and to keep the economy growing.

5.3.2 Competition among the Transportation Modes and between Telecommunication Systems Will Become Severe

It has been the case since the early 1990s that competition between air, rail and long-distance bus services for passenger transport as capacity surplus appears for some routes and the prices started to reflect the demand. At the same time, introducing more operators has eliminated the monopoly in aviation and telecommunication industries. Fare war and merging will be seen often.

5.3.3 Regional Bias for T&C Construction Will Remain

Although the privilege given to coastal cities and hence their provinces such as lower tax contribution is diminishing, the stronger economy established there so far will keep up the investment for T&C in the better-off provinces, unless special policy incentives are set to favor others.

5.4 The Major Aspects for Improvement

5.4.1 Infrastructure Management

As a common problem in developing countries, poor management of existing and newly built T&C infrastructure has been found everywhere in China, due to the lack of experience, trained managers, and proper regulations and institutional settings. For example, more-than-necessary toll-booths are found along many expressways that greatly reduce the through capacity for the entire system. The lack of standardization of telephone systems used in cities is another example. In general, the current inefficient use of T&C infrastructure means a great potential to improve the systems not by building more 'hardware', but by providing better 'managerial-ware'.

5.4.2 Awareness of the Importance of Multimodalism and Logistics

While globalized world economy has been demanding for multimodality in transport, and focusing on logistics that include all steps of supply chains rather than transport alone, very limited attention on these is found in China. For example, the railways in plan have still not considered for double-stacked container trains. Logistics centers for freight processing near ports or airports are also rarely considered. Changes shall be seen soon in these areas as more international operators for logistics and multi-modal transport go in China. However, there is still a long way to go for institutional changes to facilitate such development.

5.4.3 Planning, Orienting, and Regulating the Profit-Driven Mechanisms for Sustainable Development

From a planned economy transforming to a market economy, China has gained much economic efficiency by the profit-driven mechanisms such as privatization. However, "deregulation" in China is quite different from those in the developed world in that what demolished are those tighten up the development in a centrally planned manner. For regulations for market operations, the country had little in the first place. This means that more regulations for market operation are needed, while China is becoming a more opened economy to the world. This is particularly true to the T&C sector. T&C construction is costly, of lasting impact, and could affect regional development pattern to a large extent. How to plan, guide, and regulate T&C for an economically affordable, environmentally acceptable, and socially equitable development is a tough task for Chinese governments and planners.

REFERENCES

1. China Transport and Traffic Association. 1996. A review of development of China's transport industry during the Eighth Five Year Plan. In *1996 China Transportation Statistics Yearbook.* Beijing: China Transportation Statistics Yearbook Press (Chinese text).
2. Li, Jinghe. 1996. Highway construction in the Eighth Five Year Plan. In *1996 China Transportation Statistics Yearbook.* Beijing: China Transportation Statistics Yearbook Press (Chinese text).
3. Zhang, Baisheng. 1994. National trunk road planning. In *1994 China Transportation Statistics Yearbook.* Beijing: China Transportation Statistics Yearbook Press (Chinese text).
4. Zhang, Jishun. 1996. Coastal port construction in the Eighth Five Year Plan. In *1996 China Transportation Statistics Yearbook.* Beijing: China Transportation Statistics Yearbook Press (Chinese text).
5. Liu, Shuping. 1992. Introduction to postal and telecommunication planning in the Eighth Five Year Plan. In *1992 China Transportation Statistics Yearbook.* Beijing: China Transportation Statistics Yearbook Press (Chinese text).
6. Shi, Shanxin. 1992. Introduction to railway Ten-Year Plan and the Eighth Five Year Plan. In *1992 China Transportation Statistics Yearbook.* Beijing: China Transportation Statistics Yearbook Press (Chinese text).
7. Ni, Zubin. 1997. *A Study on the Beijing-Kowloon Economic Corridor Development.* Beijing: Climate Press (Chinese text).
8. Sun, Guoqing. 1996. Utilization of foreign investment in transportation infrastructure construction during the Eighth Five Year Plan. In *1996 China Transportation Statistics Yearbook.* Beijing: China Transportation Statistics Yearbook Press (Chinese text).
9. Wang, Jixian and Slack, Brian. 2000. The evolution of a regional container port system: the Pearl River Delta. *Journal of Transport Geography* 8: 263-275.

Chapter 7

ENVIRONMENTAL POLICY AND ENVIRONMENT OF THE REGIONS

FU Bo-jie

Professor and Deputy Director, Ecological & Environmental Research Centre, Chinese Academy of Science

LI Ya-qin

Associate Professor, Institute of Microbiology, Chinese Academy of Science

The rapid development of China's economy since the 1980s has led to substantial increase in national strength and people's living standard. However, these achievements were made with high eco-environment costs in the form of serious pollution and ecological destruction. Some of the nation's natural resources are being consumed at an accelerated rate and some are even close to exhaustion. The polluted areas have been expanding from suburban areas of a few large- and medium-sized cities to many densely populated areas. About 800,000 square kilometres of plains, deltas and valleys in the country with a total population of around 150 million, has been classified as heavily polluted by industrial and urban sewage that threatened the health of tens of millions of people. Another million square kilometres of densely populated areas were subjected to moderate water pollution. Lakes such as Taihu, Chaohu and Dianchi were seriously polluted and large rivers like Huaihe, Changjiang, Liaohe and Zhujiang were polluted to different degrees. Environmental pollution will soon damage the foundation of many productive activities, if it remains unchecked. For example, from 1984 to 1994, there is a reduction of 7.784 million ha of cultivated land due to occupation by other uses. The rate of increase of such loss showing a close correlation with GDP growth, i.e. with an r value as high as 0.98 in the economy, is the essential reason for the dramatic reduction of cultivated land. It is also closely related to the degradation of forests and grasslands, acceleration of soil and water loss, desertification, and the drop in biodiversity (National Environmental Protection Agency, 1996).

Chapters 2 to 3 have revealed the differences in regional economic growth of various regions in China. Comparisons of the eco-environment problems and economic growth of the regions show that the following types of area are normally related to ecological destruction and serious environmental pollution:

- Areas with constant growth at a high speed, such as the Zhujiang delta, Changjiang delta and areas along the Jinan-Qingdao railway line;
- Areas with a high growth rate based on resources exploitation and large-scale development of basic industries, such as parts of Henan, Anhui, Shanxi, Sichuan and Yunnan; and

- Areas of natural ecological fragility where there was irrational human exploitation and utilization such as areas bordering of Shanxi-Shaanxi-Inner Mongolia, the limestone areas of Guizhou, Yunnan and Guangxi, part of the upper reaches of Changjiang, and the middle and lower reaches of Heihe in western Inner Mongolia.

Increasingly eco-environmental problems are likely due to accelerated industrialization and urbanization, in which short-term market behavior adds to the severity of the problems, leading to increase in waste water and waste gas emission. In major cities, the total suspended particles (TSP) and sulfur dioxide (SO2) concentrations in the atmosphere have exceeded the World Health Organization (WHO) standards by two-fold. In 1996, only less than one-fifth of the country's total territory urban domestic sewage was treated and acid rain had affected 29% of China.

At the Fourth National Conference on Environmental Protection held in 1996, Premier Li Peng stressed that the situation of environmental pollution in China was still quite serious, as China was developing towards rapid industrialization and urbanization. Pollutant discharge would increase, while the extensive economic growth pattern due to a comparative backward technical and management capabilities will intensify the exploitation of natural resources. In fact, environmental pollution around the cities is aggravating and expanding towards the rural areas and the scope of ecological destruction is enlarging. Both the State and local authorities have hence put forward new ideas on environmental protection through strengthening environmental management and law enforcement. Research mostly of technical and methodologically oriented have also been carried out.

The present chapter shall take a different approach. It shall analyse the country's environmental policies, trace their implemention at the State and local levels in 1995-1996, discuss the results of their implementation, analyse investments in environmental management and ecological rehabilitation of various regions and their benefits, and point out the present status and policy measures related to problems of the regional eco-environments.

1. A New 15-Year Environmental Strategy (1996-2010)

1995-1996 has witnessed major progress in environmental protection and the deepening of reforms in environmental policy and management (National Economic Protection Agency et. al., 1996; China Society of Environmental Science, 1996; Editorial Committee of China Environment Yearbook, 1995 & 1996).

In the Fourth Session of the Eighth National People's Congress of March 1996 "Outline of the Ninth Five-Year Plan for National Social and Economic Development and the Long-Term Objectives for the Year 2000" was approved. It sets out two fundamental shifts in economic structure and mode of economic development, and laid down the two important new strategies of (1) sustainable development and (2) to reinvigorating the country through education, science and technology. It has also set the objectives of environmental protection from the Ninth FYP period to the year 2010, stressing that efforts must be made to basically control the trend of environmental pollution and ecological damage and to improve environmental quality in most cities and regions by the year 2000.

In June 1996 the Chinese Government issued a white paper on "Environmental Protection of China" and convened in July the Fourth National Environmental Protection Conference in which a general plan was drawn up. These events served as a milestone in the history of environmental development in China. In August the "Decision of the State Council on Issues Related to Environmental Protection" was drafted and measures to guarantee the realization of the five-year national plan for China's environmental protection in ten related fields were ratified. In September 1996 the State Council approved the "Ninth FYP on Environmental Protection and the Long-Term Environmental Objectives for the Year 2010". Its appendixes, the "Programme on Controlling the Total Amount of Main Pollutants during the Ninth Five-Year Plan Period" and "China's Trans-

Century Green Project (First Tranche)" lays out in detail the two important strategies for the realization of environmental protection objectives during the Ninth FYP period.

In order to implement the new sustainable development strategy, a series of policies and measures concerning environmental protection are pursued on the basis of eight management systems on environmental protection, i.e.

1) A system of environmental impact assessment
2) The "Three simultaneous system" (i.e. along aside with any planning of a construction project, the environmental protection schemes should be simultaneously designed, constructed and put to operation)
3) A charge system for pollutant discharge
4) A license system for pollutant discharge
5) A treatment system of discharged pollutant within a specified time
6) A system of concentrated control of pollution
7) A responsibility system for environmental objectives
8) A quantified checking system for urban environmental comprehensive management and control

The "Decision of the State Council on Issues Related to Environmental Protection" specifies that when important decisions on economic development, especially on production projects, are to be made, economic, social and environmental factors should be taken into account comprehensively, and environmental impact assessment should be carried out to avoid "pollution first and treatment after". Three changes in the principles were made on industrial pollution prevention and control, i.e. to bring terminal abatement into source abatement, to turn scattered abatement into integrated and concentrated abatement, and to change from control on intensity to its integration with control on total volume control.

The present legal framework on environment and resources including six laws on environmental protection and nine on resources has gradually taken shape after a lot of legislative efforts by various levels of government, i.e. the Law on the Prevention and Control of Water Pollution; Law on the Prevention and Control of Atmospheric Pollution, Law on the Prevention and Control of Environmental Pollution by Solid Wastes, Law on Marine Environmental Protection; Law on Prevention and Control of Environmental Pollution by Noise; Law on Protection of Wild Animals; Forest Law; Grassland Law; Fishery Law; Law on Agriculture; Law on Mineral Resources; Law on Land Management; Water Law; and Law on Soil and Water Conservation. Under these laws are more than 120 administrative regulations by the State Council, over 600 regulations and 300 rules by local governments, as well as nearly 400 national environmental standards drawn up by various departments concerned under the State Council. Meanwhile the enforcement of environmental laws has been substantially strengthened.

In contracting work units for environmental impact assessment a practice of bidding has been introduced to enhance its efficiency and quality alongside with the "three simultaneous" system. Charge for pollutant discharge has changed from charge for pollution exceeding a set standard to total pollutant discharge, as well as from the evaluation method based solely on concentration to consideration of both concentration and total amount. In addition, it has also changed from a single factor consideration to multifactors, with a gradually increasing charge rate. To implement responsibility system for environmental protection, a declaration and register system on solid waste, noise and pollutant discharge was pursued and a license system on water pollutant and atmospheric pollutant discharge has been put in place at the prefecture and city level throughout the country.

For four consecutive years since 1993 a nation-wide inspection was maintained to ensure the enforcement of laws and regulations on environmental protection. Law violations were investigated and prosecuted and unlawful activities were exposed and reprimanded.

Within the Huaihe drainage basin in Henan, Anhui, Shangdong and Jiangsu over 1000 paper mills using chemical pulps with an annual pulp production capacity of below 5000 tonnes were being closed in 1996. A number of other serious polluting enterprises were brought under control and pollutants of the basin have since been largely eliminated. It provides a good foundation for the realization of the objectives of bringing Huaihe River clean by the year 2000. Coordinated action was also taken throughout the country in the year for the closure of over 50,000 small enterprises that emitted serious pollutants. The "Decision on Environmental Protection" in fact demands that pollutants emitted from all industrial enterprises of the country should meet State or locally specified standards by the year 2000. Enterprises unable to meet this deadline will have to close or switch to other lines of production.

Departments in various levels of government are urged to pay close attention to the formulation and perfection of economic policy and measures to promote environmental protection, to prevent and control environmental pollution and ecological deterioration in capital construction and in their project approvals related to technical transformation, finance and taxation, banking and credit as well as the introduction of foreign investment. They must utilize more renewable natural resources and increase input for rehabilitating the eco-environment and for pollution prevention and control as a proportion in the gross national product (GNP). The proportion have to be increased substantially during the Ninth FFP period compared to the Eighth FYP period, i.e. an increase from the former 0.8% to over 1%.

To control the total amount of pollutants is another important turn for environmental enhancement under the new approach which will promote structural optimization of the economy, technical advancement and energy conservation. Cleaner production will provide an important means to meet the goals of controlling the total amount of pollutants, by deepening the license system for pollutant discharge, and by strictly charging for pollutant discharge. In addition, cross-century green project has been launched. It is a specific engineering project whose purpose is to organize national and provincial departments and enterprises such that they can concentrate financial and material resources to focus on key areas, e.g. river basins, and fundamental environmental problems for meeting requirements of international conventions. It aims at basically controlling aggravated environmental pollution and ecological destruction, improving environmental quality of important, or problem cities and regions by the end of this century and gradually realize the overall objectives of environmental protection by the year 2010.

During the Ninth FYP period, the total amount of 12 types of major pollutants will be put under controlled. The "Green Project" uses a number of practical engineering measures for realizing these objectives. It will be in three phases in 1996-2010. During the Ninth FYP period, it focuses on water pollution, particularly the treatment of the drainage basins of Huaihe, Haihe, Iiaohe, Taihu, Chaohu, and Dianchi, and atmospheric pollution in acid rain areas and sulphur dioxide polluted areas. The total number of projects involved is 1591 and the investment needed is around 188 billion yuan. Most of the investment will come from the Chinese government and US$4 billion is expected to come from international grants.

2. Environmental Pollution and its Treatment in Key Areas

Environmental pollution around China's cities is expanding outwards to the rural areas where environmental problems are increasingly restricting economic development and affecting people's health.

The major types of China's pollutions are: 1) smoke-oriented atmospheric pollution, of which dust and acid rain are the most harmful, and it is aggravating as well; 2) water-body pollution as represented by river water quality deterioration in the urban areas; and 3) noise pollution caused by road traffic. The major pollution areas include Haihe, Huaihe, Laiohe, Songhuajiang, Taihu, Chaohu, Dianchi, acid rain affected areas in southern China, the Three Gorges area, the contiguous area of Shanxi-Shaanxi-Inner Mongolia, Zhujiang River delta and Hainan Province.

2.1 Atmospheric Pollution

Atmospheric pollution is still characterized by smoke emission with suspended particles and dust widely spread throughout the country. Sulphur dioxide pollution has not been improved and vehicle exhaust emissions have increased dramatically. Nitrogen dioxide pollution has also aggravated. In Guangzhou and a few metropolises, smoke pollution and pollution caused by vehicle exhaust emissions coexist.

From 1985 to 1994, the average annual total waste gas emission had increased by 4.9%, while the rate of sulphur dioxide emission increase was 0.98%. Meanwhile, the average annual energy consumed increased by 4.9% and coal consumption by 5.2%. The total amount of emission is still increasing though the increase rate of total waste gas emission is lower than the GDP increase rate.

In 1995 to 1996, statistics of 88 cities indicated that sulphur dioxide pollution was most serious in the mining areas of high sulphur-contained coal in Southwest China which includes Guiyang, Chongqing and Yibin. In North China, sulphur dioxide pollution was serious in Shanxi and Shandong, particularly in the winter heating period. The most serious pollution resulting from the total suspended particles was found in the nine cities of Jilin, Lanzhou, Yanan, Taiyuan, Waxian, Jiaozuo, Jinan, Baotou and Baoji.

Dust pollution is common in cities of varying sizes but the total trend of pollution tends to decline, particularly so in cities of southern China. The seriously dust polluted cities include Baotou, Changchun, Tangshan, Datong, Harbin and Shijiazhuang, all are in northern China. In southern China, pollution mainly concentrates in the medium- and large-cities of Sichuan, Guizhou, Hubei, Hunan, Zhejiang, Anhui and Shanghai.

Pollution caused by vehicle exhaust emissions tends to be serious in large cities, and nitrogen dioxide has become the No. 1 air pollutant there. The annual mean value of atmospheric nitrogen dioxide concentrations over the three provincial level cities is 5-152 $\mu g/m^3$, of which the value for northern cities is 53 $\mu g/m^3$, southern, 41 $\mu g/m^3$, and that of the whole country, 47 $\mu g/m^3$. Nitrogen dioxide pollution mainly happens in large cities with a population of over 1 million. Pollution in Guangzhou and Beijing is more serious, followed by Shanghai, Anshan, Wuhan, Zhengzhou, Shengyang, Lanzhou, Dalian, and Hangzhou. Acid rain is mainly found south of the Changjiang and east of the Qinghai-Tibetan Plateau as well as in the Sichuan Basin. In the acid rain areas of Central China, particularly in Changsha, the acidity in the precipitation and its frequency tend to rise. The extent of acid rain pollution in Southwest China is next in degree of seriousness, whereas the acid rain area is widely distributed in South China and the coastal zone of East China.

In 1996, acid rain-induced pollution had generally aggravated and its distribution area had enlarged. Within the 84 State controlled network cities 43 had pH values below 5.6. In Changsha, the average pH reached 3.5, the lowest among the monitored cities. Cities with average annual acid rain pH values below 4.5 include Changsha, Xiamen, Ganzhou and Yibin. Cities with acid rain frequencies of over 60% number 24, of which Yibin, Hengyang, Changsha and Ganzhou had more than 90%. In Wuzhou, Xiamen, Huaihua, Nanchang, and Tumen, they had a frequency of more than 80% and Leshan and Guangzhou had more than 70%.

2.2 Water Pollution

China's rivers, lakes and reservoirs are polluted to different degrees. In 1996, with the exception of most tributaries and inland rivers, a general trend of aggravation was observed. Water-body pollution was serious in areas around industrially developed towns and cities and 78% of the water from river courses in cities was not fit for drinking and 50% of the urban ground water was polluted.

Compared to 1994, the total amount of waste water discharge had increased in the provincial units in 1996 except nine, i.e. Beijing, Shanxi, Liaoning, Heilongjiang, Jiangsu, Hunan, Hainan, Tibet and Qinghai. In 1985-1994, the national GDP increased by 9.8% annually and the total amount of waste water discharge increased annually by 0.73%. With the absolute increase in the amount of discharge and limited environmental capacity, China's water pollution is thus getting more serious.

Results from 2,222 monitoring stations throughout the country have revealed the seriousness of pollution in the seven major drainage systems of Liaohe, Haihe, Huaihe, Huanghe, Songhuajiang, Zhujiang, and Changjiang. The ranking order of pollution seriousness of the major fresh water lakes (from high to low) is: Chaohu (western half of the lake), Dianchi, Nansihu, Taihu, Hongzehu, Tongdinghu, Jingbohu, Xingkaihu, Bosten, Songhuahu and Erhai.

Compared to 1995, water quality conditions of the seven major drainage systems had not changed for the better but tended to get worse. Statistics of the key sections of the seven major rivers indicated that in 1996 32.2% satisfied categories I and II of the "Environmental Quality Standard for Surface Water", 28.9% satisfied category III and 38.9% categories IV and V. The water quality of the seven major drainage systems in 1995 are indicated in Table 7.1. We shall describe the situation in each of them.

Table 7.1 Water quality conditions of the seven major drainage systems (percentage of sections with different categories of standard to the respective drainage systems), 1995

Water quality / Drainage system	Category I, II	Category III	Category IV, V
Changjiang	45	31	24
Huanghe	5	35	60
Zhujiang	31	47	22
Huaihe	27	22	51
Songhuajiang, Lioahe	4	29	67
Haihe	42	17	41

Source: National Environmental Protection Agency, *Bulletin of China's Environmental Conditions 1996.*

The water quality of the Changjiang drainage system is worsening in comparison with 1994. River sections that satisfy categories I and II standards account for 38.8%, those that satisfy category III standards make up 33.7%, while those falling in categories IV and V, 27.5%. The principal pollution parameters are ammonia nitrogen, permanganate index and volatile phenol. The copper exceeds the standard in individual sections. Though as a whole water quality of the trunk stream is good, bank pollution is serious. The bank pollution belt along the trunk stream totals about 500 kilometres long.

The water quality of Huanghe River becomes increasingly serious. Sections of the entire basin that satisfy categories I and II standards accounted for 8.2%, those that satisfy category III make up 26.4%, and those that satisfy categories IV and V, 65.4%. The principal pollution parameters are ammonia nitrogen, permanganate index, biochemical oxygen demand, and volatile phenol. Water pollution tends to get worse along river banks with decrease of water flow and increase of pollutant discharge. Sewage from 1,100 enterprises along the section of Togloh to Longmen is discharged directly into Huanghe River, which makes up 5% of the daily runoff of the trunk stream.

In the trunk stream of the Zhujiang, water quality as a whole is good and pollution occurs only in sections of the tributaries. Section of the trunk stream with water quality of categories I and II accounts for 49.5%, category III, 31.2%, and categories IV and V, 19.3%. The principal pollution parameters are ammonia nitrogen, permanganate index and arsenide.

The pollution problem of the Huaihe drainage system is more obvious, where the trunk stream is seriously polluted during the low water season. The heavily polluted sections has expanded upstream, but in some key tributaries that exceed standard the extent of pollution has been dropping gradually due to treatment. Sections of the river that satisfy categories of I and II standards account for 17.6%, category III 31.2%, and categories IV and V, 51.2%. The principal pollution parameters are ammonia nitrogen and permanganate index.

Pollution is serious in the Songhuajiang and Liaohe drainage systems where the principal pollution parameters are total mercury, permanganate index, ammonia nitrogen, and volatile phenol. Total Hg pollution in the Tongjiang section is also serious and the water quality has worsened. Pollution of the Liaohe drainage system is serious during the low water period. The water quality of the rivers flowing through cities all exceeds the category V standards for surface water. Sections of the entire drainage system satisfying categories of I and II standards only account for 2.9%, category III 24.3%, and categories IV and V, 72.8%. The principal pollution parameters are ammonia nitrogen, permanganate index, and volatile phenol. Sometimes the copper, hydroxide and mercury content also exceed the standard.

The water quality of inland rivers is by far good and the total hardness and chloride content of one or two of them is high owing to the impact of physiographical conditions. The water quality of sections of these rivers satisfying categories I and II standard account for 63.5%, category III 25.4%, and categories IV and V, 11.1% of the total length.

The pollution problem of the Haihe drainage system has always been comparatively serious. Some of the important surface water sources have been polluted or threatened by pollution. The water quality of the entire drainage system, including reservoirs, that satisfies categories of I and II standard account for 39.7%, category III 19.2%, and categories IV and V, 41.1%. The principal pollution parameters are ammonia nitrogen, permanganate index, volatile phenol, and biochemical oxygen demand. With the exception of rivers of Jumahe, Tuhe and the reservoirs of Miyun, Huairou, Huangbizhuang, Panjiakou, Zhangze and Wangkuai where water quality is good, the remaining rivers, big or small, are polluted.

The water quality of the drainage systems in Zhejiang and Fujian provinces is relative good with a few river sections polluted. Water quality satisfying categories I and II standard account for 40.7%, category III standard 31.8%, and categories IV and V, 27.5%. The principal pollution parameter is ammonia nitrogen.

River water pollution is quite serious in the urban sections. In 1996, of the 138 urban river sections surveyed, 133 or 96.4% were polluted to different degrees. Sections with water quality exceeding category V standards numbered 53, category V standard numbered 27, category IV standard, 26, and categories II and III, 32. They made up respectively 38.4%, 19.6%, 18.8% and 23.2% of the total sections surveyed. The principal pollution parameters of the urban sections are petroleum related and permanganate index and the phenomenon of suspended particles exceeding the standard was very common.

After a general presentation let us examine the "Death" of Xiaoqinghe as a representative case of the nature of serious water pollution in China. Xiaoqinghe in Shandong Province is known as one of the most seriously polluted water bodies by industrial and domestic pollutants in China. Xiaoqinghe is heading for its doom and fish swimming in the water is a past scene. The dissolved oxygen content in the water of almost all river sections is around zero. Of the pollutants received the highest amount came from paper mills (19), breweries (10), chemical plants (9), pharmaceutical factories (3) and an iron and steel plant (1). The total amount of COD discharged from these principal dischargers, which made up 5% of the total, accounted for 95% of the COD in the industrial waste water discharged into Xiaoqinghe. The COD discharged from the Qilu General Petrochemical Plant alone was as high as 10%. The total COD discharged in industrial waste water from the adjacent cities of Zibo and Jinan made up around 70% of the total COD discharged.

In the past decade, the most important pollutants in Xiaoqinghe were ammonia nitrogen, BOD, COD, volatile phenol, petroleum related pollutants, and other organic pollutants. Though heavy metal pollution occurred occasionally in individual years and in one or two river sections, the annual average concentration had not exceeded standard. Analysis of concentration change in recent years showed that contents of ammonia nitrogen, BOD, volatile phenol, petroleum related pollutants, and total suspended particles (SS) in water bodies of Xiaoqinghe have been increasing while the COD content shows a declining trend (World Bank, 1997).

2.3 Control of Environmental Pollution

2.3.1 Attaching importance to technological advancement

A series of technical innovations measure were adopted by the central and regional governments to confront increasingly serious pollution. It includes the transformation of the industrial growth pattern by optimizing the industrial structure, compulsory elimination of highly energy-intensive products, or those inefficient in the utilization of resources, or are significant pollution producers in the process of their manufacturing. By October 1996, nearly 50,000 such small-scale enterprises had been closed or changed to new types of production.

These measures follow the "Three Changes" strategy as the guiding principle for attacking industrial pollution control i.e.: 1) a change from emphasising end-of-the-pipe control to pollution control of the whole production process; 2) a change from emphasising the control of concentration to focusing on the combination of control on concentration and total amount of pollutants, and 3) a change from scattered spot control to the combination of concentration control with scattered control. Governments at various levels have attached importance to the strategy for cleaner production in industrial enterprises through promoting the application of cleaner production techniques. For example, the petrol-chemical industry of the country has reduced its carbon dioxide emission by 5 million tonnes which resulted in RMB 1.47 billion yuan of economic benefits.

The "Double Enhancement" project which began in 1994 is complimentary to the new strategy. Its aim is to enhance reform in technical investment, accelerate enterprise reform, and set up technical progress mechanisms. The State has organized 57 projects with a total investment of 21.6 billion yuan. One such project involves the Jilin Chemical Industry Co. which was founded in the 1950s and is an old enterprise. The company was able to completely solve its pollution problem by striving for scientific and technological progress and in carrying out technical reforms in its production processes that consume large amounts of resources and discharge large volume of wastes.

Along with comprehensive environmental control and urban construction, a large number of enterprises with serious pollution problems have been closed, moved or put under control. This has eased the pollution situation in some areas. For example, the Beijing Municipal Government closed down the South Factory of the Capital Steel Co. because of its serious pollution problems, thus eliminating a major source of pollution in its urban area. Shanghai has also accelerated its pollution control efforts on Suzhou River, the upper reach of Huangpu River, and key parts of the city with noticeable success.

2.3.2 Increasing water pollution control in major river basins

In 1995, the Chinese Government issued the "Interim Regulations on the Protection against Water Pollution in the Huaihe River Basin" and formulated the "Programme and Ninth FYP of Water Pollution Control in the Huaihe Basin" with the aim to bring water pollution under effective control and protect the aquatic environment. The above documents have defined the targets for controlling the total amount of water pollutants discharged into river basins, as well as the maximum permissible discharge for major cities, towns, and discharge points in the Huaihe Basin. Yet they serve as models for the whole country. Thus it is useful to describe the background for their adoption (*China Environmental Yearbook*, 1995 & 1996). The Huaihe River serves as a drinking water source for a population of over 10 million along the river banks. Along its bank are over 3,000 paper mills, factories and breweries in towns and cities which directly discharge waste water into it. The total amount of such untreated industrial waste water and domestic sewage it receives annually is over 3 billion tonnes. In 1995, the water quality of only 22% of the sections of the Huaihe drainage system satisfying category III standard. The toxic level of drinking water in the basin was 10 to 20 times higher than that of the national standard. The catastrophic flood of the Huaihe in 1994 in the upper reach where there was a high concentration of waste water had caused extraordinary pollution incidents affecting the four surrounding provinces. As a consequence, the State Council set up a group for water pollution prevention and control of the Huaihe drainage basin. Compulsory measures including written commitment with local governments to bear responsibility for pollution elimination was adopted for realization of the objectives on pollution mitigation. Up to the end of 1996, 1094 paper mills had been closed or had their production suspended. The Central Government started an investment programme of a total of 6.6 billion yuan for water pollution control of the Huaihe that cover 1995-2000.

2.3.3 Control of Smoke, Sulphur Dioxide Emissions, and Atmospheric Pollution

In 1995, China promulgated the "Law on the Prevention and Control of Atmospheric Pollution". Some seriously polluted cities have adopted a permission system for air-polluting production under an experimental charge system for sulphur dioxide exhaust so as to control smoke and limit the emission of sulphur dioxide. The country has clearly delimited the acid rain and sulphur dioxide control areas and adopted strict standards for sulphur dioxide emission from industrial pits and furnaces. In some seriously polluted cities of Southwest China such as Nanchong, Yibin, Chongqing, and Zunyi, smoke pollution and acid rain problems have been eased to some degree. In order to control sulphur dioxide pollution, clean coal and clean combustion technologies have been introduced. In 1995, the smoke-elimination rate of waste gas from coal combustion had reached 88% and the purification rate of waste gas from production technology reached 69%.

2.3.4 Strengthening Control of Solid Wastes

A number of measures have been adopted to control solid waste generation and disposal. These include: to conduct experiments on solid waste registration and waste exchange; to construct facilities for urban dangerous waste disposal; to strengthen safety monitoring and management of radioactive wastes; to strictly control dangerous waste import; to increase the utilization rate of industrial "three wastes" and to develop renewable resources. In 1995, the comprehensive utilization rate of garbage had reached 39%; powdered coal reached 41%; the comprehensive utilization rate of combustive gas of large- and medium-size enterprises in the metallurgical industry averaged 39%; and the reutilization rate of water for industrial purposes in major cities reached 60%-85%. The five-years' (1991-1995) accumulated output value produced from comprehensive utilization of "three wastes" of industry totalled 70 billion yuan and the profit of 18 billion yuan generated. The total output value generated from waste material collection and utilization had reached 130 billion yuan.

2.3.5 Improving Energy Production and Consumption Structure

The implementation of both energy conservation and development has emphasised on conservation. It involves the adjustment of energy production and consumption structures; development of new and renewable energy; and the development and popularization of new and renewable energy resources such as solar, wind, geothermal, tidal and biomass energy. Efforts are also made in the development of techniques for clean combustion, in imposing constraints on direct burning of raw coal and in the adoption of a policy of encouraging residents and enterprises to use shaped coal. In 1995, shaped coal accounted for 40% of the coal used in urban households and 70% of total domestic coal consumption. Many outdated boilers and combustion facilities have been eliminated compulsorily and high-efficiency energy-saving boilers have been introduced.

3. Regional Ecological Conditions and Eco-Engineering in Rehabilitation

3.1 Major Problems and Trend of the Regional Eco-Environment

Because of impacts of natural and human factors, China's ecological situation is harsh. This is mainly reflected by the shortage of water, serious damage of vegetation, constant expansion of areas affected by soil and water loss, desertification, and a severe threat to biodiversity. China abounds in rich biodiversity. Both species of higher plants and wild plants and animals occupy around 10% of the world total. However, population growth and economic development have exerted a double threat on biodiversity. Many species have vanished, of which 200 species were higher plants. Some 15% to 20% of the animals and plants are on the brink of extinction or under threat, including 4,600 species of higher plants and over 400 wild animals. As a consequence, the productivity of the natural ecosystems of farmland, forest, grassland and rivers and lakes has been declining. In eastern China where the ecosystems are tough, land degradation and environmental deterioration has still occurred, while in the western part where the ecosystems was already fragile, their deterioration becomes even worse.

3.1.1 Shortage of Water and Deterioration of Hydrological Regimes of Rivers and Lakes

The available per capita water resources in China is only 28% of the world average. Since the 1980s, the shortage of fresh water resources and deterioration of hydrological regimes of rivers and lakes have become principal factors causing eco-environmental deterioration in many parts of China and have become essential restricting factors to socio-economic development in northern China and parts of the coastal areas.

With global warming desiccating, rapid growth of China's economy, the drastic rise of water consumption have all led to reduction of the number of medium and small rivers, shortening of river courses, decrease of water resources and even its exhaustion in some areas. Conditions in Northwest China, i.e. central and southern Liaoning, Jiaodong Peninsula, and some coastal cities, are particularly worrisome. The number of water deficit cities there has reached more than 300, and serious water deficit exists in more than 130. Due to water deficit, there is a reduction of over 2.5 million tonnes of grain because of inadequate irrigation of the farmlands.

For example, most of the drainage basin of Huanghe belongs to arid and semi-arid areas where water resource is already very scarce. As the amount of water consumed has increased, drying up of the lower reach has become more and more serious. Since 1972, natural drying up of the trunk stream of Huanghe has occurred, and its frequency has been increasing. The length of the dry section of the river has extended and the number of days when the river turns dry has also increased. In 1995, the drying up days in Lijin section reached 121, and within the estuary area, 153 days. The effects of such drying up has not only affected economic and social development and the basic ecological process of the entire Shandong Province but has extended to Henan Province for a total length of 700 km. Eco-environmental degradation has appeared in the lower reach and estuary area due to changes in the ecological succession. From 1972 to 1996, an estimated economic loss of 28.6 billion yuan in industry and agriculture was recorded, with an annual average loss of 1.4 billion yuan in the lower reach. In the 1990s, the annual loss reached 3.6 billion yuan as a result of yield reduction of 9.86 billion kg on 70.42 million mu (4,690,000 ha) of drought stricken farmland. In 1995, the drought stricken area reached 3.5 million mu (230,000 ha) in Dongying, Shandong Province, of which 2.3 million mu (153,000 ha) were unsuitable for farming. Direct economic loss registered in the Shengli Oil Field was more than 400 million yuan. Among this 91.545 million yuan was loss from agriculture and 310 million yuan in oil extraction.

Drying up and the small runoff volume have also threatened flood prevention due to silting and channel shrinkage. The latter is caused by changes in the scouring pattern. The vegetation was also damaged as a result of desertification, and salinization-alkalization in by invading sea water in the dried out channels, the flood plains. Many lakes and reservoirs are also running dry with polluted water and eutrophication, or are shrinking and disappearing. Lastly, the deteriorated water quality has led to a general drop in fish varieties, damaging the ecological balance and biodiversity.

In the upper reaches of rivers in Sichuan and Hunan, constant expansion of the reclaimed area has caused damage to the vegetation and led to soil erosion. River sedimentation of the Dongting Lake due to silt transported from Changjiang and Sishui has made it the most seriously silted lake in China. The catastrophic flood in 1996 in Changjiang caused further damage to the Dongting Lake. In both provinces, siltation make many reservoirs unable to operate effectively owing to inadequate water storage or erosion induced siltation.

3.1.2 Low Vegetation Coverage Aggravated Ecological Damage

China is one of the countries with limited forest resources. The present forest cover is about 134 million ha, or 13.92% of the country's territory. It is low compared with the world's average of 31.4%. The per capita forest area in China is only 0.11 ha and the total stored bush is 860 million m^3, which is 11.7% and 12.6% of the world average respectively. In the vast expanse of arid and semi-arid areas of western China, patches of forests have degraded and forest coverage there is even less than 1%.

In the past decades, the forest coverage in Changjiang Valley has been decreasing. In Sichuan Province which occupies 56% of the upper reach of Changjiang, the forest coverage has dropped from 20% in the 1950s to 13% in the 1980s. In its eastern part it even went down to 6%. In the entire Sichuan Basin alone, the forest coverage rate was only 4%. In history, luxuriant forests were recorded in the Three Gorges reservoir area. However, between the 1950s and 1980s, the forest area there had dropped by more than half. The present forest coverage in the 19 cities and counties in the "future reservoir area" is 19.5% and in the worst two counties, it is less than 3%. In the Boyanghu Basin of the middle reaches of Changjiang, the forest coverage is 17.8%, lower than the average level (33.6%) of Jiangxi Province, while it is 13.5% and 8% respectively in Anhui and Jiangsu of the lower reach. Overall, forest destruction in the Changjiang Valley has been widespread and serious and has led to the aggravation of soil erosion. Soil erosion occurred in 41.1% of the entire river valley which is China's largest river basin, with an area of 739,400 km^2.

Grassland is more widespread in China. It has 400 million ha, or 30% of China's total territory. 310 million ha is unusable grassland though it is also ecologically fragile, as it is mostly located in the arid and semi-arid or highly frigid areas. Because of the long history of irrational reclamation and overgrazing, the dynamics of the grassland ecosystems have degraded and they suffer from desertification and salinization-alkalization. At present, seriously degraded grasslands amount to 73 million ha, or 18.3% of the total grassland acreage in the country. The yield of the grasslands has also dropped by 30%-50%.

Grassland degradation is most serious in Xinjiang and Inner Mongolia. In Xinjiang alone, the area of seriously degraded grassland amounts to 8 million ha. Damaged areas caused by rats exceeded 13 million ha, and yield reduction due to such was over 25%. In the past 40 years, the carrying capacity of grasslands in Inner Mongolia has dropped by one-third. In 1995, an investigation carried out in Alxa of western Inner Mongolia indicated that the eco-environment there has been drastically deteriorating due to irrational exploitation and utilization. As a consequence, land there has also been desertified, vast patches of forests have disappeared, ground water mineralization has increased by three folds, and about 500,000 mu (33,333 ha) of lakes have dried up where basic survival conditions for organisms and even human beings were destroyed. Let us examine in more detail the eco-environmental Deterioration in Alxa District (Chen Yaobang, 1996). Alxa district is located in the western end of the Inner Mongolia Autonomous Region. In history it was a natural pastoral land for many minority nomadic tribes in China. The Juyan Oasis there (also named Ejin Oasis) was nurished by the Heihe River. The 800-km-long natural sacsaoul forest belt, and the dense secondary forest for conserving water sources on the Helan Mts constitute a unique ecological vegetation system in the Alxa district. They functioned as three natural windbreaks to protect the Hexi Corridor, Ningxia Plain, Hetou Plain and even Northwest and North China. However, long-term excessive exploitation of water resources in the middle reach of Heihe, substantial dam construction to intercept the river and constant increase in irrigated area have contributed to a drop of downstream discharge. In 1958, the area of water surface of east and west Juyan Lake calculated by aerial photo survey was 276 km^2 and 35.5 km^2 respectively. By 1991, the lake was completely dried up, being turned into a desert and saline-alkali beach. The well-known Swan Lake had also shrunk in 1992. Nowadays, the diversiform-leaved spruce forests lining the banks of Ejin River have decreased from 750,000 mu (50,000 ha) to only 300,000 mu (20,000 ha). The secondary water conserving forests in Helan Mts are also degrading. The deteriorated eco-environment at present experiences frequent sandstorms and natural disasters.

3.1.3 Aggravated Soil and Water Loss

China is one of the countries subject to serious soil erosion. Although some achievements have been gained in comprehensive soil and water conservation, the eroded area, erosion degree, and its harmful extent have worsened. Up to 1994, the eroded area reached 1.79 million km^2. The amount of soil loss totalled 5 billion tonne, of which 3.3 billion tonnes were cultivable soil. Siltation of lakes, reservoirs and river channels resulting from serious soil degradation related to erosion exerted adverse impacts on both production and wild life. Of the 2000-odd poverty stricken counties in the country, 87% are categorized as areas of serious soil erosion. The Loess Plateau and the Changjiang River Valley, particularly its upper reach, are the most seriously eroded areas in China.

Loess Plateau

It consists of 287 counties (cities, banners) in seven provinces with a total area of 624,000 km^2, 90% of which are situated in the middle reach of Huanghe. Their fragile eco-environment, with a long-term irrational land use practice and serious destruction of the vegetation, have resulted in extremely serious soil erosion. Soil erosion and desertification have gone hand in hand to make the Loess Plateau "the more land reclaimed, the poorer it is; the poorer it is, the more land being reclaimed". It has caused serious flooding problems lower down the river, as the river bed in the lower reach of the trunk stream is rising at an annual average rate of 10 cm, making the dikes in the lower reach "the more dangerous it is, the higher it becomes", and "the higher the dikes, the more dangerous it is".

Changjiang River Valley

It is an important economic belt consisting of ten provinces with a drainage area of 1.8 million km^2, or one fifth of the national territory. It has one fourth of China's cultivated land, and one third of the nation's population. In recent decades, forest coverage in the valley has been decreasing. Soil erosion has caused major eco-environmental problems there. In the 1980s, the eroded area of the entire valley was 739,000 km^2, accounting for 41.1% of the valley's area. At present, soil erosion is most serious in the upper reach where it affects an area of 360,000 km^2. Above the Three Gorges Reservoir area the annual amount of soil erosion is 1.56 billion tonnes every year, 680 million tonnes higher than the sedimentation in the Three Gorges. The total reservoir capacity above Yichang is 16.67 billion m^3, of which an annual average of 300 million m^3 were silted. In 1987, a "sand ridge" had formed in the Jinjiang section, causing serious damage due to blocking of the navigation channel. Within six years only 30,000 km^2 were controlled through a soil and water conservation project in the upper reach. If the projects were carried out at such a speed, it will take 70 years to fulfil the entire task of complete soil conservation.

Soil erosion is also aggravating in the middle and lower reaches of the river. In Jiangxi, the eroded area increased from 11,000 km^2 in the 1950s to 46,000 km^2 in the 1980s, and in Hunan it increased from 19,000 km^2 to 47,000 km^2. Desertification with a marked of red color in the local soil has been generally visible.

Serious soil erosion has caused land degradation and a drop in the ability in being disaster resistant. In the 1950s, in the upper reach, a serious drought occurred once every five years. In the 1970s, eight droughts occurred every ten years, while in the 1980s, a serious drought occurred almost every year. In recent years, flooding, drought and waterlogging happen nearly every year. Flooding and waterlogging frequently occurred in the middle and lower reaches where drainage on the low-lying farmland is inadequate. Geological disasters like landslides and debris flows have also increased due to ecological inequilibrium. For instance, 1,203 landslides have been recorded in the entire basin from source to the estuary. 576 landslides occurred along the trunk lines of 13 railways in its central and eastern parts and there were 271 debris flow gullies in the Three Gorges area.

3.1.4 Expansion of Desertification

Desertification is a global scale disaster, being referred to as the "cancer" of the Earth. Presently, one fourth of the land area on the globe is affected by desertification, and desertified land has reached 36 million km^2. Desertification means land degradation in arid, semi-arid and dry subhumid areas resulting from various factors, including climate variations and human activities. It is a process whereby the land is evolving towards desert-like conditions and there is a reduction of land productivity and even loss of land resources resulting from interactions between irrational human activities and the vulnerable environment. It includes land desertification, soil secondary salinization-alkalization, rocky desertification resulting from farmland degradation, forestland degradation, and grassland degradation.

China is among countries on the globe where desertification is vast in area, wide in distribution and severe in effect. In China, desertified land covers an area of 2.62 million km^2 or 27.3% of its total land territory. The desertified area is mainly distributed in 13 provinces of northwest, northeast, and north China where nearly 400 million people reside. About 60% of the poverty stricken counties are concentrated there as well. In the "Three North" (northwest, northeast, and north China) alone, 13 million ha of farmland are affected by sandy storms. The grain yield is low and unstable. 100 million ha of grassland are seriously degradated, and railway lines of over 3000-kilometre-long as well as several thousand kilometres of highways are threatened by wind-drift sand the year round. The direct economic loss caused by desertification is more than 54 billion yuan annually.

From the 1950s to 1970s, the desertified land area increased at an annual rate of 1560 km^2. From the 1970s to 1980s, the rate reached 2100 km^2 a year. Presently, the area has been expanding even faster, at a rate of 2460 km^2 annually. Northwest China is most seriously affected by sandy desertification. At present, the land area related to desert, Gobi, desertification and weathering total 1.747 million km^2, or 18.2% of the country's land area. The desertified area in Inner Mongolia makes up 6.2% of its grassland area, and in the Erdos Grassland it has reached 26.3%. Secondary salinized-alkalized land in the Kashi Prefecture of Xinjiang accounts for 57.7% of its cultivated acreage. The secondary salinized-alkalized land there accounts for another 57.7%. For the whole country the secondary salinized-alkalized land has reached 233,000 km^2, as such land, besides Xinjiang, is also developing in the Hetao Plain, Yinchuan Plain and Huanghe Delta. In the limestone area of Southwest China, rocky desertified and semi-rocky desertified areas are rapidly developing. In Guizhou Province alone, bedrock exposure has increased at an annual rate of 90,000 km^2 in the past 20 years.

3.2 Eco-Environmental Protection and Ecological Engineering

The Chinese Government has recently taken eco-environmental protection as a key for environmental protection. It has carried out ecological rehabilitation projects on forestry ecosystem rehabilitation, eco-agriculture rehabilitation, natural protected area rehabilitation and biodiversity protection. These projects have gained excellent ecological, economic and social benefits.

3.2.1 Afforestation and Forestry Ecosystem Rehabilitation

During the Eighth FYP, the focus of China's afforestation work was to eliminate wasteland by planting trees and growing grass, particularly in mountains areas and as a means of desertification prevention and control. The total growth rate of forest resources during the Eighth FYP was set to surpass the amount of consumption. The Eighth FYP was successfully carried out as the total afforested area of over 29 million ha had exceeded the set target by 40.2% of which the afforested area facilitated by "closing" of the hills reached more than 26.67 million ha. The rate of success of artificially planted trees had increased from 75% in 1990 to 89% in 1995.

In 1996, the afforested area was 4.919 million ha, and the afforested area by "closing" the hills was 3.66 million ha (Table 7.2).

Table 7.2 National Afforestation Situation during Eighth FYP and in 1996

Year	Afforested area (10,000 ha)			Afforestation by closing	Sporadic tree planted (10,000)
	artificial	aerial	total		
8th five-year plan	<2500	<400	<2900	2667	
Of which					
1995	412.9	58.6	521.5	453.8	35
1996	431.5	60.4	491.9	366.0	24

At present, the forested area have reached 134 million ha, and forest coverage has increased from 12.98% in 1992 to 13.92% in 1997. The preserved area of artificial afforestation have reached 34.25 million ha and the living stand storage reached 11.785 billion m^3. They have attained the set national plan targets. In 12 provinces wastelands have been basically eliminated by tree planting. In the cities the urban green belt coverage reached 23.8% and the per capita green belt available was 5.7 m^2.

Since the 1980s, the State has started to carry out ten macro ecological forestry projects (EFPs) focusing on the improvement of the natural eco-environment and realization of sustainable utilization of resources. According to EFPs, the planned national afforested area would be 120 million ha, and 3.57 million ha of trees would be planted in the fast growing, high-yield forest production bases.

The above-mentioned projects cover basically areas related to soil erosion, sandstorms, typhoons, and salinization-alkalization which cover about 80% of the country's total land area. They would form nation-wide ecological barriers which are rational in distribution, complete in structure and workable in function, so as to become an ecological protection network, providing a basic framework for ecological rehabilitation in China (Fig. 7.1). We shall briefly describe the ten EFPs.

Figure 7.1 Spatial scope of the macro ecological forestry projects during the Ninth FYP

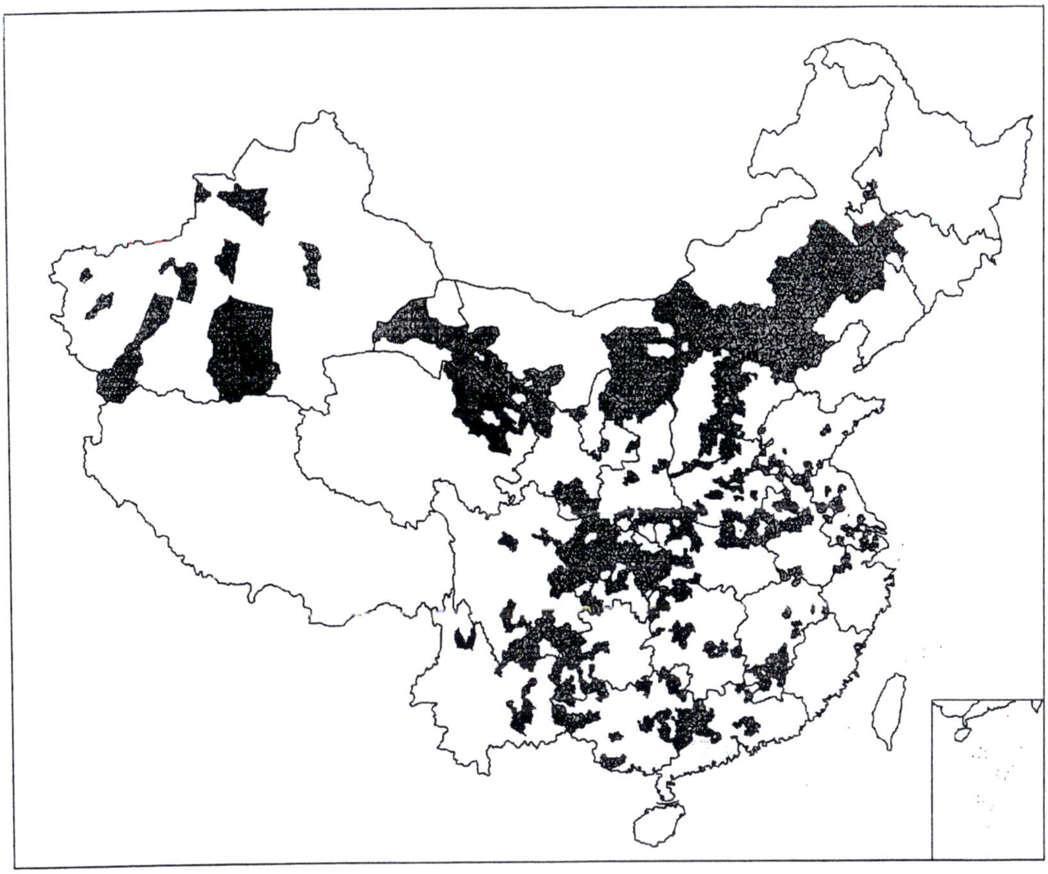

(1) The "Three North" (northeast, northwest and north China) forest-shelter system project

This is the largest EFP in the world, covering the 13 provinces of Xinjiang, Qinghai, Gansu, Ningxia, Shaanxi, Inner Mongoia, Shanxi, Hebei, Beijing, Tianjin, Liaoning, Jilin, and Heilongjiang in Northeast China, the northern part of North China and western part of Northeast China. It covers an area of 4.06 million km^2 or 42% of the country's area, or almost the entire territory of northern China. The project spans the period of 1978 to 2050. The planned afforestation area will be 35.08 million ha. The forested area will be increased to 40 million ha, and the forest coverage to 9.2%. 6.67 million ha of farmland will be brought under protection, and 200,000 km^2 of desertified land will be put under control.

During the Eighth FYP, over 6 million ha of land had been artificially afforested, 4.33 million ha afforested by "closing" the hillsides and 4 billion trees had been planted, of which, 1.33 million ha were completed in 1995, and 1.342 million ha in 1996. The first and second phase of the EFP has been completed and the third phase is underway. So far 18.51 million ha of land have been afforested, and the forest coverage has increased from 5.05% in the late 1970s to the present 8.2%. Therefore, farmland under the forest-shelter network protection reached 21 million ha and 5.2 million ha of desertified land was put under control and forest protection, of which 1.33 million ha have been transformed into farmland, grazing land and orchard. 8.93 million ha of grassland have been rehabilitated and the grass output have increased by over 20%.

In the Loess Plateau region, 2.45 million ha of land were afforeted for the purpose of soil and water conservation and nearly 8 million ha of soil eroded area have been preliminarily brought under control. Thus, the eco-environment of one-third of the counties there has started a positive ecological cycle.

(2) Middle and upper Changjiang River forest-shelter system project

In 1989, for the first time China implemented the macro EFP with a purpose of comprehensive management of river basins. This key project is of great significance to soil and water conservation of the upper and middle reaches of the Changjiang River and the eco-environment and sustainable socio-economic development of the entire basin. It involves 271 counties in 11 provinces covering the period 1989-2000. It plans to add 6.67 million ha of afforested area.

During the Eighth FYP, more than 3 million ha of land have already been afforested, of which 5.35 million ha were accomplished in 1995, 4.64 million ha in 1996. In the first phase of EFP construction, 62 counties have satisfied the planned goal.

Presently, the EFP focuses on ten key areas including the Three Gorges reservoir area, middle and lower reaches of Jinsha River, and the Jialingjiang River basin. So far 3.66 million ha of land have been afforested, and the forest coverage of the projected area was 5% higher than that in 1989. After seven years' effort, barren mountains in over 100 counties there have been eliminated by being turned into a sea of forest. The rate of sandy and rocky desertification in the mountain areas of Hunan, Hubei, and Guizhou has been slowed down. River flow is seen all year round in many previously dried up streams and wild animals unseen for many years began to increase in the forests. Soil erosion has been under check by the protection of green barriers, as is evidenced in the effects of the EFP in Gongxian County.

Gongxian in Yibin city of Sichuan province constitutes an important part of the ecological forest-shelter system in the upper reach of Changjiang and a demonstration district for the pilot project on ecological rehabilitation of the country. Primitive forests left in the county were seriously damaged during the 1950s to the 1970s. Up to the mid 1970s, the forest coverage of the county was only 8.24%. Natural disasters occurred frequently due to drastic loss of the forest. About half of the county (680 km^2) was subjected to moderate and above moderate soil erosion. In addition, hailstones, landslides and debris flows as well as drought and flood happened frequently. The output of agricultural was low and unstable.

Since the implementation of EFP in the mid 1970s, the county has undertaken a number of projects such as fast growing high-yield forest production base construction (53,333 ha, 1970s); EFP afforestation related to China's project 2606-2 of UNDP (2,000 ha, 1980s) and EFP afforestation related to the national project of the World Bank (3476 ha, 1990s). Meanwhile it also carried out economic forest projects such as the construction of fruit tree, bamboo, silkworm, and tea bases. In the past 20-odd years, 36,667 ha of artificial timber forests and 6,667 ha of economic forests involving 63 million yuan of investment have been constructed. The projects were checked and approved by provincial and prefectural level authorities in 1993. The forest area in patches has reached 41,200 ha, including 10,000 ha of afforested area by abandoning farming activities.

In 1995, the forest coverage of the county rose to 42.4%, and the living stand rose from over 200,000 m^3 to over 2 million m^3. The overall output value of forestry products totalled 92.62 million yuan in 1994, and it exceeded 150 million yuan in 1995. In 1995, the county was honored the title of "one hundred best counties" for afforestation in China. In the past years, grain production of the county gained high and stable yield due to a noticeable decrease of natural disasters that facilitated the effective functioning of water conservancy projects. In 1980, the total output of grain of the county was only 89,500 tonnes and average per mu (1/15 ha) yield was only 194 kg. In 1995, the total output reached 119,300 tonnes, per mu yield was 248 kg, and the per capita available reached 312 kg under conditions of farmland reduction and population increase. Forestry development had also facilitated the restructuring of rural enterprises. The income of farmers has increased substantially as a result of the development of timber forests, bamboo groves, fruits, silkworm, tea and medicinal plants. In addition, the establishment of the macro forestry management system also generated ecological, economic and social benefits in forestry development.

(3) Coastal forest-shelter system project

The project began in 1988 to afforest 3.56 million ha of land by the year 2010. The project covers an area 18,000 km long, starting from the mouth of Yalu River on the Sino-Korean border in the north, to the mouth of Beilunhe River on the Sino-Vietnamese border in the south. It stradles 195 counties in 11 provinces (Liaoning, Hebei, Tianjin, Shangdong, Jiangsu, Shanghai, Zhejiang, Fujian, Guangdong, Guangxi, and Haikou).

Construction started in 1991 and within the Eighth FYP, a total of 1.6 million ha had been afforested, of which 72,000 ha were completed in 1996.62 counties had satisfied the construction standard of the first phase. A primary forest-shelter belt of 13,000 km with about 30 forest parks has been built on the 18,000-kilometre-long coast line to protect the 6.56 million ha of farmland there with a disaster mitigation capacity of 30% to 40%. The forest-shelter system is rational in structure, combining forested points, patches, grids and belts. It has gained noticeable ecological, social and economic benefits. The State Council has bestowed the forest-shelter belt the status of a national special protected forest belt under special national protection.

(4) Farmland forest-shelter system project in the lowlands

There are 918 counties in China predominated by lowlands. They occupy about 15% of the country's total land area. In 1988, lowland afforestation was included into the national forest-shelter project by the Ministry of Forestry. The lowland farmland forest-shelter system takes cities and counties as units, farmland forest-shelter nets as foci, and shelter belts, inter-cropping of trees and crops, tree planting around farmlands and in areas as the main components. According to the plan, by the year 2000 farmlands in lowlands will be surrounded by forest networks. During the Eighth FYP, afforestation was accomplished in 769 lowland counties, of which 51,400 ha were afforested in 1995.45 counties met the requirement of lowland afforestation. In 1996, 36,000 ha were afforested and another 26 counties had reached the target. So far, 214,000 ha have been afforested in China's lowland areas. 34.2 million ha of farmland are now surrounded by forests, and 795 or 86.6% of the lowland counties in China have fulfilled the task of lowland afforestation.

Lowland afforestation has apparently improved the eco-environment in the farming areas. Surface wind velocity on farmlands protected by forest nets has been reduced by 20% to 30%, farmland evaporation reduced by 8% to 21%, and grain output increased by 15% to 20%. A prosperous economy, thriving domestic animals, and luxuriant forests and excellent crops can be seen in most parts of the lowlands.

(5) Re-establishing vegetation to combate desertification

China is one of the countries seriously affected by desertification and it also possesses widely distributed deserts. Its desertified land area accounts for 17.6% of the country's total land area. The project covers 599 counties in 27 provinces. According to the plan, 7.19 million ha of desertified area (mainly desertified land resulting from wind erosion) would be brought under control or will be exploited economically within 1991-2000. The focus of the project is to protect and expand forest and grass vegetation as well as psammophye, and establish a desertification combating and controlling system via the organic combination of combat, control and utilization, so as to gradually form a sand related production system.

During the first phase of the project, i.e. the Eighth FYP, 169,000 ha was afforested, and over 3.7 million ha of desertified land brought under control and exploitation. In 1996, another 867,000 ha were comprehensively controlled and exploited. Since the start of the project, more than 4 million ha of desertified land were controlled and exploited. The measures used include artificial afforestation, aerial sowing, afforestation and grass growing by closing the hillsides, reclamation and transformation of low-yield farmlands. In addition, over 600 multi-purpose management items were identified in desertified area.

(6) Taihang Mountain afforestation project

The project is of significance to the improvement of eco-environments in the Beijing-Tianjin area as well as North China, promoting national and local economic development, and facilitating farmers to become rich. It is listed in "China's Agenda 21" as a key project for global environmental rehabilitation. The project involves 110 counties in four provinces (Beijing, Hebei, Henan and Shanxi) with a total area of 12 million ha. It plans to build a comprehensive forest-shelter system with unified ecological, economic and social benefits, and with scientific allocation of multi-forest types, rational distribution of multi-tree species, and tree, shrub and grass combinations. The project covers a period of 1987-2000 for a target of 3.96 million ha of afforested area, and a forest coverage of 43.6%. Upon completion of the project, the Taihang Mountain will again become a green barrier in North China.

The project started in 1994 after a trial of seven years. During the Eighth Five-Year Plan, 1.02 million ha had been afforested, of which 427,000 ha were completed in 1995, and 402,000 ha in 1996. Up to the present, 2.151 million ha of trees, shrubs and grass have been planted.

(7) Forest-shelter system project of Huanghe's Middle Reach

The project started in 1996 in 188 counties in six provinces (Inner Mongolia, Shaanxi, Shanxi, Gansu, Ningxia and Henan). It plans to build 3.15 million ha of multi-functional forest-shelter system in 15 years, aiming at increasing the forest coverage in the project area by 10% and reducing its soil loss by 3 billion tonnes.

During the Ninth FYP, the focus of the project is to build water conserving forests in the upper reaches of all the tributaries of Huanghe, to construct bank protection and slope stabilizing forests on the banks of the trunk stream and tributaries as well as on steep slopes of gullies, and to set up forest-shelters on gentle slopes to protect farmland and pasture. In 1996, 178,000 ha had been afforested.

The project will reduce soil loss in the middle reach of Huanghe and sediment discharge into the river. It will thus improve the eco-environment and bring into play the long-term effect of water conservancy works in the lower reach of the river, particularly key water conservancy project under construction, by means of comprehensive control combining biological measures with engineering measures. Meanwhile, forest products output will also increase.

(8) Zhujiang drain basin forest-shelter system project

The project concerns 177 counties under the jurisdiction of 32 prefectures in four provinces. 1.2 million ha of forested area will be added, and the forest coverage of the project will increase from 35.01% to 37.94%, and its existing irrational tree species structure will be adjusted, while soil loss will be basically controlled and the eco-environment will be improved.

The project began in 1996 and focused first on tree species structure adjustment and the quickening of the pace of afforestation. The investment of the project for that year exceeded 100 million yuan. Of the 103,000 ha of afforested area, 38,700 ha were completed in the first tranche of 13 key counties.

(9) Huaihe-Taihu drainage basin forest-shelter system project

The State Planning Commission approved the overall planning of the project in November 1995, which includes 208 counties in seven provinces (Henan, Anhui, Jiangsu, Shangdong, Hubei, Hunan, and Shanghai), and aims at an afforested area of 1.047 million ha. Upon completion, the forest coverage of the basin will increase from the present 13.9% to 17.3%, and the total timber storage will increase from 69.91 million m^3 to 314 million m^3, and soil loss will be reduced by 28.26 million tonnes. Farmlands protected by forests-shelter will be increased by 333,300 ha, and they will expect on annual increase in grain output of 875 million kg.

The 34 pilot counties had already afforested 69,667 ha, being 3.4 times that of their trial target.

(10) Liaohe drainage basin forest-shelter system project

The project began in 1996, covering 77 counties in four provinces (Hebei, Inner Mongolia, Jilin and Liaoning). The planned afforested area is 1.2 million ha. The focus of the project is to construct a forest-shelter aiming at soil and water conservation, protection against wind, and dune stabilization. Meanwhile, economic forests and fuelwood will be developed along with timber forests.

Upon completion of the project, natural disasters such as soil erosion and sandstorms will decrease and the eco-environment of the entire basin will noticeably improve.

3.2.2 Soil and Water Conservation Projects

While constructing the EFPs, the State has also increased input on soil and water conservation. A sum of 3.5 billion yuan was invested in soil and water conservation projects during the Eighth FYP to implement the "Law on Water and Soil Conservation", the "National Programme for Soil and Water Conservation" (1991-2000), and to carry out soil and water conservation projects in the seven major drainage basins. Twenty-five major soil erosion control areas have been established at the national level. Comprehensive management of mountains, rivers, farmland, forests and roads have been carried out in the key 10,000 small river basins with serious soil erosion. A total of 71 million ha of soil eroded area were brought under control. On average soil loss was reduced by 1.1 billion tonnes a year, and the national water conserving capability increased by 18 billion m^3 (*China Environmental Yearbook*, 1995)

Soil and water conservation in the Changjiang and Huanghe drainage basins have been listed in the national capital construction plan, bringing 17.53 million ha of eroded area under control. Some 15 million ha have been controlled in the Loess Plateau region, and 30% of the eroded area had been controlled by differing degrees, and the amount of sediments carried into the Huanghe River has been reduced by 300 million tonnes.

3.2.3 Development of Eco-Agriculture

Eco-agriculture is an important way for sustainable agriculture. Thus, the State has listed eco-agriculture as the basic principle and policy target for China's agricultural development.

In recent years, progress has been made in China in demonstration projects for eco-agricultural development. Seven Chinese eco-agricultural villages or townships have been conferred the title of "Global Five-hundred Best" by the United Nations and they become a world model for sustainable agricultural development. After work carried out in over 20 years, by the end of 1995, 50 State-level eco-agricultural demonstration counties and 100 provincial-level demonstration counties had been established. By the end of 1996, 160 pilot counties and over 2,000 pilot townships and villages covering an area of 13.33 million ha had been set up. Throughout the country, there are now 69 State-level demonstration districts. At present, a development trend of taking the state-level eco-agriculture demonstration counties as foci to form a national system, of state-level with provincial- level, county-level with prefectural or urban level projects of technical demonstration has preliminarily taken shape.

Eco-agriculture has led to enhancement in agricultural development, alleviation of natural disasters, and excellent ecological, economic and social benefits as shown in the case of Northern Shàanxi (Jiang Chunyun, 1997).

Yulin prefecture in Shàanxi is situated in the transitional zone of Mu Us Desert and the Loess Plateau. In the initial period of the founding of the PRC, the forest and grass coverage in the prefecture was only 1.8% due to long-term harmful effect of sandstorms, odious eco-environment and the encroachment of shifting sand on vast patches of farmland, grazing ground and villages. The deteriorated eco-environment and extreme slow economic development made local peoples' life hard.

Since the founding of new China, particularly after the economic reform in 1978, noticeable ecological, economic and social benefits have been gained due to efforts made in dune stabilization and desert control. By 1996, the preserved area of bamboo groves in desertified areas reached 966,667 ha; forest and grass coverage, 38%; 165 patches of forests were formed in the hinterlands of deserts at an average of 667 ha each; four macro forest-shelters with a total length of 1,500 km are built; and 100,000 ha of farmland was protected by forest grids. In contrast to the 1950s, the height of sand dunes have dropped by 30%-50% on average, the annual encroachment of the decertified area have decreased from 7.7m to 1.68 m, and the annual sediment discharge into the Huanghe River was reduced by over half.

The improvement of the eco-environment has also facilitated the promotion of the comprehensive production capability of local agriculture. There are now 391,333 ha of basic farmland. The per capita available was over 0.133 ha, and the total output of grain reached 1.255 billion kg. In 1991-96, 33,333 ha of high-yield farmland had been constructed with an accumulated increase of grain of 500 million kg. A diversified economy has also been developed. In 1996, the per capita net income was 1,220 yuan in the desertified areas. Thus most of the farmers have adequate food and clothing. The scene of out-migration of people due to sand encroachment had been reduced.

The city of Yan'an in Shȁanxi is situated in a typical gully area of the Loess Plateau. In the early 1980s, the total eroded area was 28,800 km^2, or 78.4% of the city's area and the annual sediment discharge into the Huanghe River reached 258 million tonnes. In recent years, soil and water conservation work has been taken as a task of strategic importance in eliminating poverty. By 1996, the overall controlled area had reached 16,700 km^2, or 58% of the total eroded area. Forest and grass coverage had gone up to 42.9%. The per capita available basic farmland reached 0.24 mou with 0.15 ha of economic forest, and total grain output reached 890 million kg, with per capita available stabilised between 400 to 500 kg. The four leading economic products of the area of tobacco, fruits, sheep and tuber crops have reached scale production to certain degrees. The output value of the diversified economy reached 1.36 billion yuan, or 68% of the total output value of agriculture. The per capita rural net income reached 1,120 yuan, while the poverty-stricken population has dropped from 670,000 in 1988 to 236,000.

Thus eco-agriculture is an important channel to cope with the deteriorated agricultural eco-environment. As the above examples indicate, in northern Shaanxi, erosion control and construction of eco-agriculture has facilitated the apparent improvement in the eco-environment and reinforced the disaster resistant capability. In Yan'an alone, high-standard basic farmland have been built and water-saving and drought-resistant techniques have been popularized within five years since the initiation of eco-agriculture. In 1995, the output of grain was an all-time high though a rarely seen drought lasting 262 days occurred. In 1995, Liaoning was hit by a heavy flood, Changtu and Dawa, the two state-level eco-agricultural counties still got a bumper harvest.

Eco-agriculture has also promote sustainable development in agricultural production and the rural economy. The introduction of eco-agriculture in China is found to have affected a rise of grain output of over 15%, and per capita income 12% higher than the local average. In 1995, the per capita income in the 105 pilot units in Hubei Province was 32% higher than the provincial average level.

Popularization of the ecological idea has also helped to raise environmental awareness of farmers and local officials, and improve the capability in sustainable development of China's vast rural areas. These in turn have facilitated the government of various levels to attach more importance to the demonstration and popularization of eco-agriculture.

3.3 Nature Reserves and Biodiversity Protection Projects

3.3.1 Construction of nature reserves

China is among the countries having the highest amount of wild animals and plants. The establishment of various kinds of nature reserves provides an effective way to protect wild animals and plants as well the eco-environment.

Since the 1950s, China has started to set up nature reserves. Presently, a national nature reserves network has taken shape. During the Eighth Five-Year Plan, 130 forest and wild animal reserves had been established, adding an area of 48 million ha. Another ten wild animal rescue bases had also been set up. 799 nature reserves (including 99 state owned) have been designated for the conservation of a variety of species. The total acreage of existing nature reserves was 71.85 million ha, or 7.2% of the country's total territory. In 1996, the State approved the construction of another seven national nature reserves, making the total of 106, of which, 12 reserves such as the Tianmushan of Zhejiang have joined the "International Man and Biosphere Protection Network", and six were included in the "List of International Key Wetlands".

784 forest parks of various types with a total area of 7.2 million ha have been in operation. The UNESCO has listed scenic spots like Mount Huangshan, Wulinyuan, Jiuzhagou and Huanglong as components in the World Nature and Culture Heritage scheme. In 1996, China joined the "World Nature Protection Alliance".

The construction and development of nature reserves have protected many representative, typical and scientifically valuable rare and endangered species and natural ecosystems. The 258 wild animal species of state-level protection and 354 rare and endangered species have all been effectively protected and be able to carry on breeding.

3.3.2 Biodiversity protection projects

China follows the principle of combining in-situ conservation with ex-situ conservation to protect biodiversity. The establishment of nature reserves serves as an effective way for in-situ conservation whereas the establishment of zoos, botanical gardens as well as breeding centres (bases) are ways for ex-situ conservation.

By 1995, China has built a total of 175 zoos or park zoos and 227 wild animal breeding centres.

Rescue projects for endangered animals have restored many species. There are seven major rescue projects, including the giant panda rescue project, nipponia bird rescue project, Yangtze alligator rescue and development project, Hainan deer rescue project, elk rescue project, and high-nose antelope rescue project. In 1996, 25 nature reserves covering a total area of 10,690 km^2 had been built, and for the giant panda conservation project alone the investment amounted to over 7 million yuan. The number of giant pandas in the country in the wild has risen from 7 in 1981 to around 1,000 now.

In recent years, rare and endangered plant conservation has been further strengthened. Over 400 rare and endangered plant ex-situ conservation and breeding centres and material resources databases, 120 botanical gardens and tree gardens as well as 13,000 ha of seed gardens have been set up, bringing over 1,000 species of rare and endangered plants under protection, and 90% of wild plants designated for state protection under ex-situ conservation.

4. INVESTMENT BENEFITS AND POLICY EFFECT OF ENVIROMENTAL PROTECTION IN THE PROVINCES

4.1 Investment and Benefits

The severity of pollution in China has to a certain extent, corresponded to the pollution intensity of industrial enterprises. In turn there is a relationship between pollutant discharge per unit output value and the amount of investment per unit wastes. Appendix 7.1 gives the total sum used for pollution control in various provinces in 1995, including the sum for waste water and waste gas treatment, total waste water discharge and waste gas emission. Henan, Xinjiang, Jilin, Shandong, Shanxi, Hebei, Heilongjiang, Anhui, Gansu and Tianjin ranked as the first ten in terms of investment intensity in waste water treatment, with relatively higher intensity of investment in environmental protection and conservation. Xizang (Tibet), Hubei, Guangdong, Beijing, Sichuan,

Qinghai, Jiangxi, Shanghai, Hunan and Yunnan ranked the last ten (see Table 7.3), with relatively lower investment intensity in environmental protection.

Waste water treatment efficiency (WWTE) and treatment of discharge reaching standard (TERDS) can partly reflect the efficient use of the investment. In 1995, Tianjin, Beijing, Hubei, Shanghai, Shandong, Henan, Hunan, Liaoning, Guangdong and Heilongjiang ranked in sequential order the first ten in terms of industrial WWTE, i.e. all exceeding 80%. Hainan, Beijing, Ningxia, Sanxi, Shanghai, Hebei, Tianjin, Liaoning, Qinghai and Hubei ranked in sequential order the first ten in terms of industrial waste water TERDS, i.e. all exceeding 70%, with the highest of 86.7% (see Table 7.4). Henan, Shandong, Hebei and Tianjin ranked the first ten in terms of investment intensity in waste water treatment and industrial WWTE. This indicates that they have concentrated effort to cope with industrial waste water treatment issues.

Table 7.3 Investment Intensity in Waste Water Treatment in Various Provinces (1995) (WWTC 10,000 yuan)/WWD (million yuan))

Henan	33.01	Jiangsu	14.77	Yunnan	8.56
Xinjiang	32.74	Ningxia	12.78	Hunan	8.42
Jilin	30.40	Fujian	12.08	Shanghai	7.83
Shandong	29.63	Guangxi	11.16	Jiangxi	7.44
Shanxi	20.69	Zhejiang	10.98	Qinghai	6.25
Hebei	19.79	Guizhou	9.81	Sichuan	4.30
Heilongjiang	19.25	Hainan	9.49	Beijing	4.20
Anhui	18.24	Liaoning	9.02	Guangdong	4.07
Gansu	16.13	Shaanxi	8.68	Hubei	4.01
Tianjin	15.67	Inner Mongolia	8.61	Tibet	1.02

WWTC--waste water treatment cost;wWD--waste water discharge.

Table 7.4 Industrial Waste Water Treatment Efficiency and Treatment of Discharge in Various Provinces in 1995 (%)

Region	Treatment efficiency	Treatment of discharge reaching standard	Region	Treatment efficiency	Treatment of discharge reaching standard
Beijing	87.7	82.9	Henan	84.3	45.8
Tianjin	90.4	75.8	Hubei	75.2	70.5
Hebei	88.5	76.8	Hunan	83.2	53.4
Shanxi	77.1	81.2	Guangdong	81.1	56.6
Inner Mongolia	72.2	42.2	Guangxi	69.8	45.3
Liaoning	82.5	71.1	Hainan	58.4	86.7
Jilin	71.6	56.1	Sichuan	55.6	49.6
Heilongjiang	80.6	47.4	Guizhou	76.6	54.3
Shanghai	85.9	80.4	Yunnan	63.3	44.7
Jiangsu	73.5	56.0	Tibet	2.2	0.0
Zhejiang	76.0	56.3	Shaanxi	77.9	67.4

Anhui	69.2	52.7	Gansu	69.2	56.2
Fujian	64.1	46.6	Qinghai	45.0	71.0
Jiangxi	69.6	48.3	Ningxia	66.7	82.7
Shangdong	84.8	64.5	Xinjiang	58.1	50.3

Comparing industrial WWTE with TERDS, several types of provinces can be recognized. Beijing, Shanghai, Tianjin, Liaoning, and Hebei belong to type I, for both WWTE and TERDS in them are relatively high. Hainan, Ningxia and Qinghai belong to type II, where the WWTE is low but TERDS is relatively high. The financial resource of these provinces are adequate for coping with industrial waste water treatment issues. Hubei and Shanxi belong to type III, where both WWTE and TERDS are at a moderate level.

As the industrial structure of various provinces varies, resulted pollution also differ. Even within the same industrial sector, the pollution intensity in different provinces vary to certain degrees. Case studies of Liaoning, Beijing, Shanghai, Guangdong and Sichuan indicated that waste water chemical oxygen demand (COD), atmospheric SO_2 and total suspended particles (TSP) have been dropping since 1988. This is particularly obvious in Guangdong and Sichuan. However, differences in pollution intensity still exist in the provinces. The most serious pollution is produced by enterprises in Sichuan and the least in Shanghai. It is found that changes in industrial pollution intensity are mainly dependent on environmental management, local residents' pressure and awareness as well as the industrial enterprises themselves.

According to the Environmental Status Bulletin of Shandong, total investment in pollution control and environmental protection there in 1996 was 3.264 billion yuan, an increase of 50.8% compared with the previous year. The implementation rate for environmental impact assessment and the "three simultaneousness" reached 99.8% and 91.8% respectively. The five largest cities in the province had carried out a total of 133 control projects. The largest project in the province's history for environmental protection and water conservancy began in 1996. This is an integrated control of the trunk stream of Xiaoqinghe. With constant increase in economic growth, the aggravated situation in environmental pollution and ecological destruction has been checked to a certain degree and the province's environmental quality has been improved.

4.2 Analysis of Policy Effect

4.2.1 Economic reforms promoted environmental protection

China's economy has developed rapidly in the recent ten years. Its industrial structure has also been transformed and its international trade has developed at a marked speed. Meanwhile, its environmental policy has been universally implemented. Under such circumstances the environmental quality of China has changed positively in the following three aspects.

(1) The expansion of production scale in industry, as seen in the proportion of large-scale enterprises had gone up from 30% in 1988 to 37% in 1994. The growth rate of Shanghai is particularly remarkable; it rose from 41% to 58% in the corresponding period. As the pollution intensity per unit output value of large-scale enterprises is low, increment in the production scale has resulted in less harm to the environment.

(2) The proportion of state-owned enterprises has dropped gradually. Changes in this aspect are also remarkable in Shanghai where state-owned enterprises had dropped from 70% in 1988 to only 54% in 1994. As non-state-owned enterprises are more eager to reduce pollution discharge, the decrease in the number of state-owned enterprises has led to a marked drop in pollution intensity.

(3) Rapid economic growth has facilitated readjustment of the industrial structure. The proportion of industrial enterprises that discharge vast amounts of TSP in the country has dropped

whereas enterprises that discharge substantial amount of SO_2 and COD has remained unchanged. In Beijing, Shanghai, Guangdong, Liaoning and Sichuan, the "clean" environment effect in their reformed industrial economy is quite obvious. For example, the proportion of water pollution-intensive industrial enterprises of these provinces has dropped. As for atmospheric pollution, with the exception of Liaoning (SO_2 emission), the proportion has also dropped. The greatest change is witnessed in Beijing where there are industrial enterprises previously emitting vast amount of COD, and enterprises in Guangdong and Shanghai which have formerly emitted a lot of SO_2 and TSP.

Trade reforms have equally exerted positive effects on the protection of the eco-environment in China. The effect works through increased production efficiency of export-oriented enterprises as well as through their quick absorption of new and clean production techniques from the developed countries. These will compensate the adverse impact on the environment due to China's inadequate environmental management and its use of low cost, pollution-induced raw materials.

There is evidence that China is shifting slightly away from a pollution-intensive industrial structure. The proportions of the five major pollution-induced industries such as chemical, pulp and paper making, non-ferrous metal, ferrous metal, and nonmetal refinery (cement industry in particular) have gone down in the national total. Another indicator is the net import volume (imports minus exports) of these five industries. If pollution-intensive industries tend to increase in China, then there will be an increase in the net import of pollution products. In fact, since 1987, the volume of actual net export of these industrial products has increased.

In a word, with the development of foreign trade in China, output of less polluting products has been on the increase. All industrial sectors, including the pollution-intensive industries have been developing very fast. However, non-pollution industrial enterprises, no matter measured in domestic production scale or in terms of their international trade have been developing faster than polluting industries.

4.2.2 Actual drop in charges for pollutant discharge without touching major pollution producers

The "Actual charged level" was practised as a basic environmental management means and has gained remarkable effects. Studies on the relationship between pollution intensity and pollution protection strength have indicated that control by means of actual charge has played a critical role in controlling industrial pollution intensity in the provinces: an increase of 1% of the fees collected leads to a drop of COD pollution intensity of 0.8% (unit output value COD discharge). For atmospheric pollution, an increase of 1% of the fees would lead to about 0.3% drop in SO_2 pollution. Pollution intensity is apparently lower in provinces where high fees were collected. It is vice versa in areas of low fees. Since 1987, pollution intensity has also dropped in places where the fees collected have increased gradually. Thus different efforts in environmental management will produce different effects.

The fees collected for atmospheric pollution of the five provinces studied have all dropped since 1987. The amplitude of Beijing and Shanghai was surprising, though the harm of atmospheric pollution there has constantly been rising during this period. The fees collected for pollutant discharge in general has also generally declined except in Liaoning and Shanghai where it rose rapidly (Sichuan rose relatively slowly). Though the practice of fees charge can affect pollution intensity of industrial enterprises, there is yet no sign that as a result polluting industrial enterprises have shifted to other products.

Since 1987, the drop in fees collected for atmospheric and water pollution in Beijing and Shanghai may have stimulated the rise of pollution there. Hence, the actual drop in pollution intensity recorded is likely due to the impact of economic reforms and other management measures, such as industrial scale, and changes in ownership of enterprises and the industrial structure. As for atmospheric pollution, this conclusion is also suitable to the three provinces of Liaoning, Guangdong and Sichuan where the fees collected have also been reduced. For water pollution, the drastic drop of pollution intensity of the above mentioned provinces has been caused by the interactions of two factors: the drastic rise in the fees collected for waste water and the shift of industrial structure towards a low pollution type.

4.2.3 Preliminary effect of the new move in environmental protection

In 1996, the State Council drafted the "Decision on Issues Related to Environmental Protection" which specified ten decisions on the realization of objectives related to environmental protection by the end of 2000. The "Decision" has produced some initial effects:

- Prior to September 30, 1996, paper mills with an annual production capacity below 5000 tonnes, tanneries with are annual production capacity equivalent to or less than 30,000 cowhides, dyestuff plants with an annual production capacity of less than 500 tonnes as well as enterprises with backward and out-moded equipment for making coke and sulphur were outlawed. Enterprises with indigenous methods to smelt arsenic, mercury, lead and zinc, refine oil, sort gold, produce insecticides, bleaching and dyeing, electroplating, as well as those that produce asbestos and radioactive products were ordered to close or suspend production. These are the so-called 15 types of small enterprises which yet are seriously polluting. In 1996, the 15 types of small enterprises with serious pollution numbered a total of 62,000 firms were either closed or suspended production (with a rate of closure of 85.5%). This has made marked progress in promoting structural readjustment of the industries and products and facilitated the shift of the economic growth pattern.
- Work on water pollution control of the three rivers of Huaihe, Liaohe, and Haihe and the three lakes of Chaohu, Taihu and Dianchi has been unfolded. Up to June 30 of 1996, 1,111 paper mills with an annual pulp production capacity of less than 5,000 tonnes had been outlawed in the Huaihe drainage basin. By September 30, 1996, another 3876 small enterprises of serious pollution had been outlawed. Many enterprises succeeded in switching production to non-polluting sectors after production suspension or enterprise closure. For example, of the 47 small paper mills closed or suspended production in Tongxian county, 21 had their production shifted to other sectors which are related to new technology, less environmental pollution, high economic efficiency and having potential for future development. However, as only a few of the 1,562 enterprises with daily waste water discharge exceeding 100 tonnes have reached the waste discharge standard, it is deemed difficult for the entire basin to realize the target of industrial pollutants reaching the discharge standard before the end of 1997.

5. Appraisal and Conclusion

At the Fourth National Conference on Environmental Protection Premier Li Peng pointed out that we should keep sober-minded that the situation was still quite serious in environmental protection in China. As China was at a development stage towards rapid industrialization and urbanization, increase in pollutant discharge will result. The intensification in natural resources exploitation plus China's extensive nature in economic growth, with a comparative backward technical and management capability would make the situation of environment worse.

In the country as a whole, environmental pollution around the cities is still aggravating and is expanding towards the rural areas. The scope of ecological destruction is still enlarging. In some regions, environmental pollution and ecological destruction have hampered healthy development of the economy, and has even caused direct threat to people's health. In the next fifteen years, the demand for resources will be increasing at a faster speed as there will be continuous growth of population and the economy. Great pressure would be exerted on resources, the environment and the ecology. With increasing demand of the people for better environmental quality, the solution to these problems are requiring seriousness and urgency of treatment. As both the old environmental pollution problems left over in history and new environmental problems created in the process of development have to be controlled for sustainable, rapid and healthy development of the national economy, environmental protection is thus an arduous work in China.

5.1 The Objectives for China's Environmental Protection have not been Achieved

As the basis for the future fifteen years' work on environmental protection, the "Outline of the Ninth Five-Year Plan for National Social and Economic Development and the Long-Term Objectives for the Year 2010" stressed on efforts to basically control the trend of increasing environmental pollution and aggravation of ecological damage, and to improve environmental quality in some cities and regions by 2000. They also pledge to basically change the situation of environmental deterioration and to improve environmental quality in most cities and regions by 2010 as China's cross-century environmental protection objectives.

In terms of the standards for environmental protection, China has set a low level target; yet compared to China's present environmental protection efforts and environmental conditions, there is still a long way to go to even achieving such a target (Box 1).

Box 1 Objectives for China's Environmental Protection

China has formulated a set of clear-cut quantified objectives system. For water pollution, the objective concerns the control of waste water discharge increase rate to 14% and COD discharge to zero growth rate during the Ninth Five-Year Plan period. For air pollution, the Central Government hopes to basically maintain the present TSP and SO_2 discharge level.

	1995 level	Objectives for 2000
Water		
Waste water discharge (billion tonne)	42.2	48
Industrial waste water	27.1	30
Domestic sewage	15.1	18
COD discharge (million tonne)	22.3	22
industry	10.9	16
domestic sewage	11.4	6
Atmospheric pollutant emission (million tonne)		
Smoke	17.4	17.5
Industrial dust	17.3	17.0
SO_2	23.7	24.6
Others (million tonne)		
Solid waste produced		1110
Industry		930
Domestic garbage		180
Industrial solid waste disposal amount		60
Petroleum pollutants (tonne)		83100

Objectives for water environment required that industrial waste water treatment rate be increased from 63% to 74%, urban domestic sewage treatment rate raised from 5% to 25%. According to estimation by the State, the total investment for the next five years for pollution control would be 450 billion yuan, half of which would be used for atmospheric pollution control and about 40% for controlling waste water discharge.

Source: National Environmental Protection Agency and others, the "Ninth Five-Year Plan on Environmental Protection and the Long-Term Objectives for the Year 2010", 1996.

5.2 Improvement in Environmental Protection Policy and Measures are Needed

Development of China's market-oriented industrial enterprises has provided opportunities for more effective utilization of market-based pollution control measures. However, if no reasonable guiding principle is available for improving the present management system, negative effect on environmental management will result. Once market mechanisms are introduced, existing administrative control mechanism should also be strengthened. The effective implementation of pollution control measures have to rely on local governments. It is therefore necessary to adopt the following changes in the State-level environmental policies:

(1) Redesign environmental impact assessment (EIA) and the "three simultaneous" system. A new management method should be adopted to combine EIA and the "three simultaneous" system with measures for controlling the total amount of pollutants to make them effective stimuli for industry in applying clean production techniques. EIA and the appraisal basis for the "three simultaneous" system should be gradually expanded to include the self-purification capacities of local and regional atmospheric/water body environments rather than to merely focus on whether the pollutant discharge standard has been satisfied. The combination of EIA and the "three simultaneous" system with a market-based pollutant discharge license system will encourage enterprises to adopt pollution measures to minimize charges.

(2) Reform on the fees system for pollutant discharge. To realize the objectives for controlling pollutants, it is necessary to substantially increase fees for pollutant discharge. The National Environmental Protection Agency has proposed to increase the fees collected for atmospheric pollutants by ten times. Studies of the World Bank indicated that the ideal sum charged for SO_2 emission in Zhengzhou should be 50 times that of the present level. The severity of pollution at present is obvious, and a vast amount of investment should be made to improve atmospheric environmental quality of the cities in China.

It is also necessary to improve the charge design and its collection efficiency. The present fees for pollutant discharge are fixed on the basis of a ceiling for pollutant discharge. When the amount of discharge is below the ceiling, it would be free for pollutant discharge. The penalty is also directed to the single predominant pollutant without considering the cross effects among various pollutants. A more practical approach in waste gas and water discharge control is to adopt a complete charge system capable of evaluating a particular pollutant and all the impacts it can cause. The establishment of a charge system based on the total amount of pollutants will be helpful to its organic combination with the pollutant discharge license system.

To reform the management of the fees collected for pollutant discharge is essential to the overall realization of the potentiality of the charge system in pollution control. A portion of the fees collected is returned to the enterprises in the form of cheap loans for pollution control investment. This will ease the difficulties in the collecting of fees for pollutant discharge. Moreover, financial supervision should be exercised to guarantee rational use of this fund. At least this money should be concentrated for reallocation to enterprises based on the principle of maximum marginal pollution reduction effect. Secondly, the appropriation of the fees should be transparent and be taken as a portion of the budgetary revenue. Thirdly, fees collected for pollutant discharge should not be used for administration or as revenue of the implementing departments.

(3) There should be serious attention to the rapid increase in pollutant discharge from township and village enterprises. If township and village enterprises were put under the scope of control by environmental management departments, then many of these enterprises would not be so bad at present. Market mechanisms should be depended upon to solve the environmental problems of these enterprises which are huge in number and are sensitive to market prices. The Central Government should start studying and trying out a set of control systems established on the monetization of natural resources and environmental taxes, and constantly improve and strengthen the existing control measures.

In order to consolidate previous environmental protection achievements, China has to carry out sustained economic reform. Economic reforms and the resulting changes in technical and structural improvements of the economy, ownership and market size can sufficiently remedy the impact of pollution caused by the rise of total output. Sustainable reform in the future decade will also produce similar mitigation effects. Rapid industrial growth will cause more serious environmental pollution if no further transformation in the ownership and production scale are to be effected.

A stimulation-based pollutant discharge license system will lower pollution control cost and help to realize local and regional pollution control objectives. China should implement measures for controlling the total amount of pollutants as soon as possible through the introduction of such a pollutant discharge license system. Its organic combination with the existing pollution control system would constitute an important task confronting the environmental control departments in the coming years.

5.3 Institutional Reform

In China the overstaffed implementation organizations have hampered the effective implementation of various environmental protection systems. It was particularly so for environmental protection systems which need to be carried out in coordination by various government agencies sharing the same resources (such as water) or facing the same pollution issues (such as acid rain). Besides, inadequate public participation is also a major defect in China's environmental management system.

It is suggested that the State should separate the three major aspects of protection, i.e. construction, development, and operation from government departments for commercialized management and operation. The functions of government departments will then concentrate on supervision of the eco-environment, resources protection and exploitation activities that do harm to the eco-environment and resources, including supervision based on laws and regulations, checking up, licensing, resources fee collection, propaganda, education, and crack down on law enforcement. Assessment indexes used should mainly refer to improvement of the eco-environment, the maintenance and rise in value of ecological assets and resources, overall improvement in the awareness of ecological and resources protection, and the introduction and implementation of various protection measures.

It is necessary to establish national natural resources management organizations. There are many precedents abroad. For example, unified management is performed by the U.S. Department of Interior on water resources, land resources, and the conservation of wild animals and plants for the whole country. Unified management can change the long existed phenomena of 'too many cooks spoil the broth'. As the power of such an organization is higher than other ministries and committees, it can perform unified assessment, planning, standard formulation and implementation by coordinating various departments. To vest the organization with high authority can drive it to perform good management against the huge interests of the market economy. Such an organization can be named as "natural resources management committee", composing of the functional departments on natural resources management such as the State Bureau of Land Administration, Ministry of Water Conservancy, Ministry of Forestry, and National Environmental Protection Agency. Natural resources management and protection can be practised at different regional levels according the natural drainage systems and physical geography of the country.

The present task is to improve eco-environmental protection of the central management organizations, to set up and improve local management organizations; to reform irrational division of labour in current management organizations, to reinforce cooperation among departments, and to set up coordinative management and feedback mechanisms.

Taking wetland as an example, it is imperative to set up a wetland protection leading group led by the Ministry of Forestry with the involvement of relevant ministries and commissions for the sake of strengthening linkages and coordination among various departments, so as to gain common understanding on the major issues and take integrated action, to intensify responsibilities and obligations of the management organizations, and to establish networks for coordinated cooperation. According to conditions in China, it is still necessary to adopt administrative methods. A decentralized management system cannot do much good in the management of wetland resources. It is thus necessary to proceed from the overall interest of the country and carry out management of key regions and trans-provincial (regional) drainage management based on administrative means in formulating and implementing principles, policies and laws and regulations, and the division of functions at the national and regional level.

5.4 Financial Reform - Establishment of National Eco-Environmental Protection Financial Support System

So far there has been no financial support system for environmental protection in China. In addition to government allocation to large projects on eco-environmental protection at the provincial and prefectural levels, China depends mainly on international support from the World Bank, Asia Development Bank, and loans of foreign governments. There are several dozens of loaned projects operating every year. As China will soon be unqualified for preferential loans from international organizations, it is necessary for China to set up its own eco-environmental protection financial support system. The functions of the system should include:

1) Provide preferential loans to projects related to eco-environmental protection.
2) Use centrally allocated preferential loans as start-up fund to promote matching funds from local government or enterprises.
3) To change budget allocation from the Central Government to that of preferential loans from banks, and to change the present almost sole effort of the government to economic behaviour of private financial organizations for the sake of effective and scientific use of funds, and for convenient fund management and supervision to accord with international practices.

The eco-environmental protection financial support system can either be a newly established national development bank or a new department inside the present national development bank. Because the fund is invested for public welfare and for sustainable development, it should be allocated separately from other funds. The initial sum should at least be greater than the sum of input for eco-environmental protection projects China receives from the World Bank.

5.5 To Increase Public Participation and Awareness

Public participation and awareness on environmental issues in China is generally weak and is much worse in environmental decision-making and environmental ethics. It is suggested that the State should raise the national awareness in resources conservation and environmental protection through education, propaganda and ideological and political work, and take them as specific requirements for patriotism and list it as an important aspect for the construction of socialist civilization. Regarding regional resources assessment, resources planning and management as well as resources exploitation, public participation can be specified in law to encourage the public to care and participate in environmental protection. This is of great importance to democratic decision-making and the rule of law in environmental protection as well as in the building of an honest and clean government.

REFERENCES

1. Bulletin on China Environment Conditions. 1996. 1997. *China Environmental Daily*. June 7.
2. Chen Yaobang. (ed.) Textbook on Sustainable Development Strategy. Beijing: China Planning Publishing House (Chinese text).
3. China Society of Environmental Science. 1996. *Theory and Practice for Controlling the Total Amount of Main Pollutants*. Beijing: China Environmental Science Publishing House (Chinese text).
4. Division of Society's Work, China Association for Science and Technology. 1995. *The Yangtze River - Development for 21st Century*. Beijing: China Surveying and Mapping Publishing House (Chinese text).
5. Editorial Committee of "China Environment Yearbook". 1995. *China Environment Yearbook*. Beijing: China Environment Yearbook Publishing House (Chinese text).
6. Editorial Committee of "China Environment Yearbook". 1996. *China Environment Yearbook*. Beijing: China Environment Yearbook Publishing House (Chinese text).
7. Integrated Survey Team of Chinese Academy of Sciences on Loess Plateau. 1992. *Overall Programmes and Studies of Major Issues Concerning Comprehensive Management of Loess Plateau Region*. Beijing: China Economic Press (Chinese text).
8. Jiang Chunyun. 1997. An investigation report on soil and water loss in northern Shaanxi province and rehabilitation of its eco-agriculture. *People's Daily*, September 3 (Chinese text).
9. Li Jie. 1997. Drying out of the yellow river: a new task for harnessing yellow river. *People's Daily*, May 24 (Chinese text).
10. Li Yucai. (ed.) 1997. *Competing Desertification in China*. Ministry of People's Republic of China (Chinese text).
11. Li Yucai. (ed.) 1997. *Making China Green*. Ministry of the People's Republic of China (Chinese text).
12. National Environmental Protection Agency. 1996. *Meeting the Challenge of the New Century--Collected Papers on Theories of Environment and Development*. Beijing: China Environmental Science Publishing House (Chinese text).
13. National Environmental Protection Agency. *et al.* 1996. *China's Trans-Century Green Project* (First Tranche, 1996-2000). Beijing: China Environmental Science Publishing House (Chinese text).
14. World Bank. 1997. *China's Environment Towards 21st Century* (Chinese text).

Chapter 8

AGRICULTURE AND FOOD SUPPLY POLICIES

GU Shuzhong
Associate Professor and Deputy Head, Department of Resource Economics,
Commission for Integrated Survey of Natural Resources,
Chinese Academy of Science
Becky P.Y.LOO
Assistant Professor, Department of Geography,
University of Hong Kong

1. BACKGROUND

Given the huge and expanding population, which has skyrocketed from 541.7 million in 1949 to 1223.9 million in 1996, the problem of food grains has always been a major issue on the government agenda ever since the foundation of the People's Republic of China (PRC). In the pre-reform era, agricultural production has often been emphasized, except during the relatively short spans of the Great Leap Forward (1957-1959) and the late 1970s when political ideology and national security were the predominant considerations.

No wonder, the reforms of the Open Policy were first introduced in the rural sector in the form of the Household Responsibility System. The prototype of the Household Responsibility System was first introduced in Guangdong, i.e. the "wuding yichan", that is, fixed labour, plot size, costs, production and work points, and bonus for higher-than-quota production and penalty for lower-than-quota production. The re-establishment of links between production outputs and income has greatly stimulated the initiatives of the farmers. The 1980s, therefore, witnessed the rapid increase in agricultural production of the country and has generated a picture of initial euphoria on the possibility of meeting the demand for food grains of the population and even some surplus for export to the international market to earn foreign exchange. The optimism has, however, died down in the mid-1980s when the hectic pace of industrialization and urbanization following the Open Policy began to exert pressure on the cultivated land, food supply and agricultural sector. Under such changes, Brown (1995)'s book *Who will Feed China* has alarmed not only the Chinese government but also the international community on the seriousness of the problem and its possible trend to continue or even to aggravate (Liang, 1996; Sun, 1996).

In China, food problem is not confined to the agricultural sector and the supply of food grains to feed the population alone but to the entire national economy (Chen, 1995). In fact, high inflation since the early 1990s has, to a large extent, been caused by declining agricultural production and, hence, increasing prices of agricultural produce. From this point of view, the agricultural policy of the government would be a major factor affecting the overall future economic development of the nation. One of the consequences as well as policy responses is seen in the changing spatial pattern of grain production. It too becomes an important facet of regional development in the country. We shall first review the national picture. Then, the regional trends of growth and decline will be examined as a basis to assess the national policy of food grain production in different regions.

2. Declining Agricultural Production Growth Rates

Table 8.1 shows the year-to-year growth rates of grain production in China. Relatively high fluctuations ranging from 9.49% in 1990 to -6.92% in 1985 exist. In per capita terms, the grain production is generally on the increase. During 1981-1995, the average annual growth rate was 1.05% and grain output per capita had 2 peaks in 1984 and 1990, reaching 393 kg. Such levels are still very far away from the official goal of producing at least 400 kg per capita. Towards the 1990s, that is, the Eighth FYP (1991-1995), the average grain per capita has stagnated at around 370-380 kg.

Table 8.1 Grain Production in China

	Grain			
	Total Production (in 10 thousand tonne)	**Year-to-Year Growth Rates (%)**	**Production Per Capita (kg)**	**Year-to-Year Growth Rates (%)**
1981	32502	1.39	324.77	0.02
1982	35450	9.07	348.73	7.38
1983	38728	9.25	375.99	7.82
1984	40731	5.17	390.29	3.80
1985	37911	-6.92	358.14	-8.24
1986	39151	3.27	364.16	1.68
1987	40298	2.93	370.27	1.68
1988	39408	-2.21	359.65	-2.87
1989	40755	3.42	367.69	2.24
1990	44624	9.49	390.30	6.15
1991	43529	-2.45	375.82	-3.71
1992	44266	1.69	377.79	0.52
1993	45649	3.12	385.17	1.95
1994	44510	-2.50	371.38	-3.58
1995	46657	4.82	385.21	3.72

Source: State Statistical Bureau. *A Statistical Survey of China*. Beijing: State Statistical Bureau, 1996.

After the Open Policy, domestic grain consumption does not just rely on domestic production but is supplemented by net imports. The historical trend shows that China is still not strongly dependent on imports (Fig. 8.1). In 10 out of the 15 years in Figure 8.1, China has net imports. Yet, the amount of net imports was quite low ranging from 4.19 million tons in 1982 to -1.72 million tons in 1993, less than 5% of the total domestic production.

Figure 8.1 Relationship between Domestic Grain Production and Net Imports

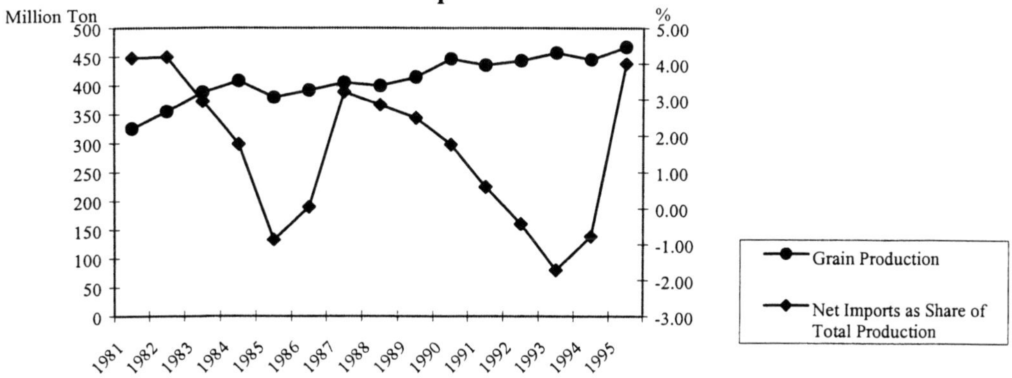

In general, China's agricultural policy can be categorized into four major aspects. They are the procurement of agricultural produce, land use policy and the supply of production inputs, financial policy and loan policy. China's agricultural policy has undergone 2 major phases of changes after the Open and Reform Policy. In Phase One, steps were taken to break away from the constraints of the planned economy by gradually promoting rural economic system reforms. Attempts were taken to increase production autonomy for farmers in the hope that this would release their "suppressed" production initiatives, while the traditional ways of production and management have not been changed. The limited market reforms have always gone into conflicts with the centrally planned economy and this had given rise to an odd "planned commodity economy".

In Phase Two, which began in 1992, the local governments and farmers were given operation autonomy with regards to their production. The major reforms introduced included the lengthening of the land tenure to 30 years. A grain protection system was also set up with the establishment of a grain security fund and savings. Major moves were made not only to designate but also support "Chief Grain and Cotton Production Regions". In practice, 500 large counties producing commodity grain, 150 large counties producing good-quality cotton were selected for exclusive loans and higher capital construction investments. In addition, the Provincial Head Responsibility System was implemented to ensure balance in the supply and demand of food grain. Besides, a plan for a system of comprehensive agricultural resource balance and chemical fertilizers reserves was drawn up; and a Special Agricultural Product Tax was introduced to support agricultural development since the Eighth FYP. In the transition to market economy, the Chinese government has passed a series of important laws and regulations on agricultural production.

Official attention on agricultural production cumulated in the promulgation of the White Paper on "Grain Problem in Our Country". The White Paper elucidated the need for diversification of agricultural production and increase in both grains and non-grain production potentials. More importantly, an official target of a grain self-sufficiency rate of not lower than 95% was also put forward. The basic principle of mainly relying on domestic resources, effecting basic self-sufficiency in grain but not eliminating the possibility of utilizing foreign resources as a supplement was established. Furthermore, it states that the role of imports is only to fulfill one of the following functions: widening the choice of agricultural produce, extending the price range and addressing domestic regional imbalance. The Paper concludes that the country would prove that it is capable of not only feeding itself but also improving the quality of life of its people gradually. Thus, China would not become a threat to the world food security. Instead, it would make increasingly significant contribution to world food development.

3. Changes in Cultivated Land and Regional Variations

Agricultural land is one of the indispensable resources for grain production. Thus, any decline in the size of cultivated land or its regional imbalance is a major issue in any serious research on the agricultural sector and regional development in China (Zhou, 1996a).

3.1 Dynamics of Regional Balance

The PRC may be divided into two large North and South regions by roughly following the Changjiang River. The former includes Beijing, Tianjin, Hebei, Shanxi, Inner Mongolia, Liaoning, Jilin, Heilongjiang, Shandong, Henan, Shaanxi, Gansu, Qinghai, Ningxia and Xinjiang; and the latter includes 15 provinces and municipalities of Shanghai, Jiangsu, Zhejiang, Anhui, Fujian, Hubei, Hunan, Guangdong, Guangxi, Hainan, Sichuan, Guizhou, Yunnan and Tibet. From 1991 to 1995, the regional share of the national agricultural land has not changed much with the North region falling slightly from 39.8% to 39.2%. Similarly, the official tripartite division of East, Central and West has not shown much changes with the East declining marginally from 32.5% to 32.0% in 1991 to 1995. The share of the West was basically unchanged. The Central has increased slightly from 43.8% to 44.3%.

These figures suggest that major decline in the Eighth FYP was confined mainly to the Southeast. This part of the country is where most of the Special Economic Zones, Open Cities, and Open Economic Zones were located. When Shanghai, Jiangsu, Zhejiang, Fujian, Guangdong, Guangxi and Hainan are taken together, the reduction in the size of agricultural land was sizable. From 1991 to 1995, net drop in agricultural land area there was as large as 470,000 hectare, or about 70% of the total national decline. By 1995, cultivated land in the region has dwindled to 12.92 million hectares. This declining trend has caused a major concern of the government and the international community.

In view of such changes in the Eighth FYP, 15 areas were designated for the exploitation of abandoned land and land regeneration as listed in Table 8.2.

Table 8.2 Areas Designated for Increasing Cultivated Land

Area	Method	Size (hectares)
Inner Mongolia	Exploitation of Abandoned Land	320,000
Liaoning	Land Regeneration	67,000
Jilin	Exploitation of Abandoned Land	130,000
Heilongjiang	Exploitation of Abandoned Land	600,000
Jiangsu	Land Regeneration	67,000
Zhejiang	Development of Marshy Area	20,000
Jiangxi	Exploitation of Abandoned Land	34,000
Shandong	Exploitation of Huai-Huang Delta Bay Area	667,000
Hubei	Exploitation of Land at the River Estuary	100,000
Hunan	Exploitation of Abandoned Land	133,000
Guangdong	Exploitation of Abandoned Land	43,000
Tibet	Exploitation of Abandoned Land	15,000
Qinghai	Land Regeneration	34,000
Ningxia	Exploitation of Abandoned Land	---
Xinjiang	Exploitation of Abandoned Land	467,000

If these official goals are achieved, the size of cultivated land in the PRC will be increased by 28 million hectares. 57% of the increase will be found in Liaoning, Jilin, Heilongjiang, Inner Mongolia and Xinjiang in the Northeast and the North. Thus, the entire national grain supply will likely to become increasingly dependent on the North. Another obvious characteristic of the land exploitation policy is the large increase in the number of major areas designated for increasing cultivated land compared to the Eighth FYP. This policy of continual exploitation of new land cannot be sustainable ecologically and can lead to severe loss of soil nutrients by erosion and runoff, desertification or the destruction of ecological balance.

The control or even reversal of the decline in agricultural land area in the largest agricultural provinces is also very important. The need to arrest continually declining is particularly urgent in Hebei, Shanxi, Anhui, Henan, Hubei, Hainan, Sichuan, Guizhou, Gansu, Liaoning, Jilin, Shandong, Yunnan, Shaanxi and Jiangxi. In provinces where the demand for non-agricultural land has been particularly strong, the priority needs to be set for reserving land for grain production. These provinces included Zhejiang, Fujian, Guangdong, Guangxi, Beijing, Shanghai and Tianjin.

3.2 Dynamic Balance of Agricultural Land

Based on the agricultural trend and its related policy, it is possible to classify the provinces/municipalities into three categories. The characteristics of these three types of areas are summarized in Table 8.3.

Table 8.3 A Typology of the Dynamic Balance of Agricultural Land

Type	Examples and Locations
Dynamic Balance with Possibility of Increase	*Examples* • Heilongjiang, Inner Mongolia, Tibet, Gansu, Qinghai, Ningxia and Xinjiang *Locational Characteristics* • All, except Tibet, are all in the Northeast or Northwest
Dynamic Balance	*Examples* • Hebei, Shanxi, Liaoning, Jilin, Anhui, Shanxi, Shandong, Henan, Hubei, Hunan, Hainan, Sichuan, Guizhou, Yunnan and Shaanxi. *Locational Characteristics* • Mainly scattered throughout the country
Decreasing	*Examples* • Beijing, Tianjin, Shanghai, Jiangsu, Zhejiang, Fujian, Guangdong and Guangxi *Locational Characteristics* • All eastern provinces

3.3 Changes and Regional Characteristics of Sown Area in the Eighth FYP

Similar to the cultivated land, the sown area has not undergone much change in spatial focus following the North-South divisions. In terms of the East-Central-West division, the fall in the share of the East can again be identified. Most of such decline has been made good by increases in the West.

According to the agricultural resources of different areas, the country can be divided into major potential agricultural development regions.

3.3.1 Northeast Plain

The favourable and sustainable factors include rich reserve of agricultural land and other resources, e.g. relatively rich water resources, relatively less serious agricultural pollution and higher agricultural labour productivity. The major unstable factors are relatively few agricultural products, loose operation, cool summer, dry spring and frequent flooding in the summer. The major development strategy should therefore be the commercialization of agricultural production, diversification of the agricultural management system and suitable exploitation of reserve resources without disrupting the ecological balance.

3.3.2 Huang-Huai-Hai Plain

The major favourable factors there include flat relief, good geographical location, sufficient sunlight and diversity of plant species. The major unfavourable factors include dry spring, "saltiness of land", high ratio of low to middle-high land, lack of water resources and low agricultural labour productivity. The major development strategy is on irrigation and transformation of low to middle-high yield land.

3.3.3 Southwest Hilly Region

The sustainable factors there include a wide variety of relief and natural resources, special species cultivation and relatively sufficiently rainfall. The major non-sustainable factors include the lack of cultivated land, low quality of land productivity, low water utilization rate and severe loss of land and water in some regions. The major strategy for development are terracing and the utilization of underground water or river.

3.3.4 Xinjiang Region

The major sustainable factors there include relatively sufficient cultivated land reserve and sufficient sunlight. The major non-sustainable factors include the lack of water resources and severe desertification. The major development strategy is suitable exploitation of cultivated land reserves, irrigation, dry farming and oasis farming.

4. Regional Balance of Grain Supply and Demand

Affected by the above and other agricultural factors, there are obvious regional characteristics in grain supply of China. Population distribution and economic development characteristics are also showing very prominent regional characteristics. These factors have laid the foundation for differences in the grain supply and exerted a persistent influence on the agricultural and grain goals and policy in different regions.

4.1 Changes in the Spatial Pattern of Grain Production

Again, if the North-South division is followed, the trend for the shift of focus to the former is obvious in the Eighth FYP. The share of the South has declined from 55.4% to 54.0%. When the absolute amount is taken, the fall in the supply is particularly obvious. Following the tripartite division of East-Central-West, the Eighth FYP witnessed the declining shares of East and Central and the accompanying increase in the importance of the West. Nonetheless, the former two still remain relatively important in national grain production as they shared 33.7% and 44.7% respectively in 1995. If one follows the division of the 7 major regions, the Southeast's contribution to national food production has dropped from a high of 21.0% in 1990 to only 19.9 in

1995. The most notable is the absolute decline in grain production of Shanghai, Jiangsu, Zhejiang, Guangdong and Guangxi.

4.2 Per Capita Grain Production

Regional disparity in the average grain production per capita can be calculated by the Regional Disparity Index (RDI), by the sum of the difference between regional and national grain per capita divided by the sum of the regional grain per capita. Following this formula, the RDI in 1990 was 14.49% and in 1995, it dropped slightly to 13.43%.

If one uses the average provincial grain production as a yardstick, 18 provinces and municipalities have average grain production levels lower than the average in 1990. By 1995, the number has fallen to 16. The two provinces which have risen out of this lower-than-average category were Henan and Hebei.

The successful implementation of the "responsibility system" by provincial heads is a major contributory factor to falling regional disparity in grain. This system is quite strict and, to a certain extent, has reverted to the planned economy practice of indicative planning. In brief, specific "grain sown area" and "grain production" targets are set for different levels of government. As there were drastic falls in grain production in 1992 and 1993, the State Council promulgated the "Notice about Deepening Reforms in Food Procurement System" (1994/5/9). In the Notice, the State Council demanded the implementation of a responsibility system in each province, autonomous region and municipality to ensure the balance of regional grain production, and to stabilize cultivated land and sown area, food production and food reserves. Therefore, the "Opinions of the Central Committee of the CCP the State Council about Improving Agriculture and Rural Villages Works" (1995/3/11) further emphasizes that the government at every level, particularly at the provincial level should shoulder the responsibility of ensuring a general balance of food supply and demand and "actually" implement the provincial head responsibility system. The system was expounded later in the "Notice of the State Council about Deepening Reforms of the Grain Procurement System, Cotton and Chemical Fertilizers" (1995/4/2). Specifically, the provincial head has to shoulder the responsibility of ensuring local food supply, including agricultural produce and the balance of chemical fertilizers and other production factors. Furthermore, he/she has the responsibility of stabilizing price. Basically, this system has four major advantages: increasing the initiatives of agricultural production at the local level, maximizing the locational advantages under the planning of the government and achieving the balance of agricultural produce.

4.3 Major Types of Food Deficit Areas

4.3.1 Classification Criteria

Grain security is a major component of food security. It has been brought to the attention of the government when serious grain deficit prevailed in the late 1980s. Grain security includes the general security of total volume, quality, type, reserves and a secure system of demand and supply. In most situation, the security of total amount is the most important. Within the PRC, there are two commonly-used yardsticks for defining grain security. They are the average grain output of 400 and 320 kg per capita. In view of the fact that by now the average grain output in China has not yet reached 400 kg, the use of the latter as a yardstick seems more realistic.

We set the range of 320 to 400 kg as the basically balanced zone and 400 kg as the minimum for classifying grain surplus areas. According to these criteria, there are altogether 13 grain deficit provinces/municipalities. These deficit regions can be further divided into 3 types, that is, deficits of total amount, deficits because of structural imbalance and deficits in fodder and industrial crops.

Guizhou, Gansu, Qinghai, Shaanxi and Tibet fall into the first group. Shanxi and Yunnan fit the second group and Beijing, Tianjin and Guangdong as the third group.

Conversely, there are 10 grain surplus provinces, which have traditionally been labeled the "agricultural provinces" as they have richer resources in cultivated land, cultivated land reserves and agricultural technology. The roughly balanced provinces include Jiangxi, Liaoning, Sichuan, Henan, Ningxia, Zhejiang and Guangxi. The pressure on agriculture in these provinces is less than the others.

4.3.2 Relief

In the PRC, the composition of the population living in plain, hills and valley areas is 40%, 18% and 34% respectively. The rest of the population (18%) live in basin and other regions. The average grain consumption per capita in the 3 areas is 453,408 and 333 kg respectively.

In other words, food security in the hilly areas is the most unstable. Of the population with an average grain output of 200-400 kg, 41%, 29% and 30% live in the hilly areas, valleys and plain respectively. In other words, the major focus in ensuring grain security should be the hilly regions, as they are also where agriculture, pasture and forestry are all taking place and the loss of soil nutrients by runoff and erosion are severe. In these regions, the problems of sustainable development are quite serious and solutions are not readily available. The fluctuations in agricultural output and population pressure are also major reasons behind the problem of food insecurity in these areas.

4.4 Agricultural Policy and Planning in the Ninth FYP

Following the large-scale drop in grain production from 1992 to 1994, which was not just confined to individual regions but affecting many regions, the policy of setting grain production goals for individual provinces and municipalities was once again implemented (Quanguo nongcun jingji fazahan "jiuwu" jihua he 2010 nian yuanjing guahua bianweihu, 1996). Table 8.4 lists the grain output targets for individual provinces and municipalities.

Table 8.4 Grain Output Targets at the Provincial Level, 1997

Region	Grain Development Targets
Beijing	• Stabilize total production at above 2.5 million tons
Tianjin	• Stabilize total production at above 2.25 million tons • Achieve basic self-sufficiency
Hebei	• Generate surplus after meeting local needs • Improve the variety mix • Improve the quality
Shanxi	• Balance in total volume
Inner Mongolia	• Fulfill the government assignment of 7 billion of grains
Liaoning	• Basically achieve balance • Generate surplus after meeting local needs
Jilin	• Further increase marketable ratio • Further increase grain production efficiency

Heilongjiang	• Further raise the net marketable ratio • Improve the variety mix • Raise efficiency
Shanghai	• Control and slow down the rate of grain output decline
Jiangsu	• Achieve balance in total volume • Generate grain surplus after meeting local needs
Zhejiang	• Achieve basic self-sufficiency • Adjust local deficits and surplus
Anhui	• Further increase 10 billion jin of grain
Fujian	• Increase grain output gradually
Jiangxi	• Net inter-provincial grain "exports"
Shandong	• Increase output by 10% or more • Increase internal reserves of grains
Henan	• Maintain the situation of net grain "exports"
Hubei	• Increase by average annual rate of 1.47% • Average grain of 430 kg per capita
Hunan	• Stabilize output • Improve quality • Secure efficient system of demand and supply
Guangdong	• Local self-sufficiency rate of 70% • Self-sufficiency in staple and seed grains
Guangxi	• Achieve internal balance
Hainan	• Achieve basic self-sufficiency
Sichuan	• Continue to be achieve balance in quantity
Guizhou	• Achieve basic sufficiency in rural staple • Raise self-sufficiency rate
Yunnan	• Achieve basic self-sufficiency in grain, meat and oil
Tibet	• Gradually achieve basic self-sufficiency in grain, meat and oil by farmers
Shaanxi	• Achieve self-sufficiency in grain • Try to create surplus after fulfilling local needs
Gansu	• Try to be self-sufficiency
Qinghai	• Achieve a self-sufficiency rate of 83%
Ningxia	• Increase the marketable rate
Xinjiang	• Increase the marketable rate • Try to be the "Grain Warehouse" in the West

Based on the targets set by the government, individual provinces have also worked on their own specific agricultural development strategies. These strategies are listed in Table 8.5.

Table 8.5 Grain and Agricultural Development Strategy at the Provincial Level, 1996-2000

	Grain and Agricultural Development Strategies
Beijing	• Achieve scale economy • Improve quality • Agricultural diversification
Tianjin	• Water utilization
Hebei	• Dry farming • Improvement of species • Water irrigation strategy of "reduce taking from wells, encourage diversion and improve drainage"
Shanxi	• Increase multiple cropping index • Practice dry farming
Inner Mongolia	• Exploitation of 0.32 million hectares of abandoned land • Expand irrigation area by 0.4 million hectares • Practice dry farming
Liaoning	• Land rehabilitation by 60,000 hectares • Reduce in usage of grain for farming • Usage of ocean resources -- "Liaoning on sea"
Jilin	• Exploitation of 0.13 million hectares • Improve 2 million hectares of low-to-medium yield farmland
Heilongjiang	• Exploitation of land by 0.6 million hectares • Improve 1.73 hectares of low-to-medium yield farmland • Further develop grain and fodder crops processing
Shanghai	• Use of technology • Commerce-industry-agriculture • Rural economic unification
Jiangsu	• Land Rehabilitation by 60,000 hectares • Improve 1 million hectares of low-to-medium yield farmland • Set up a agriculture risk protection system • Raise labour productivity • Improve self-sufficiency in agricultural inputs

Zhejiang	• Stabilize farmland area • Mainly aimed at a single species • Improve quality • Dual structure of staple and economic crops • "Reclaiming" 0.3 million hectares of farmland from sea • Transform 5 million hectares of low-to-medium yield farmland • Reform the household registration system • Achieve scale economy
Anhui	• Stabilize farmland area • Commercialize agricultural outputs • Make use of technology
Fujian	• Transform 0.2 million hectares of low-to-medium yield farmland • Increase multiple cropping index • Mitigate effects of natural disaster
Jiangxi	• Exploitation of 30,000 hectares of new land • Improve 0.64 hectares of low-to-medium yield farmland • Reduce 0.94 million hectares of land affected by loss of soil nutrients by runoff and erosion • Develop fruit gardening
Shandong	• Increase multiple cropping index • Stabilize of farmland area • Transform 0.3 million hectares of low-to-medium yielding farmland • Exploit ocean resources -- build "Shandong on sea"
Henan	• Increase the base of commodity grain • Reducing tobacco-growing • Processing of agricultural products
Hubei	• Stabilize farmland and grain production area • Terracing • Transform 0.3 million hectares of low-to-medium yield farmland • Exploit 0.1 million hectares of "river estuary"
Hunan	• Exploit 0.13 million hectares of new farmland • Transform 0.3 million hectares of low-to-medium yield farmland
Guangdong	• Protect farmland • Stabilize farmland area • Protect agriculture (price, reserve and insurance) • Family-based farm

Guangxi	• Stabilize farmland area • Plant high-quality rice • Special agricultural produce
Hainan	• Stabilize farmland • Develop tropical crops • Promote Hainan-Taiwan co-operation • Improve selling network of agricultural products
Sichuan	• "Stabilize grain output increases and work towards a stable situation" • Exploit the west
Guizhou	• Stabilize grain production area • Comprehensive development by terracing
Yunnan	• Stabilize grain production area • Increase multiple cropping index • Develop the southwest and Jinshajiang exploitation
Tibet	• Exploit 15,000 hectares of new land • Transform 0.1 million hectares of low-to-medium yield farmland
Shaanxi	• Exploitation of new land in part of the province • Utilize different sources of agricultural funds
Gansu	• Transform low-to-medium yield farmland • irrigation plan • Dry farming
Qinghai	• Rehabilitate 30,000 hectares of land • Transform 0.12 million hectares of low-to-medium yield farmland
Ningxia	• Exploitation of new land • Water diversion • Irrigation
Xinjiang	• Exploitation of 0.46 million hectares of new land • Four major belt of cultivation

From this table, it is clear that the strategies have been relatively vague in Tianjin, Shanxi, Anhui and Shaanxi. These local governments are relatively unwilling to pin down specific agriculture and grain production strategies. In the following analysis, it becomes clear that these are areas of relatively low production areas.

4.4.1 Authoritativeness and Feasibility of National Agricultural Goals

After the Open Policy, there was a trend for the government to reduce the scope of indicative planning targets. Nonetheless, the national agricultural goals set by the central government still have a certain degree of binding effects on local governments and individuals. This is particularly the case for staple production. However, the results/policy outcomes have not all met national goals. As local governments began to set goals for themselves, comparison of the national targets and the sum of the provincial/municipal targets show that there have been substantial differences between the two. In particular, the target for meat was below the national target by as much as 22%.

4.5 Comparisons of Agricultural Production Potentials across Regions

In making comparisons of the situation in different regions, we attempted to look into the 3 major aspects, i.e. recent agricultural production plan, long-run agricultural production potential and locally-determined agricultural growth rates.

4.5.1 Trend of Agricultural Production

Agriculture embodies cultivation, forestry, husbandry and fishery. The agricultural production index (API) in China is calculated by expressing the value of the agricultural production at the end of the period over the base year. In other words, if there is no increase of agricultural production during the study period, the API is 100. This index can also be calculated at the provincial level. During the Eighth FYP (1991-1995), the national average agricultural production index is 107.5. Taking this national average as a yardstick, spatial variations in agricultural production trend can be examined by plotting the differences between the provincial API and the national average. Figure 8.2 shows the situation during the Eighth FYP. From the Figure, one can see that Fujian, Hainan, Guangxi, Shadong, Henan, Jiangsu and Liaoning are having the highest growth rates of agricultural production. The ability in generating higher value added agriculture-related industries, such as the wine industry, in Shandong and the success of tropical and local food production in Hainan and Guangxi are noteworthy. Moreover, one can observe that the performances of the coastal provinces have not been unsatisfactory. In contrast, provinces like Fujian, Shangdong and Hainan were having the greatest increase in agricultural production. If one just considers cultivation, the cultivation production index (CPI) can be calculated and they are also displayed in Figure 8.2. For cultivation, the national average was 105.7. Again, provinces with better performances are Fujian, Hainan, Xinjiang, Hebei, Guangxi, Jiangsu and Shandong. Provinces which lag behind include Qinghai and Jiangxi. Hebei, Shangdong and Jiangsu in East China represent a contiguous belt of coastal provinces having relatively high growth of cultivation production from 1991 to 1995.

Further breaking down the average annual growth rates by year, 5 categories are produced. Table 8.6 summarizes the major characteristics of these 5 types.

Table 8.6 Grain Production Trends by Type

Type	Examples and Characteristics
Steady Increase in Grain Production	*Examples* • Hainan, Tibet, Anhui, Henan, Yunnan, Guizhou, Xinjiang, Shandong, Hebei *Major Characteristics* • Larger areas of cultivated land and more medium- and low-producing fields
Steady Increase in Grain Production	*Major Characteristics (Con't)* • Steady but slow increase in Tibet and Guizhou
General Increase with Small Fluctuations	*Examples* • Fujian, Tianjin, Heilongjiang, Jiangxi, Hubei, Hunan and Ningxia.

	Major Characteristics • Natural-disaster prone • Relatively poor agricultural production basic infrastructure
Relatively stable production	*Examples* • Inner Mongolia, Shanxi, Jilin, Jiangsu, Sichuan and Gansu
Large Fluctuations	*Examples* • Liaoning and Guangxi *Major Characteristics* • Possible response to the fluctuations in the nearby coastal areas
Steady Decrease	*Examples* • Shaanxi, Qinghai, Guangdong, Beijing and Shanghai *Major Characteristics* • Those having enough money to supplement local deficiencies • Those do not have enough money to supplement local deficiencies

5. OVERALL ASSESSMENTS

Based on the earlier analysis, the different provinces and municipalities are given an overall agricultural potential index (OAPI) based on the following 3 criteria:

i) Average annual growth rate of total agricultural production in the past decade;
ii) value of fixed assets in agriculture and ability to resist natural disasters and;
iii) targets of agricultural growth rates set locally.

For each of these criteria, the top 20% of the thirty provinces and municipalities in China are given a score of 5, the next 20% are given a score of 4 and so on. The OAPI is calculated by summing up these three scores. Table 8.7 lists all the scores and OAPI of the provinces and municipalities. Moreover, the scores of the average annual growth rates of farming and grain outputs in the last decade for each province/municipality are also shown.

Based on the values of the OAPI, the country can be divided into five major categories to reflect the overall conditions of agriculture in different areas of the country. These five categories are “good”, “relatively good”, “average”, relatively poor” and “poor”. Areas having OAPI of 13 or above have “good” agricultural potentials. Provinces and municipalities in this group include Tianjin, Liaoning, Shangdong, Fujian, Guangxi and Hainan. In addition, seven provinces, that is, Heibei, Heilongjiang, Guangdong, Zhejiang, Henan and Anhui, have OAPI of 11 or 12. Their indices are relatively high and they may be classified as having “relatively good” potentials for agricultural production. Areas having OAPI equals to 9 or 10 have “average” conditions. These localities are Jiangsu, Jiangxi, Xinjiang, Hubei, Hunan, Shaanxi, Jilin, Beijing, Sichuan and Shanxi. Areas having relatively poor conditions are Inner Mongolia, Shanghai, Yunnan, Tibet, Gansu and Ningxia. These localities have OAPI equals to 7 or 8. Within this “relatively poor” category, all areas, except Shanghai, are relatively backward areas. Lastly, provinces having OAPIs below 7 are classified as having "bad" potentials for agricultural production. These localities include Guizhou and Qinghai. They are relatively backward areas. The above categories can then be used as a basis for the government to formulate the most suitable agricultural policy to suit local circumstances. Besides, it is an important basis for understanding the potentials and problems for the local economy, and an important element in regional development in the country.

Table 8.7 OAPI in Different Provinces and Municipalities in China, 1997

Region	Agriculture (a)	Farming (b)	Grain Output (c)	Fixed Assets in Agriculture (d)	Resistance to Natural Disasters (e)	Long-Term Agricultural Potential ($f=\frac{d+e}{2}$)	Target Growth Rate Set Locally (g)	API (h=a+f+g)
Beijing	3	3	1	5	5	5	1	9.0
Tianjin	4	4	4	4	5	4.5	5	13.5
Hebei	4	5	5	4	4	4	4	12
Shanxi	2	2	3	2	4	3	4	9
Inner Mongolia	2	3	3	4	3	3.5	2	7.5
Liaoning	5	4	2	3	3	3	5	13.0
Jilin	3	2	3	4	3	3.5	3	9.5
Heilongjiang	4	3	4	5	4	4.5	3	11.5
Shanghai	3	2	1	2	5	3.5	1	7.5
Jiangsu	5	5	3	3	4	3.5	2	10.5
Zhejiang	4	3	1	5	4	4.5	3	11.5
Anhui	4	3	5	4	3	3.5	4	11.5
Fujian	5	5	4	3	3	3	5	13
Jiangxi	4	1	4	2	2	2	4	10
Shandong	5	5	5	3	5	4	4	13
Henan	5	4	5	4	4	4	2	11
Hubei	4	3	4	1	3	2	4	10
Hunan	4	3	4	1	2	1.5	4	9.5
Guangdong	4	3	1	4	4	4	4	12
Guangxi	5	5	2	2	4	3	5	13
Hainan	5	5	5	3	4	3.5	5	13.5
Sichuan	3	3	3	3	3	3	3	9
Yunnan	3	4	5	4	3	3.5	2	8.5
Guizhou	3	4	5	1	2	1.5	2	6.5
Tibet	4	3	5	5	2	3.5	1	8.5
Shaanxi	4	4	1	2	2	2	3	9
Gansu	3	3	3	2	1	1.5	4	8.5
Qinghai	1	1	1	4	3	3.5	1	5.5
Ningxia	2	3	4	1	3	3	3	8.0
Xinjiang	4	5	5	3	5	3	3	10.5

REFERENCES

1. Brown, Lester A. 1995. *Who Will Feed China?* Worldwatch Institute.
2. Chen Yaobong. 1995. *Zhongguo liangshi fazhan wenti yanjiu* (Research on Development Problems in China). Zhonghua gongshang lianhe chubanshe.
3. Liang Ying. (ed.) 1996. *Can China Feed Herself?*. Beijing: Economic & Technology Press (Chinese text)
4. Sun Honglie. 1996. Utilization of different resources to meet the increasing demand for food in China. *China Technology Press*, 11th October (Chinese text).
5. The Editorial Committee. 1996. *Long-Term Planning of Rural Economic Development in the Ninth FYP and 2010 Plan*. Beijing: China Association of Industry & Commerce (Chinese text).
6. Zhou Jiahua. 1996. Necessity and feasibility of maintaining dynamic balance in cultivated land. In *China Land* 7 (Chinese text).
7. Zhou Yuchuan. 1996b. Measures for maintaining dynamic balance in cultivated land. In *China Land* 7 (Chinese text).

Chapter 9

POVERTY STRICKEN AREAS AND POVERTY RELIEF

LIU Hui
Associate Professor, Institute of Geography,
Chinese Academy of Science

Poverty is a serious challenge facing the world today. There are about 1.2 billion poor people in the world, 300 million of which were added in the past five years. In March 1995, a summit conference on social development was held by the United Nations at Copenhagen, Denmark, in which the year 1996 was defined as the International Poverty Relief Year and October 17 as the International Poverty Relief Day. In 1994, the Chinese government promulgated a 7-year National Poverty Relief Program aimed at concentrating manpower, material and monetary resources to solve the basic survival problems of food and clothing for the 80 million poor people in China's rural areas (State Council, 1994). The National Poverty Relief Program is aimed to resolve the rural poverty problem, reduce the wealth distinction between the East and the West, achieve the goal of common prosperity. It states that from 1994 to 2000, the government will concentrate manpower, material and monetary resources, and will mobilize all social strength, to solve the basic problems of food and clothing for the 80 million poverty-stricken people in the rural areas of China. The details of the program is as follow: -

1. Goals

(1) to achieve a per capita net income for poverty-stricken family of no less than 500 yuan according to the fixed price of 1990

(2) to establish the essential conditions for poor families to solve their food and clothing problems.

(3) to improve infrastructure: to basically solve the problem of water supply for both men and live-stock, to ensure that most of poverty-stricken towns and poor areas that have markets or commodity products have highways, and to eliminate the no-electricity counties, so that most of poverty-stricken towns will use electricity.

(4) to change the backwardness in culture and education of these areas, make elementary education universal, develop professional technological education and training, improve medical condition, and stringently practise family planning.

2. Policies and Measures

(1) to emphasize the development of agricultural, breeding and other related primary industries, and establish the means by which products are transported and traded.
(2) to make vigorous effort to develop rural enterprises that use local resources and intensive labor.
(3) to speed up the exploitation and utilization of uncultivated land, waste mountains, waste slopes and waste waters.
(4) to export and resettle labor.

3. Fund Management

The current poverty relief fund will be used continuously until the year 2000. In 1994, the central government increased the fund for providing work as a form of relief and the poverty alleviation loan. From now on the government will continue to increase the poverty relief fund.

Besides, the government has worked out 9 preferential policies on loans, import and export, etc. in this program. At the same time, it also assigned poverty relief tasks to each department of the government and social groups. It seeks also to encourage international cooperation on poverty relief.

According to the target of the Program by the year 2000, abject poverty will be dispelled from China. This is one of the most important tasks of regional development in China by the end of the last century.

From the mid-1980s, initiated by the State Planning Commission, the Institute of Geography of the Chinese Academy of Sciences studied the poverty-stricken areas in China and their corresponding physical and economic characteristics. The study divided the poverty-stricken areas into six categories. Since then, there had been a lot of research on the topic and they have considerable achievements. Among these, Kang Xiaoguang (1994) had analysed the causes of poverty in China and evaluated anti-poverty strategies of the 1990s. Yet, he failed to analyse the implementation of such policies. There were also Jiang Dehua (1989) who concentrated on the classification of poor areas in China and Guo Laixi (1992, 1996), who studied the poverty problem on a regional scale. Other than these, Chen Junsheng (1994) studied the new conditions arising after the implementation of the National Poverty Relief Program. In a continual battering against poverty, the World Bank (1990) used some developing countries as examples in their study of the relationship between poverty and economic development, and pointed out that no other task had been more important for policymakers in the 1990s than poverty relief.

This chapter will describe the standards used for poverty determination in China and the principal characteristics of poverty-stricken areas. It will make a systematic exposition of the national and regional poverty relief policies and their implementation and it illustrate with examples their effects on solving the basic survival problems, i.e.: decreasing the size of the poverty-stricken population, promoting local economic development and improving the environment. It will also reveal the initial results of the Chinese government's poverty relief policy of developing the individual poverty-stricken households or persons, and analyse the varying effects of poverty relief in different regions and at the same time evaluates the latest poverty relief policies of the central government and the local provinces.

This chapter will therefore focuses on poverty relief policies as a positive step and offer poverty relief plans for these areas to achieve their poverty relieve goal.

1. DISTRIBUTION OF POVERTY-STRICKEN POPULATION

During the Eighth FYP (1991-1995), the number of Chinese living in poverty decreased from 9.7 million to 6.5 million (Chen, 1994). In 1996, the country successfully lifted 7 million rural residents out of poverty and the size of the poverty-stricken population consequently decreased to 5.8 million. Although the percentage of poor people to total rural population is not very high, the distribution of poverty-stricken population is highly uneven. Most of them live in plateaux, mountainous and desert areas under harsh physical conditions with little access to transportation and communication.

1.1 Poverty Stricken Areas and their Determining Standards

There had not been any standard for distinguishing poverty-stricken areas in 1949-1978. Towards the end of the 1970s, the government determined 18 regions[1] as poverty-stricken areas. In order to solve the food and clothing problems of the poverty-stricken population, the central government distinguished, on the national level, three groups of poverty-stricken counties.

The first group was published in 1986, which includes 331 counties. Their determining standards are: (1) annual per capita net income in 1985 of less than 180 yuan for general counties, (2) annual per capita net income in 1985 of less than 200 yuan for minority autonomous regions, (3) annual per capita net income in 1985 of less than 300 yuan for the Old Liberated counties, (4) an annual average per capita net income from 1984-1986 less than 300 yuan for the purely pastoral counties or less than 200 yuan for the semi-pastoral counties.

The second group was published in 1991, which added 236 'special loan' counties as poverty-stricken counties on the basis of the first group, leading to a total of 567 counties. The standard adopted was that the annual per capita net income in 1990 was less than 300 yuan.

The third group was published in 1993. Considering the rise in the general price index and the cost-of-living index, the standards determined in the 1989s were found incompatible with the living demands of the 1990s. So, when the government formulated the National Poverty Relief program in 1993, the standards were readjusted. Counties with a per capita net income of no more than 400 yuan in 1992 were categorized as poverty-stricken counties, whereas if the poor counties determined in 1986 had achieved a per capita net income of more than 700 yuan in 1992, they would no longer be recognized as poverty-stricken counties. As a result, the number of China's national-level poverty-stricken counties increased from 331 in 1986 to 592 in 1993 (Table 9.1).

---

[1] The 18 poverty-stricken regions are: Qinba mountains, Wuling mountains, Wumeng mountains, mountains of the south-west of Yunnan province, Hengduan mountains, Taihang mountains, Luliang mountains, mountains of the north-west of the Guangxi Zhuang autonomous region, Jiuwanda mountains, Luluerhu mountains, Xihaigu regions, Dingxi regions, Tibet, the old liberated south-west and north-east Fujian, north of Shannxi, Jinggang mountain, south of Jiangxi and of the Yimeng mountain

Table 9.1 Changing Number of Poverty-stricken Counties

Province	Number of poverty-stricken counties in the Seventh FYP	Number of poverty-stricken counties in the Eighth FYP		Number of poverty stricken counties in the national poverty relief program	Change
		Total	Increase		
China	331	567	236	592	+25
Hubei	14	32	18	39	+7
Shanxi	14	31	17	35	+4
Inner Mongolia	16	32	16	31	-1
Liaoning	3	9	6	9	
Jilin		6	6	5	-1
Heilongjiang		11	11	11	
Jiangsu					
Zhejiang	3	3		3	
Anhui	9	16	7	17	+1
Fujian	14	14		8	-6
Jiangxi	17	20	3	18	-2
Shangdong	9	9		10	+1
Henan	15	22	7	28	+6
Hubei	13	25	12	25	
Hunan	8	10	2	10	
Guangdong	4	7	3	3	-4
Guangxi	23	29	6	28	-1
Hainan	3	6	3	5	-1
Sichuan	21	34	13	43	+9
Guizhou	19	46	27	48	+2
Yunnan	26	70	44	73	+3
Tibet				5	+5
Shaanxi	34	48	14	50	+2
Gansu	31	44	13	41	-3
Qinghai	10	13	3	14	+1
Ningxia	8	8		8	
Xinjiang	17	22	5	25	+3

1.2 Distribution of the Poverty-Stricken Areas and Poverty-Stricken Population

The number of poverty-stricken people of the 592 poverty-stricken counties is more than 70% of the total poverty-stricken population in China. So the statistics of these 592 counties basically reflect the distribution of abject poverty by administrative division. (Table 9.2, Fig. 9.1, 9.2, 9.3).

Table 9.2 Distribution of Poverty-stricken Population in China (1993)

Province	Number of poverty-stricken counties	Persons with annual per capita income less than 500Yuan (in 10,000)	Persons with annual per capita income less than 300 Yuan (in 10,000)	Rate of poverty occurrence (%)*	Share of poverty in China (%)
China	592	12142	5496	11.3	100
Hubei	39	827	314	11.7	6.8
Shanxi	28	1054	397	15.2	8.6
Inner Mongolia	10	156	30	5.9	1.3
Liaoning	9	70	17	7.1	0.6
Jilin	5	49	29	10.9	0.4
Heilongjiang	11	47	24	17.9	0.4
Jiangsu				3.6	
Zhejiang	3	64	23	1.3	0.5
Anhui	17	665	163	8.3	5.4
Fujian	18	202	47	6.4	1.7
Jiangxi	25	797	323	7.0	6.5
Shangdong	10	393	186	7.7	3.2
Henan	8	13	2	2.1	0.1
Hubei	3	24	2	1.1	0.2
Hunan	5	30	14	3.8	0.2
Guangdong	28	503	304	18.8	4.1
Guangxi	43	1372	535	11.2	11.2
Hainan	48	1344	860	17.4	11.0
Sichuan	73	1791	1119	18.3	14.7
Guizhou	5				
Yunnan	31	281	115	20.8	2.3
Tibet	35	411	102	14.8	3.3
Shaanxi	50	887	396	18.4	7.3
Gansu	8	140	74	18.4	1.1
Qinghai	41	863	335	30.4	7.1
Ningxia	14	83	34	21.8	0.7
Xinjiang	25	157	72	17.6	1.3

* World Bank, Poverty Relief Strategy of China in 1990s, 1993.

Fig. 9.1 Distribution of Poverty Population in China

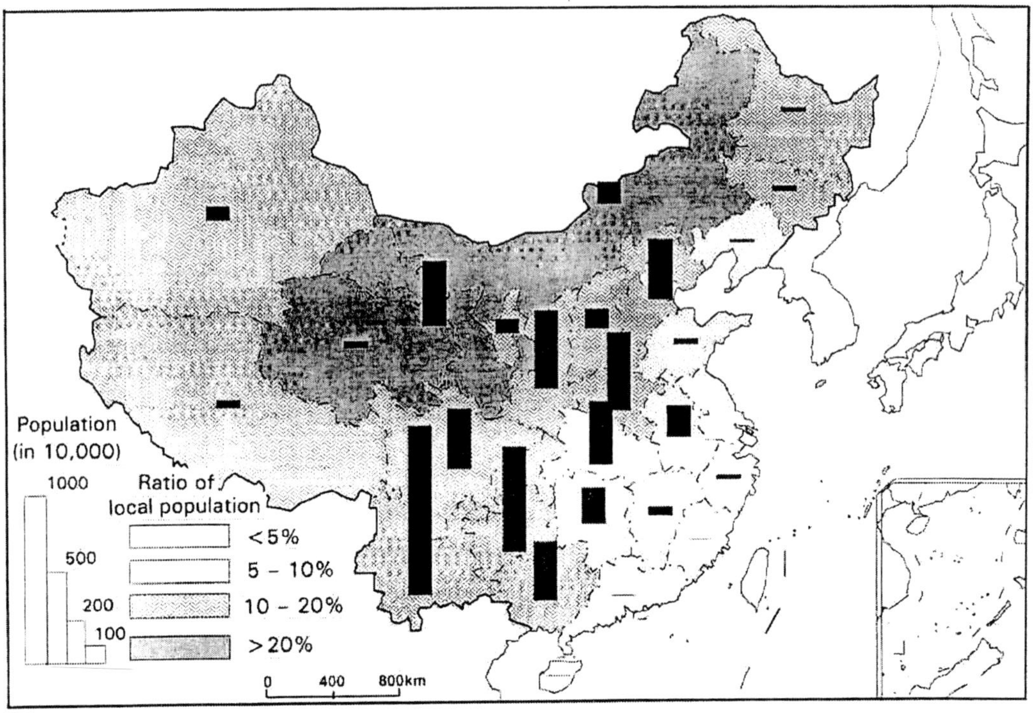

Fig. 9.2 Degree of Poverty in China

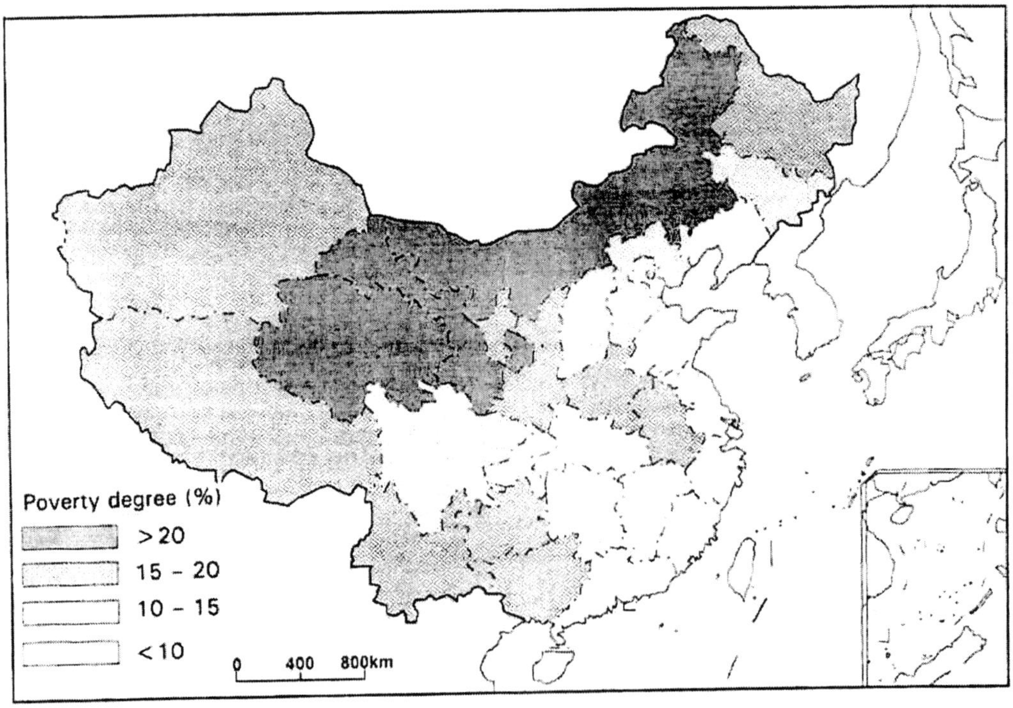

Fig. 9.3 Distribution of Poverty Counties in China

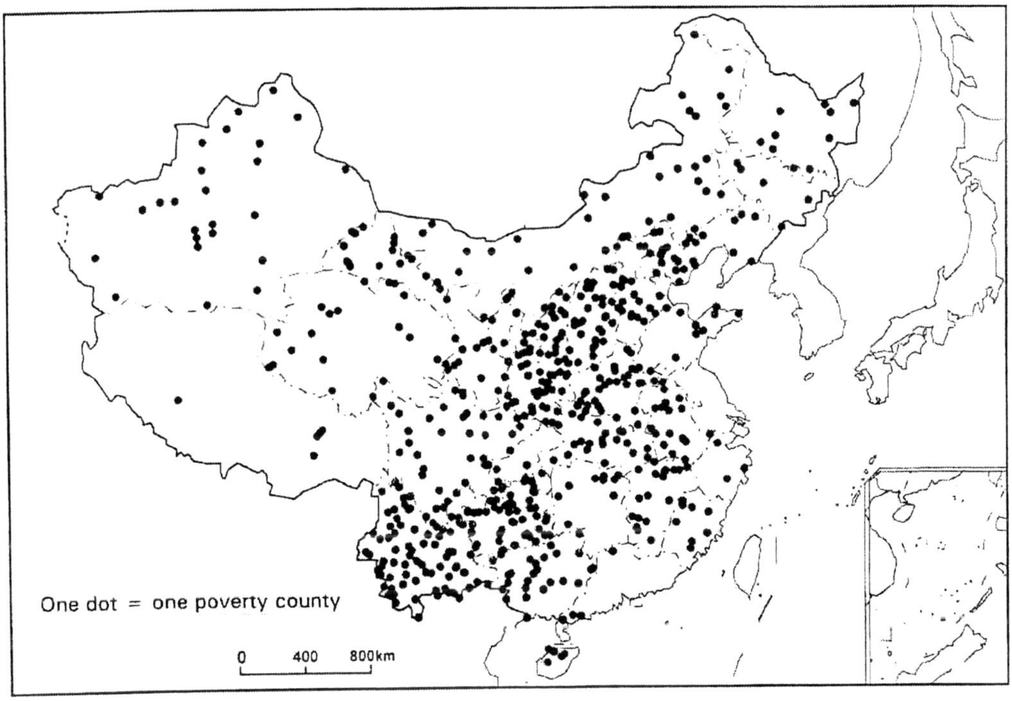

The distribution shows the following characteristics:

(1) the poverty-stricken areas are extensively distributed. Almost all province in China (except Beijing, Tianjin, Shanghai and Jiangsu), has one or more national-level poverty-stricken county(ies).
(2) the poverty-stricken population are mostly concentrated in Sichuan, Guizhou, Yunnan and Henan provinces, in which the number of poverty-stricken people sums up to 46% of the total poverty-stricken population of the country.
(3) the poverty-stricken people concentrate mainly in the Central and Western provinces, such as Yunnan, Guizhou, Sichuan, Shannxi, Ningxia, Gansu, Henan, Hebei, Hunan and Anhui. There are 437 poverty-stricken counties in the Central and Western China, which make up 82% of the country's total poor counties.
(4) provinces with a large poverty-stricken population often have a higher properisity of being poor. Therefore, poverty relieve is most difficult in these provinces as local resources are limited, such as Yunnan, Guizhou, Shannxi, and Gansu.

Jiang Dehua (1989) has summarized the distribution of the poverty-stricken population according to China's physical geography characteristics. It indicates that the percentage of the poverty-stricken population living in mountainous areas is 91.3%, deserts 7.3% and arid areas 1.4%. That is to say, most of the poverty-stricken population in China live in the mountainous area with harsh physical conditions. Thus physical geography is an important factor in the distribution of the poverty-stricken population.

Another spatial dimension of poverty distribution is the rural-urban differences. According to the Chinese government, the entire population of abject poverty in 1993 resided in the rural areas. However, with the introduction of market economy and the readjustment of the industrial structure, unemployment in the urban areas has increased. The problem of urban poverty has emerged in China as well. According to Kang Xiaoguang (1995), there are about 12 million poverty-stricken people in the urban areas, which accounts for 13% of the total poverty-stricken population in China. The rate of poverty occurrence in the urban and rural areas is 3.6% and 9.4% respectively. In addition, there are 5.3 million registered unemployed workers within the urban areas. If we include those who depend on social relief as within a loose definition of poverty, then the total number of poverty-stricken population in China's urban areas may amount to 20 million, or roughly 20% of the total working population in the urban areas.

1.3 Changes in the Size of the Poverty-Stricken Population

With the expansion of China's economy in the 1980s, the size of the poor population has greatly decreased. The number of Chinese living in poverty was 250 million in 1978 -- then 26% of the China's total population and a quarter of world's poor population, to 58 million in 1996 -- 4.8% of the total Chinese and 5% of the world total poor population (Fig. 9.4; Table 9.3 & 9.4). Although the size of the world's poor population increased by 25 million every year, the actual number of China's poor has registered continuous decline according to three discreet stages.

Fig. 9.4 Number of Poverty Population in China

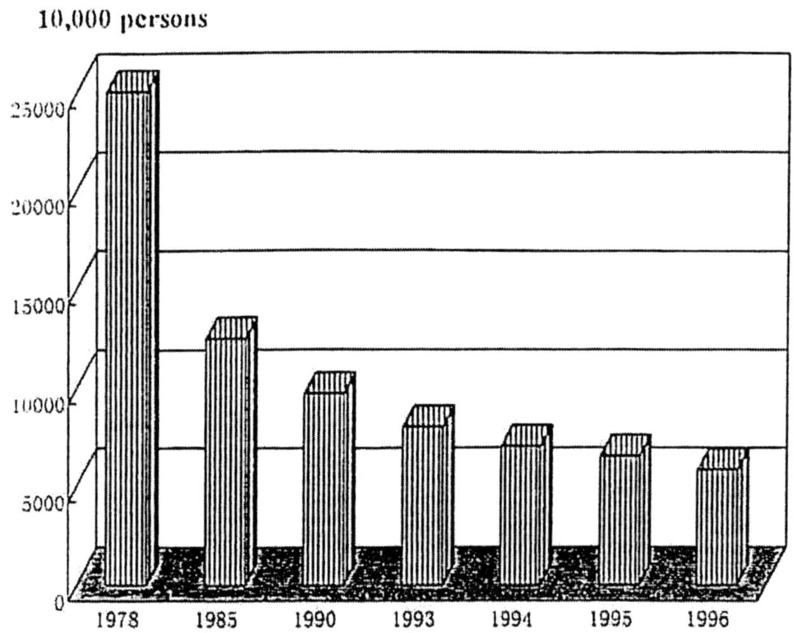

Table 9.3 Change in Number of Poverty-stricken Population

Year	Number of poverty population (in million)	Rate of rural poverty occurrence (%)
1978	250	30.7
1985	125	14.8
1990	97	10.8
1993	80	8.9
1994	70	7.8
1995	65	7.1
1996	58	6.3

Table 9.4 Poverty Population in Major Poverty-stricken Provinces

Regions		China	Guangxi	Guizhou	Shaaxi	Ningxia
Number of Poverty-stricken Population (in 10,000)	1993	8000	800	1000	500	139.8
	1994	7000	706			128.8
	1995	6500	606	789	393	132.8
	1996	5800	486	629	301	76.0
Annual rate of reduction (%)		8.5	13.1	12.3	24.1	15.2

First stage: from 1978 to 1985, with the implementation of rural economic reform, the number of poverty-stricken people was halved from 250 million to 125 million due to much increased rural productivity.

Second stage: from 1986 to 1993, it was a stable decreasing stage. The number of poverty-stricken population decreased from 125 million to 80 million. In this stage, despite the implementation of favorable poverty relief policies, and increased government assistance on poverty relief, the speed of poverty relief slowed down. The country only lifted 6.43 million people out of poverty each year.

Third stage: from 1994 to 2000, it is expected to be the most arduous poverty relief stage. In order to materialize China's poverty relief goal by the year 2000, 11.42 million poor people are to be removed from poverty each year. The speed of channeling people out of poverty will need to be nearly doubled as compared to the second stage. However, at the end of 1996, there were still 58 million poor in China. Between 1994 and 1996 the average decrease of the poverty-stricken population is 7.3 million a year.

During the Eighth FYP (1991-1995), the country successfully lifted 32 million poor people out of poverty, with a rate of 6.4 million a year. In 1996, there were 7 million poor removed from poverty, more than the average in the Eighth FYP. Yet it is still far from the goal of China's poverty relief program.

2. National Poverty Relief Policies and their Implementation

During the Eighth FYP (1991-1995), the central government provided financial assistance and special loans totalled 39.4 billion yuan to the poverty-stricken areas. The poverty relief funds came from various sources and it has been increasing (Table 9.5). In 1994, the State Council promulgated the National Poverty Relieve Program (1994-2000). In 1996, the Central Committee of the Communist Party of China (CPC) and the State Council publicised their decision to solve the food and clothing problems for the poverty-stricken population in the rural areas as soon as possible, in which the government further defined the fundamental policy of poverty relief by the method of development, and pointed out that the related activities should target towns, villages and individual households (Chen, 1994).

Table 9.5 Poverty Relief Funds from the Central Government

Source of fund	Amount (Yuan)	Year beginning	Name of fund	Remarks
Ministry of Finance	800 million annually	1980	Fund for assisting undeveloped regions	(1) The Central government have added 1 billion Yuan for providing work as a form of relief and 1 billion Yuan special discount loans for poverty relief every year since 1994. (2) The Central government will increase 1.5 billion Yuan of financial allocation and 3 billion Yuan of special loans for poverty relief from 1997. (3) The management of poverty relief funds are changed from the People's Bank, Agriculture Bank, Industrial & Commercial Bank to China Agriculture Development Bank after 1994. (4) Fund of providing work as a form of relief was provided in the form of materials before 1996 by the State Planning Commission. Since 1996 it has been provided in the form of money by the Ministry of Finance.
Ministry of Finance	200 million annually	1983	Special fund for "Three West Regions" construction	
People's Bank	1 billion annually	1984	Loans for the old-revolutionary bases, the Minority and the Border Areas	
People's Bank	400 million annually	1988	Loans for rural enterprise program in poverty counties	
Agriculture Bank	300 million annually	1985	Loans for helping poverty areas' economic development	
Agriculture Bank	1 billion annually	1986	Special discount loans for poverty relief	
Agriculture Bank	50 million annually	1988	Special discount loans for poverty relief in pastoral areas	
Agriculture Bank	500 million annually	1991	Special discount loans for poverty relief	
Agriculture Bank	100 million annually	1991	Special discount loans for border poverty farms	
Agriculture Bank	400 million package	1992	Special discount loans for farmland capital construction in poverty counties	
Agriculture Bank	150 million package	1992	Special discount loans for construction of water conservancy works in poverty counties	
Agriculture Bank	100 million	1992,1993	Special discount loans for the poor disabled	
Agriculture Bank	500 million package	1993	Special loans for poverty relief	
Industrial & Commercial Bank	200 million annually	1988	Special loans for industrial program in poverty counties	
Construction Bank	100 million annually	1984	Special loans for industrial program in poverty counties	
State Planning Commission	28.3 billion	1984	Provide work as a form of relief	

Since the implementation of the National Poverty Relief Program, the per capita net income and per capita grain owned have increased a lot in the 592 national-level poverty-stricken counties. In 1993-95 the per capita income of farmers in poverty-stricken counties had increased by 70.6 per cent, and the per capita grain owned by 4.9 per cent. They indicate that there had been success in the poverty relief policies. We shall review the poverty relief policies of the three different periods, with emphasis on those adopted after the mid-1980s.

2.1 Poverty Relief by "Transfusion of Blood" Before 1980

Since the Liberation in 1949, poverty in China has been given a clear definition. Abject poverty only occurs in rural areas as various welfare services are provided in the urban areas. Before 1978, poverty relief was mainly in the form of urgent relief programs administered by the central government's Civil Administration relief system. Because the number of abject poor then was very large and the national economy still weak, thus the government was the only channel of relief resource allocation. Poverty relief then included social relief, natural disaster relief and special care to the disabled and family members of revolutionary martyrs. It was a simple method of "transfusion of blood" which temporarily solved the difficulties for the individuals, but it could not help the development of the poverty-stricken areas and root them out of poverty. Thus, before the 1980s, poverty relief by succor was a major poverty relief policy. It was carried out by the Civil Administration from central government to local government. The major content is as follows:

(1) Social relief – It was mainly for disabled people. For example, it provide relief regularly to widowers, widows, disabled people, and orphans. Most of them lived in urban areas. In the rural areas, social relief was given only to the "five ensure household" (guarantee food, living, medicine, burial and education for orphan).
(2) Natural disaster relief - It provided urgent relief grain, clothing etc. to people in natural disaster areas in order to maintain their living.
(3) Special care – It provided special care to disabled servicemen and to family members of revolutionary martyrs.

2.2 Poverty Relief by Promoting Economic Development in Poverty-Stricken Region (1991-1995)

Since 1980, the Chinese government changed its poverty relief policy from 'blood transfusion' to 'developing the local economy'. In 1984, the CPC and the State Council sent out a 'Notice On Helping Poor Areas To Change Their Backwardness Conditions As Soon As Possible', which defined the poverty relief policy by development and marked a fundamental change in China's poverty relief strategy.

In this period, the tendency of government's poverty relief effort concentrated on enhancing the economic development of the poor regions. Conceptually, it takes regional economic development as a means to bring along poverty relief. Such 'poverty relief' thus targeted whole counties instead of individual rural households. The central and provincial government provided poverty relief funds to poverty-stricken counties according to a fair principle and these counties had to use the funds according to an agreed efficiency principle.

The local governments managed the poverty relief funds, and used them in programs which could solve local financial difficulties, such as setting up rural enterprises. However, in reality, the speed of poverty relief had not matched the speed of local economic development. Table 9.6 shows that the new poverty relief strategy by developing the local economy had played a very important role in improving the local economy, especially in increasing the local government's revenue. Yet it had done little in ameliorating the living condition of the abject poor. In short, it has moderated the local poverty problem in poverty-stricken areas, but increased the degree of abject poverty.

Table 9.6 Statistics of Poverty-stricken Counties in China (1989-1994)

	1989	1996	Change (%)
Rural per capita social gross products (Yuan)	834.0	1376.5	+65.04
Rural per capita net income (Yuan)	389.8	413.0	+5.96
Rural per capita grain production (Kg)	357.0	351.0	-1.68
Number of towns with no highway (%)	16.1	2.5	-13.60
Number of towns with no electricity (%)	22.2	14.4	-7.80
Rate of rural poverty occurrence (%)	11.4	8.7	-2.70

2.3 Poverty Relief Policies Targetting At the Poor Households Since 1996

Since 1990, the characters of the poor population in China have changed markedly. The poverty relief strategy of the 1980s, by developing the local economy, had solved the food and clothing problems of most of the spatially concentrated poverty-stricken population. Yet the rest of the poor population in remote mountains or isolated areas were unaffected by the county's economic development. Poverty relief activities should therefore target at the individual poverty household, and measures of poverty relief should change from those of large-scale industrial development to projects that individual households can participate directly and be benefited.

On 23 October 1996, the CPC and the State Council announced a new poverty relief strategy based on the development of poor households. Besides increasing the poverty relief funds and the range of preferential policies for poverty-stricken areas, the government also adopted a series of poverty relief measures (Table 9.7), which included the provision of work as a form of relief, poverty relief through the innovations of science and technology, fixed-site poverty relief by organizations, and regional and international cooperation.

2.3.1 Providing Work as a Form of Relief

Providing work as a form of relief belongs to the category of succour, different from simple hand-out relief as it combines relief with construction. It provides relief funds or materials to the poverty-stricken people to allow their participation in public projects, including infrastructure construction. Instead of simply 'transfusing blood' to poverty-stricken areas, it improves their ability to develop. A good example is developing agriculture by providing work as a form of relief in Hongqiao Village. The Hongqiao Village was a long-famous poverty-stricken village in Yulin, Shannxi. The condition there was bad for agricultural practices. It has sandstorm hazardous and little access to means of transportation and communication. The 1989 per capita grain production was 228 kg, and the per capita net income was 287 yuan. In 1990, the central government decided to implement agricultural development programs in the desert areas north of Yulin by providing work as a form of relief. Hongqiao Village implemented these programs by carrying out construction of agricultural infrastructure, with water conservation facilities as the emphasis, and by readjusting the structure of agriculture. They have achieved remarkable success. By the end of 1994, the per capita grain production went up to 1,200 kg, and the per capita net income 1,550 yuan. All households there had improved their living above the poverty standard and it became an example of agricultural development in Yulin.

During the Eighth FYP, the total amount of money spent by China in providing work for poverty relief added up to 24.7 billion yuan. In its 1996 budget, the central government allocated

9.85 billion yuan for poverty relief, 4 billion of which, or 40% of the total, was spent on work provision.

Since the implementation of the work provision measure in 1984, there were seven batches of projects, six of which were carried out during the Eighth FYP, four in 1995 alone, and two in 1996 (Table 9.8). We may cite one example of work provision which involves a transport project in Jinlong county. Partly as a result of the limited access to the means of transportation, the 1984 per capita net income of farmers in Jiulong county, Sichuan, was only 194 yuan, and the revenue of the county was only 530,000 yuan. Since 1985 the local government had built 8 roads of 230 km long, thus allowing the exploitation and development of the abandoned copper ore and the forest resources in the area. It had brought about advances in the local economy. In 1993 the county's revenue reached 9.56 million yuan and the per capita net income 661 yuan, which were 18 times and 3.4 times that of the level in 1984 respectively.

Table 9.7 Principal Measures of Poverty Relief

Measures		Year beginning	Inputs or effects	Administrative organization
Provide work as a form of relief		1984	28.63 billion Yuan (total)	State Planning Commission
Poverty relief by science & technology		1986	70 million Yuan of poverty relief loan and 400 million of "spark plan"	State Science & Technology Commission
Fixed site poverty relief by organizations		1987	947 million Yuan of direct input and 990 million Yuan of induced funds in 1995	Central government, State organizations, enterprises and institutions
Regional cooperation	East-West cooperation of rural enterprises Developed regions help poor regions	1995 1996		State Council and provincial government
Social poverty relief	Glorious facilities	1994	2.74 billion Yuen total	Association of Industry & Commerce of China
	Well-being project	1995	20 million Yuan	Well-being Project Organization Committee
Export of laborers		1994	25 million laborers (1994-1995)	Labor Department
Resettlement of poor farmers			Nearly solved 1 million people's food and clothing problems	Various local governments
	Projects in South-West	1995	4.23 billion Yuan, of which US$247 million loaned by the World Bank	Chinese government
	Project in Qinba Mountains	1997	3.32 billion Yuan, of which US$200 million loaned by the World Bank	Chinese government

Table 9.8 Details of Projects of Providing Work as a Form of Relief

Project	Implementation time	Contents	Regions
First	1984-1987	Help poverty-stricken areas to construct roads, water channels and small water conservancy facilities by providing works as a form of relief through the payment of 5 billion Kg of grain, 100 million Kg of cotton and 500 million meters of cloth	25 Provinces except Beijing, Tianjin, Shanghai and Jiangsu, the major regions are 18 concentrated poverty-stricken regions
Second	1989-1991	Besides construction roads and water conservancy facilities, stressed water supply facilities through the payment of middle and low-grade industrial goods worth 600 million Yuan	All poverty-stricken areas
Third	1990-1992	Carry out comprehensive development of "mountain, water, forest, field and road through the payment of industrial products worth 1509 million Yuan	Central and Western part of China, some part of the Eastern regions, especially in minority areas
Fourth	1991-1995	Practices provide work as a form of relief through the payment of 200 million Kg grain each year. It takes agricultural infrastructure as the key, especially basic farm land construction to improve fields of medium and low level productivity, consummate conveyance system of large water conservancy facilities. In the mountainous areas, terracing on the slopes was the principal activity, with water conservancy facilities, roads and water supply facilities	Mainly in the West, ethnic minority areas, remote mountainous areas, plateau and frigid areas, and areas where the people eat "resold grain"
Fifth	1991-1995	Using grain and industrial products worth 2 billion Yuan each year as payment for work to harness large rivers and large lakes, and help disaster areas to repair infrastructure destroyed by flood and to resume production and rebuild their home	Involving 30 Provinces, especially in the lower reaches of rivers and serious disaster areas
Sixth	1993-1997	Each year using grain, cloth, cooking oil, middle and low-grade industrial products worth 2 billion Yuan as payment for work to develop capital construction of water conservancy facilities, and water supply facilities to plant orchard and trees, develop livestock raising, and to construct rural transportation and communication facilities	Mainly in poverty areas in the West, including areas that have not yet solved food and clothing problems and areas which have a high rate of poverty in Hebei, Liaoning, Fujian and Henan provinces
Seventh	1994-2000	Using goods kept in stock worth 1 billion Yuan each year, emphasizing on road construction and water supply facilities	Mainly in 592 poverty counties of 21 Provinces defined in China's Poverty Relief Program

The decade of implementing 'provision of work as a form of relief' has achieved remarkable success in improving the conditions of the poverty-stricken areas, enhancing production and the living conditions, and promoting economic development there (Table 9.9).

Table 9.9 Input and Effect of Provide Work as a Form of Relief

Program	1st batch	2nd	3rd	4th	5th	6th	Special fund for poverty farms in border areas	7th
Period	1984-87	1989-91	1990-92	1991-95	1991-95	1993-97	1991-95	1994-2000
Regions involved	25 provinces	26 provinces	26 provinces	18 provinces	30 provinces	23 provinces	Xinjiang, Yunnan, Guangxi, Heilongjiang Inner Mongolia	26 provinces
Input (billion Yuan)	2.7	0.6	1.5	5	10	10	0.5	7
Types of materials	Grain, cotton, and cloth	Industrial goods	Industrial goods	Grain	Grain and industrial goods	Grain, cotton, cooking oil and industrial goods	Grain and industrial goods	Grain and industrial goods
Roads build (10,000 km)	12	1.1	1.3					
Bridges build number/length (Km)	7200/162	722/31	1000/38					
Solving water supply problem (million) People/ livestock	14.04/9.71	6.47/3.85	8.75/6.98					
Increased irrigation land (million Mu)*	13.9	0.38	20.82					
Water & soil conservation area (million Mu)	17.01		71.84					
Terraces on slopes (million Mu)			64.77					

* 1 ha = 15 Mu

Source: Department of Agricultural Economy of the State Planning Communission, "Provide Work as a Form of Relief and Benefit People", 1996, p. 541.

As of today, the seven batches of work provision has been a successful measure of poverty relief which fares fairly well in the Chinese context. It have played an important role both in improving the infrastructure in poverty-stricken areas and in increasing the income of the poor.

2.3.2 Poverty Relief through Innovations in Science and Technology

Since 1986, under the initiation and coordination of the State Planning Committee, the State Commission of Science and Technology has carried out poverty relief through innovations in science and technology in 55 counties of the Tabie Mountain, Jinggang Mountain, north of Shannxi, and some ethnic minority areas (*Guangming Daily*, 7/7/1997). In 1986-1995, it has been successful in increasing per capital income and local revenue of the poverty-stricken areas, i.e. from 266 to 1,097 yuan and 510 million to 1,500 million yuan, from 3.4 million in 1990 to 1.7 million in 1995 and in decreasing the size of the abject poor.

The aim of poverty relief through innovations is to introduce advanced and useful scientific techniques to poverty-stricken areas by implementing large-scale scientific demonstration programs, constructing scientific demonstration sites, and providing scientific training, such that farmers might be instructed and encouraged to develop agriculture with the back-up of science and technology. It is hoped that this will improve the level of resource exploitation and labor productivity, promoting economic development, and accelerating the speed of poverty relief. One example is the planting of bamboo shoots for poverty relief in Zhijin county, Guizhou. Bamboo shoots in Zhijin is a valuable product. Before 1985, 1 kg of wild bamboo shoots was worth the price of 50 grams of gold in Hong Kong. Aware of the value of bamboo shoots, scientists of Zhijin began their research in 1983. Large-scale planting began in 1989. In 1995, the number of households planting bamboo shoots had increased (from 150 in 1989) to 70,000, the planting area (from 1,050 square meters in 1985) to 1.7 million square meters, and the total output (from 125 kg) to 204,000 kg. During the Eighth FYP it had created a total output value of 200 million yuan i.e. a per capital income of 250 yuan. All 32 villages in this county benefited from this in different degrees. There were more than 500,000 farmers working on it, with the result of 60,000 households or 250,000 individuals left 'poverty', which accounted for achieving 50% of the total poverty relief target in the county. Bamboo shoots became the 'first-pillar industry' in the county and played a major role in poverty relief.

In 1986-1996, the State Commission of Science and Technology provided special funds of 70 million yuan for poverty relief through innovations in science and technology and 400 million yuan for poverty alleviation loan. As a result, more than 500 programs of resource development and science-and-technology popularization, had been implemented. More than 1,500 scientific demonstration sites had been set up, with 1.3 million farmers trained in one or two practical techniques. The results have proven that employing innovations in science and technology in poverty relief and equipping farmers with the required implementation techniques are highly effective measures for promoting sustainable growth in poverty-stricken areas, thus increasing farmers' income.

2.3.3 Fix-Site Poverty Relief by Government and Party Organizations

Since 1987, all central government departments or party organization have each been assigned one fixed poverty-stricken area which they have to look after. Every year they have to send a working groups there to help to relieve poverty. As of today, 122 state organizations, state enterprises or government and party institutions have assisted a total of 330 national-level poverty-stricken counties, or 56% of the total poverty-stricken counties in the country. In 1995 alone, these organizations directly provided materials and funds amounting to 947 million yuan, and injected 990 million yuan cash to their 'adopted' sites. In doing so, they have initiated and implemented the 'Poverty Relief Plan by Transportation', 'Water Supply Program' and 'Poverty Relief by Electricity'.

The government of provinces and autonomous regions had also organized county governmental organizations to carry out fix-site poverty relief in accordance with the guidelines of the State Council. Poverty relief by county organizations is mainly targetted at poverty-stricken villages. These organizations must keep in touch with the villages until they have graduated from poverty. In 1996, the government of Hunan created 60 working groups of professionals from 60 provincial organizations, to carry out fix-site poverty relief in 60 abject poverty-stricken villages. The project was very successful. The per capita net income of these villages was increased by 200 to 400 yuan (*Guangming Daily*, 15/1/1997).

2.3.4 Regional Cooperation

In early 1995, China launched the 'demonstration project of cooperation' of rural enterprises between the East and West, to assist the poverty-stricken areas in the West to develop their rural enterprises and promote rural economic development. The central government plans to set up 100 such rural enterprise cooperation demonstration sites, and 1,000 industrial demonstration projects by 2,000 with 1,000 new techniques and products introduced, and 100 pairs of brethren cooperation counties created. By promoting cooperation, the comparative strength in the East can be made complimentary to that of the West, such that the differences in wealth between the East and West can be reduced. In 1996, the State Council arranged to have 9 provincial units, and 4 municipalities to assist 10 provinces in the western region as fixed partners to relieve poverty (*People's Daily*, 8/1/1997) (Table 9.10). This project is currently making good progress. The first group of poverty relief programs, with more than 30 development projects, have already begun. One example being that Qingdao, Shenzhen, Ningbo and Dalian municipalities assist Guizhou to relieve poverty. They have injected 183 million yuan of development fund and contributed other resources to Guizhou.

Table 9.10 Pairing between the East and West for Poverty Relief

Developed provinces or cities in the East	Poverty-stricken provinces in the West
Guangdong	Guangxi
Shanghai	Yunnan
Liaoning	Qinghai
Shenzhen, Dalian, Qingdao, Ningbo	Guizhou
Fujian	Ningxia
Tianjin	Gansu
Jiangsu	Shaanxi
Beijing	Inner Mongolia
Shangdong	Xinjiang
Zhejiang	Sichuan

2.3.5 Labor Export, Resettlement of Poor Farmers, Private Sector Participation and International Cooperation

In 1994 and 1995, the poverty-stricken counties exported to the rest of the country 25 million laborers, of which 10 million came from Sichuan. Every year the exported Sichuan laborers contribute about 5 billion yuan to the province in the form of postal remittance. Sample surveys reveal that if a household exports one laborer, the whole family can be removed from poverty. Labor export not only can raise the household income, it also raises labor quality by raising the skill level, and exposing the labour to new concepts and new environments.

Resettlement of poor farmers living in harsh physical conditions where there is a lack of resources and basic living and production conditions to areas of better potentials in compliance with their wishes is another new government measure. Nearly one million poverty-stricken people have been taken out of poverty as a result of such a resettlement operation. 600,000 had been relocated from the poorest areas of Dingxi, Gansu, and the Xihaigu areas of Ningxia Hui autonomous region.

The best known effort of private sector participation is 'Glorious Enterprise'. It was sponsored by privately-owned enterprises in 1994, which is a form of civilian activities to fight poverty. In April 1994, Mr. Liu Yonghao, director of a famous private enterprise, the Hoping Group of Sichuan, and ten other private enterprises proposed to all private enterprises in China to 'join the glorious enterprise of poverty relief'. The 'Glorious Enterprise' jointly proposed by this united front and the Chinese Association of Industry and Commerce then came into being. It combines the financial resources, technology, talents and management techniques of private enterprises with the comparative advantages of physical and human resources the poverty-stricken areas possessed in poverty relief. By June 1996, more than 3,000 private enterprises in 29 provinces participated in the 'Glorious Enterprise'. As a result 2,735 million yuan was put into poverty relief, 699 poverty relief programs carried out, and 20,000 persons in poverty-stricken areas trained.

International assistance is an important means through which the problem of funds deficiency in China can be remedied. The first World Bank program of south-western poverty relief and the second World Bank relief program of Qinba Mountain poverty relief are examples of international cooperation on poverty relief (Table 9.11). The third World Bank program is currently under negotiation. Should this program be agreed, the total World Bank assistance fund for poverty relief will amount to 650 million US dollars. Including the conveyancing funds, the total inputs of these programs will add up to 10 billion yuan, which can solve the problems of food and clothing for 100 million poverty-stricken people. It certainly will help to speed up China's poverty relief progress. Nonetheless the procedures involved in obtaining these funds are complicated, and often local governments lack the necessary matching funds.

3. Results of National Poverty Relief Policies of 1995-1996

1995 and 1996 are the second and third years of the implementation of the Chinese Poverty Relief program. In 1995, five million people had been removed from poverty, and another seven million in 1996. However, there are regional discrepancies in such an achievement due to different conditions of poverty and subsequent differences in the focus of poverty relief policies implemented (Table 9.12).

In the East region which had greater economic strength, with limited poverty-stricken areas, lower degree of poverty, and a higher living standard of the poor (with the exception of Guangxi) the poor there are better off than those in the West. Hence poverty relief in the East is largely self-dependent. The eastern regions has a high speed of poverty recovery, and is able to achieve the National Poverty Relief Goal in advance. For example, the Fujian government uses fix-site assistance and resettlement to relieve poverty. The size of the poverty-stricken population in the province decreased from 800,000 in 1993 to 500,000 in 1995. With the 420,000 people removed from poverty in 1996, it is predicted that by 1997, all poverty-stricken people in Fujian would have been relieved from poverty. Another example is Guangdong where the government proclaimed that it would solve the problem of food and clothing for 600,000 abject poverty-stricken people in the province by 1997. It employed the measure of making large companies construct development bases upon which households could develop agricultural projects and was able to achieve poverty relief three years ahead of the intended deadline.

Table 9.11 World Bank Poverty Relief Programs in China

	Program of South-West	Program of Qinba Mountains
Start time	July, 1995	July, 1997
Total input	4.23 billion Yuan, of which US$247 million from World Bank	3.32 billion Yuan, of which US$200 million from World Bank
Regions involved	1798 villages in 290 towns of 35 poverty counties in Yunnan, Guizhou and Guangxi	3231 villages in 461 towns of 26 poverty counties in Shaanxi, Sichuan and Ningxia
Objectives	To solve farmer's food and clothing problems, promote social and economic development	Using integrated methods to concentrate funds and investment on a certain large area, and to solve food and clothing problems for the people in this area until eventually it can achieve a good circle of economic development
Project selection	835 projects in 8 types, including agriculture, the secondary and tertiary industry, exporting of labors, infrastructure, education, hygienic, monitor of poverty, and construction of organizations etc.	854 projects in 8 types, including land and household development, rural enterprises and markets, rural infrastructure construction, education and hygienic, export of labors, construction of organizations, monitor of the project and poverty, and small loans to poor farmers, etc.
Implementation time	6 years	20 years
Expect results	When the program is finished, per capita net income will increase by 100%. The per capita net income will achieve 690 Yuan according to 1990 fixed price. It will solve food and clothing problems for 3.5 million poverty-stricken population.	By the year 2010, the per capita net income of the regions covered will achieve 509 Yuan according to 1990 fixed price, an increase of 298.4 Yuan compared with 210.6 Yuan in 1995. The average annual increase rate is 9.31%. It is expected that when the program is finished, food and clothing problems in the abject poverty-stricken regions will be basically solved.

Table 9.12 Regional Difference in Poverty Relief Results

Region	Province	Major measures	Annual reduction in number of poverty population (1994-1996)	Annual rate of poverty population reduction	Time to achieve the goal of poverty relief
East	Fujian	Fixed sites cooperation, resettlement	310 thousands	39%	1997
	Guangdong	Developing modern agriculture	All poverty population were out of poverty in 1997		1997
	Guangxi	Initiate economic group, resettlement	1.047 millions	13.1%	Hard to be realized by 2000
Central	Shanxi	Develop dominant industry, cadre effort in fixed site poverty relief	860 thousands	23.6%	Can be realized by 2000
West	Guizhou	Terracing on slopes, developing new farm land	1.24 millions	12.4%	Hard to be realized by 2000
	Shaanxi	Stress on key projects with concentrated effort	640 thousands	24.1%	Can be realized by 2000
	Gansu	Underground water cellars, plastic sheet plant cover	413 thousands	8.2%	Hard to be realized by 2000
	Sichuan	Exporting of labors	2.3 millions	19.5%	Hard to be realized by 2000
	Ningxia	Underground water cellars irrigation, resettlement	213 thousands	15.2%	Hard to be realized by 2000
	Tibet	Developing productivity forces	61.5 thousands	13.5%	Hard to be realized by 2000
	Yunnan	Infrastructure, developing superior industry	770 thousands	10%	Hard to be realized by 2000

Inner Mongolia is an example in Central China which has successfully lifted one million poor out of poverty. Its speed of poverty relief is the fastest in China. By the end of 1995, Shannxi, also in Central China, has wiped poverty off 4 counties completely, and had 14 counties close to achieving the poverty relief goal. Its per capita income in 1995 amounted to 748 yuan, whereas the number of poverty-stricken people decreased from 3.81 million in 1993 to 2.3 million in 1995. An extra 1.06 million people were removed from poverty in 1996.

Western China is the major and most difficult area needing poverty relief. All western provinces have adopted various poverty relief measures according to their needs and have achieved considerable results. For example, in the past few years, Guizhou has implemented important poverty relief measures including terracing on slopes and developing wasteland. As a result, 910,000 poverty-stricken people were removed from poverty in 1994. Subsequently, there were 1.2 and 1.6 million people relieved from poverty in 1995 and 1996. In Shannxi, a detailed annual poverty relief plan was drawn up. Even during a two-year continuous drought (1994 and 1995), the local government was able to lift one million poor out of poverty. An additional 920,000 were removed from poverty in 1996. Its per capital income in 1995 reached a high level of 547 yuan. In Yunnan, the government developed local industries such as tobacco, sugar cane, tea and rubber growing, as well as tourism, and attempted to relieve poverty by making long-term infrastructural inputs. In 1996 alone, there were about 1.2 million poor people out of poverty.

However, the speed of poverty relief in the 1997-1998 had slowed down in comparison to that of 1994 to 1996. It is unlikely that, with the exception of Shannxi, the Western provinces will be able to accomplish the poverty relief goal by 2000. Poverty relief remains an arduous task to be completed in these regions.

4. Measures and Results of Poverty Relief in Major Poverty-Stricken Areas

Poverty-stricken areas in China fall into different categories owing to their varied physical and economic conditions. In order to achieve the poverty relief goal by 2000, every province has to plan its provincial poverty relief program in accordance with the national poverty relief policy, yet targeted at specific problems of the locale. The following is a summary of these poverty relief measures that have adapted to suit the most concentrated and difficult poverty areas of the country.

4.1 Poverty Relief in the Karst Regions of South-eastern China

The Karst mountains in south-eastern China cover part of the Yunnan-Guizhou Plateau and its border mountain areas, namely parts of Yunnan, Guizhou, and the Guangxi Zhuang Autonomous Region. In 1994, the three provinces had 25 million poverty-stricken people, or 31% of the national poverty-stricken population, one of the areas where the poor are most concentrated. Poverty relief there is made very difficult as a result of the high mountains, thin and poor soil, and a lack of water supply. Despite such challenges imposed by nature, clever measures have been carried to combat poverty in an attempt to achieving the national poverty relief goal.

4.1.1 Terracing on Slopes and Wasteland Development

Guizhou is an ethnic-minority area located near the national border. It is mountainous with little flat land, and the per capita farmland is only 0.26 acre, which explains why that for a long time, even basic survival problem of food and clothing of the people could not be resolved. There are currently 6.29 million poverty-stricken people in the province, or 11% of the total poverty-stricken population in China. It is one of the provinces with the highest concentration of the abject poor, who make up about 40% of the provincial population. Poverty relief is thus difficult. A poverty relief plan was adopted in 1994 by the provincial government to implement the National Poverty Relief Program. Since then, 3.71 million people have been lifted out of poverty, of which 1.6 million were removed in 1996 alone. The per capita income and per capita grain owned in the province have also been increased, i.e. from RMB 355 to 700 and 204 kg to 309 kg in 1993-96.

As a major province needing poverty relief, Guizhou receives poverty relief funds from various sources which averaged 250-370 million RMB annually, or 6-7 per cent of China's annually total poverty relief fund. Besides the policy of work provision, there are 11 other types of poverty relief funds granted to Guizhou (Table 9.13). Its total annual poverty relief funds amount to 1,400 million yuan. Besides injection of money to needy areas and projects, the Guizhou government also uses slope terracing and wasteland development to create new farmland, thereby alleviating poverty. In August 1991, the provincial government passed a resolution to develop more cultivable land by work provision. During the Eighth FYP, Guizhou had formed 2.95 million acres of new farmland through terracing, 77% of this was the effort of workmen from the poverty-stricken counties. Thus statistics have revealed that terracing on slopes can increase grain production by 300 million kg every year. In 1996, there were 48 poverty-stricken counties, 20 of which had attained self-sufficiency in grain. In 1997, an extra one million acres were formed by terracing. In Qiannan Prefecture, there are 50,000 acres of land in the form of terraces. In 1997, the Prefecture attained a level of 53.3 kg grain and 60.6 yuan income per mu. The annual income of an average family working on the improved soil had increased to 149.1 yuan.

Table 9.13 Funds of Poverty Relief in Guizhou Province

Name of fund	Amount of fund (Yuan)	Year of providing
Comprehensive agriculture developing fund	294 million	1990-1992
	270 million	1993-1995
Green industry loan	400 million	From 1997, annually
World Bank and international loans	1.0788 billion	From 1994
Special discount loans	100 million	From 1986, annually
Funds for assisting undeveloped regions	31 million	Annually
Development fund for poverty relief	20 million	From the Central Government, annually
	20 million	From Shenzhen, annually
Low interest loans for the old-revolutionary bases, minority and border areas	60 million	Annually
Loans for enterprises in poverty counties	33 million	Annually
Development fund for ethnic minority areas	3 million	Annually
Circulation fund for poverty relief	3 million	Annually
Discount loans for poverty-stricken villages	15 million	Annually

In addition, the Guizhou government has undertaken to develop wasteland and local industries as a means to relieve poverty. Such a measure is important in raising the income of the poverty-stricken population. At the same time, it makes possible the formation of a number of locale, upon which industries are built. Examples include a base for medicine, raw material sources for chemical industries and other local industries. The development of these industries has enhanced economic development in the poverty-stricken areas, and lays a foundation upon which problem of food and clothing can be solved and poverty relieved. Developing wastelands and closely monitoring the exploitation of resources also have positive ecological effects and can contribute to sustained and continuous development of poverty-stricken areas.

4.1.2 From Provision of Relief to Development

Guangxi is another example of a large number of concentrated poor in a karst landscape. In 1993 there were 49 poverty-stricken counties, taking up 56% of the total provincial area. The number of poverty-stricken people amounted to 8 million, about one tenth of the total population of poverty-stricken in China. In April 1995, the Guangxi government drew up a poverty relief plan, the aim of which is to solve the problem of food and clothing of the 8 million poor by 2000. To attain that goal, the method of altering the pattern of economic growth, i.e. to attain poverty relief through development has been used. During the Eighth FYP, the agricultural output value of the province was increased by 9.6% each year. The number of poverty-stricken people had decreased from 10 million in 1990 to 4.86 million in 1995.14 poverty-stricken counties were removed from poverty, and the poverty occurrence rate was decreased to 12.8%. In 1995, all indices on the provincial economy and the per capita income had increased and the living condition and production have improved (Table 9.14). In 1996, an additional 1.2 million people were lifted from poverty, and the problem of water supply was solved for 385,000 people and 300,000 heads of livestock.

Table 9.14 Economic Situation in 49 Poverty Counties, Guangxi, 1995

GDP (billion Yuan)	37
Gross Production of Agriculture and Industry (billion Yuan)	49.24
Compare with last year	+38.6%
Farmer's per capita net income (Yuan)	1008
Compare with last year (Yuan)	+276
Number of poverty-stricken population (million)	6.06
Compare with last year	-1

Although Guangxi is located in a karst landscape, it has achieved remarkable results in poverty relief. Besides closely related to the policy of development as mentioned above, such achievements are also related to efforts to adjust the local industry structures, create poverty relief economic groups' and carry out resettlement programs. These latter measures are briefly described below:

(1) Adjusting the structure of local industry. The government put grain production as the first priority in the economic structure. All land suitable for cultivation is used to produce grain first, thereby increasing the per unit area yield and solving the problems of food in the poverty-stricken areas. At the same time, the government has brought in and utilized other local resources on top of the increased grain yield to adjust the agricultural structure e.g. agriculture and forestry are developed simultaneously in the development in poverty-stricken areas.

(2) Creating economic groups for poverty relief. In the poverty-stricken areas, owing to the poor physical conditions for production, low level of technology and skills, and low ability in risk-management, self-development is difficult if households work separately instead of pooling their strength together. The Guangxi government has helped the forming of economic groups, which draw together efforts of individual households to assist the relieve poverty with much greater efficiency. Economic groups created by the

Guangxi government for poverty relief bears the characteristics of being multi-form. There are several forms in this project, including: an organization plus farmer households, an economic group plus farmer households, and competent individuals plus farmer households. It uses economic measures, scientific management techniques, and advanced production technology to organize large amount of households under an intensive and unified system of production. For example, there are 1,065 economic groups in Daxing county. They have helped to lift 11,000 households and 63,000 people out of poverty. The Tianyang county authorities have created three mango farms and thus removed 6 villages and 20,000 households, 34% of the total poverty-stricken population in the county, out of poverty.

By 1995, there were 2,537 economic groups for poverty relief in Guangxi. They were provided within 334 million yuan of poverty relief funds and had carried out a total of 5,000 development projects, which lifted 125,000 families and 666,500 people out of poverty. In 1996, the per capita income of those lifted out of poverty had gone up to 1,300 yuan.

(3) Resettlement. Most of the poverty-stricken people in Guangxi are distributed in mountainous areas with adverse physical conditions unsuitable for living and production. The provincial government has thus decided to resettle 200,000 of the poor who are living under such conditions within a period of five to eight years from 1993.

By 1995, a total of 275 million yuan was injected into the resettlement projects. As a result, 122 resettlement sites were constructed, and 16,000 households and 87,000 poor from 22 poverty-stricken counties were moved into the new settlements. These projects work alongside with comprehensive agricultural development. By so doing, most resettled families had their problem of food and clothing resolved within a year of the resettlement, and some were even being lifted from poverty altogether. In 1996, by improving the original resettlement sites, the provincial government resettled a further 21,000 poor. A total of 108,000 people had thus far been resettled.

In addition, the provincial government has utilized foreign capital in relieving poverty, most notably through the World Bank program. By 1996, 226 million yuan, 31% of the total investment of the program, were used within a period of two years for the construction of 86,700 square meters of schools and clinics. The also helped to solve the problem of water supply for 26,000, constructed 103 km of highway, planted 28,000 hectares of forest, orchard and farmland, bred 39,000 heads of livestock and helped to export 44,000 laborers. The number of households benefited from this program has exceeded 180,000.

4.2 Measures on Poverty Relief in the Loess Plateau

The poverty-stricken areas in the Loess Plateau are mainly found in Shannxi, Shanxi, Ningxia and Gansu. In 1994, the number of poverty-stricken people there amounted to 18 million, or 22.5% of China's poverty-stricken population. Water and soil erosion are the major physical problems in this area. Drought, together with ecological degradation, has rendered the Xihaigu region in Ningxia the poorest area in China. However, after ten years of unceasing and comprehensive development and poverty relief, the Xihaigu regions have developed a series of effective poverty relief measures with good success.

In 1994, the Ningxia government drew up the 'Ningxia Autonomous Region Poverty Relief Plan' for one million poverty-stricken people in 100 villages by 2000. In order to achieve the target of 3 mu of farmland per capita by the end of the century, the government proposed the full utilization of the three different sources of water: rain water, surface water, and underground water. It also proposed to construct three kinds of farmland: land on the river bank, terraces, and irrigated land. During the Eighth FYP, there was a serious drought which lasted for five years. Many who had already been removed from poverty had fallen back to their original poor condition, i.e. 104,000 returned to the situation of poverty in 1995, raising the poverty population from 1.29 million in 1994 to 1.33 million in 1995. Thus, there were only 70,000 people lifted from poverty in 1994 and 1995. In 1996, with the combined efforts of the increased poverty relief funds, implementation of various poverty relief measures, and favourable climatic conditions, 568,000 poor people in Ningxia had their problem of food and clothing resolved. However, there are still 760,000 poverty-stricken people in 144,000 households in the province.

The Xihaigu region in Ningxia is a typical poverty-stricken area in the Loess Plateau. Poverty in this area is closely related to the lack of water supply characteristic of the region. The provincial government has addressed this directly using the principle that 'if there is water, maximizes its use; if there is none, focus on economic activities that do not require water; otherwise find other means to sustaining the economy'. A series of unique and feasible measures has been established after years of dedicated work, as will be described below.

4.2.1 Irrigation using Water-Storage Pits

The five-year drought in the region mentioned earlier had inspired people to an ingenious method of irrigation, using man-made water-storage pits which was declared as a 'remarkable achievement' specialists from the World Bank. Between 1995 and 1996, the provincial government injected 20 million yuan into the constructing of these water-storage pits. Together 81,000 pits were constructed in eight poverty-stricken counties. Consequently 126,200 mou of farmland were irrigated. Complementary to the pit construction, the government also introduced a new technique of minimizing water loss by covering the maize-growing land with plastic sheets. In 1996, the eight poverty-stricken counties employed this technique in cultivating their 258,400 mou of land, and obtained a yield of 120 million kg of maize. Thus, by utilizing only 3.7% of the province's cultivated land, the counties were able to produce a yield equivalent to 16% of the province's total. Zhou Huicheng, a farmer in Haiyuan county, has dug the largest pit in the county, which has a storage capacity of 380 cubic meters. Using his pit for irrigation, Zhou has obtained a high yield of 625 kg per mu on his 11-mou land, and has his family lifted out of poverty. A study reveals that farmer households in this area with pits have an average income of no less than 2,700 yuan per year, and the highest has more than 8,000 yuan and that their per capita income is no less than 600 yuan in this area.

Using water-storage pits for irrigation is an inexpensive yet effective means of resolving water deficiency. It is therefore worth promoting to other poverty-stricken areas that face a similar problems. However, considering that poverty-stricken areas should target at long-term development instead of short-term benefits, the new technique of maize cultivation should not be upheld as an example for other poor regions.

4.2.2 Irrigation using Water from the Yellow River and Collective Resettlement

Irrigating the drier mountainous areas with water from rivers in adjoining valleys is another successful poverty relief measure. Irrigation works can also be carried out along side with resettlement. The provincial government has therefore pumped up water from the Yellow River on one hand, and resettles people from poverty-stricken areas to these new irrigated areas. Each family participating in the resettlement plan was given 400 yuan to supplement the cost for building a new houses. In this the authorities and the resettled farmers worked hand-in-hand, i.e. the government provided infrastructure such as roads and canals, whereas the farmers cultivated the land. Since the Opening and Reform, twelve pumping stations have been constructed and nearly one million farmland irrigated. In total, 21 resettlement sites have been built, and 260,000 poverty-stricken people resettled. The achievement is remarkable. Resettlers usually would take about one year to build their houses, two years to prepare a piece of land for cultivation, and three to five years to achieve a certain level of affluence. For example, the Locaowa Resettlement Village in Jingyuan county has only 10.7% of the county's total cultivated land, but it alone is responsible for 32.7% of the county's total grain production. Its per capita income was doubled by 1996.

The Huaxi Resettlement Village was another key project in 1995 and 1996. It resettled 3,000 poverty-stricken people from 500 families in the Xiki, Haiyuan, Guyuan and Tongxin counties. The project included the development of 3,500 mu new farmland, utilization 1,500 mu of wasteland, planting 50,000 trees, and building 1,000 houses. This village has now a stablized food production and is serviced by electricity, water, highways and postal communication. Secondary and tertiary industries also begin to develop, while eight industrial enterprises have already been established. The Huaxi Resettlement Village has become one example of poverty relief by resettlement.

In 1996, the central government attempted to extend the national key construction projects into the Central and Western regions. The Yellow River pumping project (the '1236 Project') was approved as a major component in poverty relief and was listed as one of the national key projects for the Ninth FYP. 3,000 million yuan will be used in a period of 6 years to develop 2 million mu of newly irrigated farmland, with 1 million poverty-stricken people around it will be lifted out of poverty. This project will play an important role in relieving poverty in the Xihaigu region.

4.2.3 Construct the Irrigated Farmland on Slopes

Terracing has been a fundamental measure in developing farmland in the Xihaigu region. It cost around 100 yuan per mou to convert slopes into cultivated land by terracing, but 100 kg grain per mou can be produced on such land. In 1996, the government divided the valleys into units and combined terracing with the development and management of the mountains, rivers, farmland, forests, and government roads. A total of 215,000 mou of high-standard farmland were so opened up. As of today, 184 small river valleys in Xihaigu and some 4,800 square km of eroded land there have been brought under control. There, 4,584,000 mou of new farmland was developed, and the per capita farmland in the region has risen to 2.2 mou. At the same time, an area of 2,720,000 mou of newly planted forests and another 1,460,000 mou of reafforestation land, 6,480,000 mou cultivated grassland, and 2,500,000 mou of conserved land have been achieved, which generated great economic and ecological success.

4.2.4 Combining Individual-Household Poverty Relief with Local Industries and Agricultural Industrialization

On the basis of concentrating the pool of poverty relief funds on major projects, Xihaigu moved closer to the poverty relief goal in 1996. The various poverty relief projects were directed towards improving the livelihood of the poverty-stricken villages and households. This is most noticeable in the choice of local industries. The provincial government inclined to develop industries which would involve the participation of, and benefit, the mass of poor.

Potato is the chief agricultural product in Xihaigu. Potato processing, at the same time, is the key industry for poverty relief. There are some 70% potato surplus every year in the 8 poverty-stricken counties. Hence the provincial government established a potato processing industry as the 'pillar industry' there, and provided it with financial assistance.

The Xiye Village in the Xiji county has 47 households with a total of 264 inhabitants, two-third of whom lived below the poverty line. Since 1996, 95.7% of its inhabitants became involved in the potato processing industry. With the help of this industry, the per capita income of the village has shot up to 1,300 yuan. The villagers also use waste water from the industry as fertilizers to improve the yield of grain and potato, while the residuals is used as fodder for cattle. As a result, the per capita grain owned by the village in 1996 was raised to 878 kg and the per capita income 2,067 yuan. The county also established two enterprises, one selling starch and the other vermicelli made from starch. These are sold in many coastal cities, as well as exported to overseas markets.

Meanwhile, labor exporting is also developed to improve the farmers' income. In 1995 and 1996, 459,000 and 517,000 laborers were exported. In return, the county received an additional income of 315 million and 360 million yuan respectively in the two years. Another poverty-stricken county was assisted by a provincial coal-mining factory to export 600 workmen of the factory in 1995. Upon their return to the village, 98% of them become independent and self-reliant.

In addition, since 1996 the provincial government has taken vigorous actions in seeking overseas support for its poverty-relief effort. It was able to obtain funds and assistance from the World Bank poverty relief program, which help to accelerate its speed in achieving the poverty relief goal.

5. CONCLUSION: POVERTY RELIEF QUESTIONS TO BE ADDRESSED IN THE FUTURE

During the Eighth FYP, the central government continued its policy of using development as poverty relief. Since the drafting of the National Poverty Relief Program by the State Council in 1994, increasing attention is given to poverty relief by both the central and the provincial governments. As a result there is continuous increase in the input of funds and policy efforts, and poverty relief in China has achieved remarkable success. The size of the poverty-stricken population was reduced from 97 million in 1991 to 65 million in 1995, at a rate of 6.4 million per annum. The per capita net income of farmers in all poverty-stricken counties has increased on average by 70.6%, i.e. from 483.05 yuan in 1991 to 824 yuan in 1995. In 1996, the central government convened a conference on poverty relief, in which it further directed poverty relief as to target individual villages and households. At the same time, the government implemented various related measures, such as the fixed-site poverty relief projects, and the introduction of large-scale development projects in poverty-stricken areas. 7 million people were lifted from poverty in 1996, the rate of increase has exceeded that of the Eighth FYP. Nonetheless, it has not yet met the targets of the National Poverty Relief Program.

According to the National Poverty Relief Program, each province was to draw up its poverty relief plan, with measures adapted to the local conditions. However, as there are differences in economic structure and degree of poverty between the provinces, their degree of success in poverty relief also differ. In 1993, there were 592 poverty-stricken counties in China, of which 105 were located in the Eastern, 130 in Central, and 307 in the Western region, which account for 17.7%, 30.4%, and 51.9% respectively of the national total. In 1995, there were only 119 counties where per capita net income was below 500 yuan, of which 1 (0.8%) was in the East, 13 (10.9%) in Central China, and 105 (88.3%) in the West. The West out-proportioned the rest of the country in terms of the number of poverty-stricken counties, and the figure was found on the increase. Poverty is most concentrated in south-western China, especially in Yunnan, Sichuan and Guizhou. The National Poverty Relief Program predicts that the East will be able to achieve the poverty relief goal in advance, the Central region will probably achieve it in time if it works hard on it, whereas the West, with the exception of Shaanxi, will have much difficulty in achieving it on time.

The central relief strategy of enhancing development means combining governmental assistance with the strength and competence of the poverty-stricken people to increase their ability to self-develop. The goal is to put an end to poverty once and for all. Work provision has proven to be a successful measure in attaining this goal. It plays an important role by laying a foundation upon which self-perpetuating development is made possible.

The use of external assistance to relieve poverty and to enhance economic development is another effective means of tackling poverty. During the Eighth FYP, various organizations of the central government were involved in relieving poverty in fixed-sites. In 1996, regional cooperation between the coastal, developed provinces and the western, poverty-stricken provinces, as well as the World Bank program in south-western China, have played important roles in improving the livelihood of the poverty-stricken population. The resettlement projects in Guangxi and Ningxia had also achieved significant results.

Despite the success, many problems still exist in the implementation of China's poverty relief policies. We shall outline the main problems below:

(1) The Neglect of Birth Control

The central government's fundamental policy is to combine poverty relief with family planning. It states that population growth must be controlled in order to help the poor to escape the vicious circle of population growth breeding more poverty. However, in some poverty-stricken areas, especially where ethnic minorities are found, family planning policy is not rigoriously carried out. For example, the family planning policy in Ningxia Hui autonomous region is that of '1, 2, 3', i.e. birth control varies with ethnicity. The Hans can only have one child, where the Huis on the plains can have two, and those in the mountains, three. This policy has resulted in a high population growth rate of 2.186% in the mountain areas.

According to population census of the autonomous region, 90,000 resettled poor were lifted from poverty at the commencement of the Eighth FYP. However within that time, 110,000 new babies were born in the original villages, which exceeded the number of the resettled population. The Yellow River Pumping Project intends to resettle one-tenth of the Xiji county population and move them out of this poverty-stricken area. However, with increasing population growth, it will not take long to make up the loss, and the problem of poverty remains.

(2) Ignoring the Poverty-Stricken Areas' Perpetuating Development.

In order to achieve the poverty relief goal on schedule, provinces have drawn up different plans with regard to removing the immediate problems on food and clothing, and on increasing the income of households. Inevitably these efforts are important in solving the problem of poverty, but many such measures simply have ignored the importance of conserving the environment that would allow these areas to further develop in future.

(3) Excessive Investment and Scale in Resettlement Plans

Where there is deficiency in farm land, water supply, means of transport, and electricity in a poverty area, resettlement is a successful measure for relieving poverty. However, the investment so involved is huge. In Guangxi, 1,500 yuan per head is needed to relieve poverty by developing the original place of residence, whereas resettlement would require 5,000 yuan per head. Currently, too much emphasis is placed on resettlement, and its scale is expanding. As a result, provincial governments become incapable of providing even basic utilities and infrastructure to the resettlement areas. More consideration, planning, and precision in design should be made to these resettlement plans.

(4) Insufficient Governmental Funds for Poverty Relief

According to statistics of the Poverty Relief Office of the State Council, it takes 1,500 yuan per poverty-stricken person to attain the poverty relief goal. The total 1996 poverty relief funds summed up to 10 billion yuan only. Given that all of them were devoted to relieve poverty, every poverty-stricken person would receive 153.8 yuan, which was far from the expected amount. In 1995, funds devoted to Ningxia amounted to 224 million yuan, of which 165 million yuan was used in the Xihaigu region. This means every farmer in Xihaigu received no more than 128 yuan. In 1996 one billion yuan was granted to Guizhou for poverty relief, and the poverty relief fund per head was still just 126.7 yuan. Even with the matching funds from the provincial government the amount was still far from the expected goal. Hence there is a need to increase the amount of funds provided by the central government.

(5) Some National Policies Render the Poverty-Stricken Areas More Unfavourable than Before

At the commencement of Opening and Reform, the central government granted many favourable policies to the coastal areas to encourage their development. Although the emphasis of development begins to shift to the West, the government has not yet adjusted its policies accordingly to encourage development in the West. It has increased the amount of grant given to the West, while fiscal and management measures various policies, including taxation, credit and land management, remain uniform throughout the country. The poverty-stricken areas do not have the finance necessary to match with the demands of the central government, and this deficiency has seriously impedes the implementation of poverty relief projects and local economic development.

It is therefore apposit to learn from experiences in the implementation of the National Poverty Relief Program and the positive and negative lessons from the provinces for amending existing policies and measures so that the goals of poverty relief will be attained and that there will be less regional disparity in a longer term and sustainable manner.

REFERENCES

1. Chen Junsheng. 1994. *Poverty Relief Problems of China*. Chongqing: Three Gorge Press of China (Chinese text).
2. *Guangming Daily*. Dec. 30, 1996 (Chinese text).
3. *Guangming Daily*. Jan. 7, 1997 (Chinese text).
4. *Guangming Daily*. Jan. 10, 1997 (Chinese text)
5. *Guangming Daily*. Jan. 25, 1997 (Chinese text).
6. *Guangming Daily*. May 22, 1997 (Chinese text).
7. *Guangming Daily*. June 7, 1997 (Chinese text).
8. Guo Laixi. 1992. *Poverty – A Problem Facing Human Beings*. Beijing: Science and Technology Press of China (Chinese text).
9. Jiang Dehua. etc. 1989. *Types and exploitation of poverty-stricken areas of China*. Beijing: Press of Tourism and Education (Chinese text).
10. Kang Xiaoguang. 1995. *Theory of poverty and anti-poverty of China*. Nanning: Guangxi People's Press (Chinese text).
11. *People Daily*. Jan. 8, 1997 (Chinese text).
12. *People Daily*. March 28, 1997 (Chinese text).
13. *People Daily*. Dec. 26, 1997 (Chinese text).
14. Research group of anti-poverty. 1996. *Go Out of Poverty - Study on Anti-Poverty Agricultural Construction in Xihaigu Region*. Ningxia: People Press of Ningxia (Chinese text).
15. The State Council. 1994. *National Poverty Relief Program (1994-2000)* (Chinese text).
16. World Bank. 1990. *World Development Report in 1990*. Beijing: Finance and Economy Press of China.

Chapter 10

IMPACT OF FOREIGN INVESTMENT AND FOREIGN TRADE ON GROWTH AND REGIONAL DEVELOPMENT

Victor F.S. SIT
Professor, Department of Geography,
University of Hong Kong

CAI Jianming
Associate Professor, Institute of Geography,
Chinese Academy of Science

The oil crises of 1973 and 1978 had caused a world-wide economic recession and initiated profound economic restructuring in the More Developed Countries (MDC) which led to a shift of their traditional manufacturing industries into the Less Developing Countries (LDC). Facilitated by the rapid development of computer and information technologies, this shift has hence gone faster and deeper. As a consequence, the Old International Division of Labor (OIDL) marked by trading of products and goods between nations is being replaced by the New International Division of Labor (NIDL). The latter is characterized by Foreign Direct Investment (FDI) and its related production and services. Thus, economic globalization has taken place (Sit & Yang, 1997). Since 1990 as the "Cold War" ended, this process has been accelerated, with the rapid increase of FDI of Multinational Corporations (MNCs) as its main driving force.

From the late 1970s, partly as a result of recognizing the importance of globalization on economic development, China began to open her door gradually to the world and undertook a series of reforms, some of which are aimed at attracting international capital and new technologies. She has since then increased her connection with the world and experienced an ever faster growth in the economy for nearly two decades. What are the impact of foreign investment and foreign trade on China's economic development? What has been their effect on regional disparity in recent years? As a late comer what strategies should China take to catch up with the advanced nations in the process of integration with the world economy? This chapter shall explore these questions.

1. Patterns of Utilizing Foreign Direct Investment (FDI)

1.1 FDI and its Impact on Regional Development

1.1.1 MNCs become major players in FDI outflow

Before the 1980s, international investment was conducted mainly through government aids, loans, bonds and other types of indirect investment. Since the late 1980s, FDI has increased dramatically. 1992 could be regarded as a turning point when for the first time, the share FDI inflow to LDCs surpassed 50 per cent of their total international capital inflow. This trend has continued since (Fig. 10.1)[1].

FDI mainly comes from MNCs, especially the large MNCs. According to the *World Investment Report 1996* (UN, 1996), the world's largest 100 MNCs (excluding banking and financial institutions) have roughly $1.4 trillion worth of assets abroad and accounted for around a third of the global FDI stock.

Figure 10.1 Composition of foreign capital inflow into Developing Countries, 1986-1994

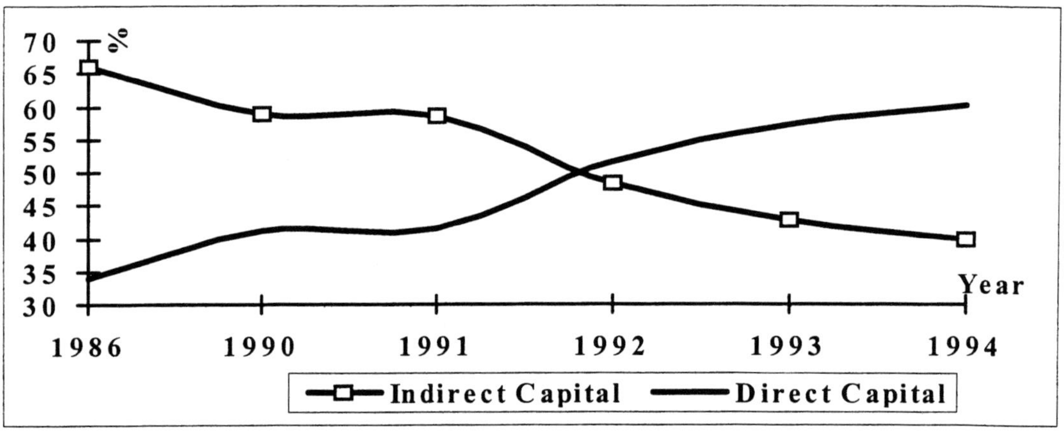

MNCs have different behavior from other international investment institutions (Dunning, 1993). To pursue maximum profit, MNCs have adopted different strategies in their global operations: (1) applying absolute and relative ownership advantages to acquire maximum regional market share; (2) to utilize internalizing advantages with arbitrary pricing for their semi-products for cost and tax-effective division of labor between the parent company and affiliates; (3) to exploit locational advantages in different countries/markets to create a diversified enterprise culture and new image to promote its competitiveness internationally; and (4) through necessary cooperation while competing with other MNCs to appropriately disperse R&D risks in investment.

[1] Data source: Ministry of Foreign Trade and Economic Cooperation of China (ed.), *Yearbook of Foreign Economic Trade of China, 1985, 1986, ..., 1997*, China's Economic Press, Beijing.

Note: The same data source used in figures and tables in this chapter will not be noted again.

1.1.2 Impacts of FDI on the development of Host Countries

Whether FDI by MNCs are desirable or not on the development of the host LDCs is still debatable. There are two polar attitudes regarding this issue (Souza & Stutz, 1994).

Views in favor Scholars holding this view believe that the MNC is probably the most efficient social, economic and political institution ever devised and it exerts a positive impact on the LDCs through FDI. Their impact works through: (1) enlarging capital investment and effectively reallocating regional resources; (2) creating new organizations with standard international management; (3) innovating, adopting and transferring advanced technologies; (4) enable LDCs to enter the international market more easily through MNCs' existing networks; (5) enhancing the quality of labor of the host country through regular personnel training programs and providing host countries with overseas elites; (6) serving as a major source of local savings and taxes; (7) facilitating the creation of vertical organizations which allow for the smooth, reliable and effective progression of local goods from one stage of production to another; and (8) probably the most important one, providing LDCs with new growth poles to drive the corresponding development of local up/down-stream industries.

Views in disfavor It comes mainly from the Dependency theory based on empirical studies of Latin American cases. FDI is seen to be negative to LDCs because: (1) foreign enterprises in most cases are located in enclave areas with limited and fragile economic connection with the local economy, thus their roles as growth poles are not effective; (2) since most of the profits earned by MNCs are remitted back to home countries, it aggravates the balance of payment and international debts problems of host countries; (3) the local economy may be controlled and manipulated by MNCs, thus the sovereignty and independence of host countries may be threatened; and (4) the increase of FDI may further enlarge the gap and inequality between the Developed and Developing countries.

However, empirical studies of Asian cases have drawn different conclusions (Sit & Yang, 1997). Therefore the real impacts of FDI on the development of host countries needs further research.

1.2 China as the Biggest Foreign Investment Recipient among LDCs Since 1992

Since 1978, China's actual utilized foreign investment has been increasing remarkably, especially after 1992 when Deng Xiaoping delivered his talks in his Southern tour (Fig. 10.2). The utilized foreign investment in China was only US$0.26 billion in 1978, while in 1992, this figure rose to US$16.2 billion. In 1996, it further jumped to US$54.8 billion. With an annual increase rate of 37 per cent, much higher than the growth of any economic sector, China's accumulated utilized foreign investment during 1978-1996 reached US$271.4 billion. As a consequence, China has since 1992 become the world's largest net foreign investment recipient among LDCs.

Figure 10.2 Utilized foreign capital in China, 1978-1996

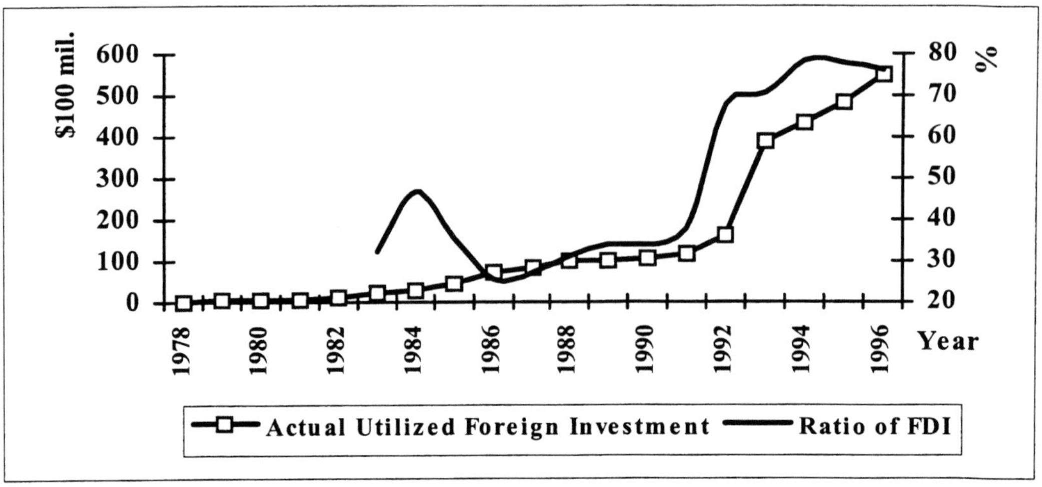

China's share in the total foreign capital inflows of the world to LDCs has also risen quickly in the last ten years. The net foreign capital inflows to LDCs in 1986 was US$74.8 billion, in which China's share was 9.7 per cent. In 1994, the corresponding figures rose to US$183.8 billion and 23.5 per cent (Fig. 10.3).

Figure 10.3 China's share in the world's net foreign capital inflows to the Developing countries

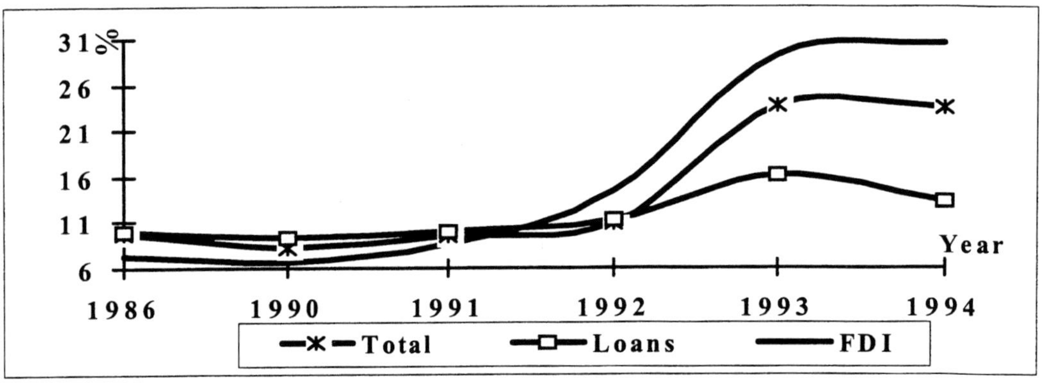

1.2.1 FDI has Become the Main Form of Utilized Foreign Capital

Changes in foreign investment utilization in China follow identically the changes of different international capital origins. In the 1980s, international capital was mainly in the form of various government development assistance, so China's utilized foreign capital was mainly composed of loans, while FDI only accounted for 30-40 per cent. After 1992, as MNCs dramatically increased their direct investment to host countries, the composition of international foreign capital inflow has changed completely. Accordingly, the proportion of FDI in China's total utilized foreign capital has increased from 37.8 per cent in 1991 to 67.9 per cent in 1992. It has remained between 76 and 78 per cent since (Figure 10.2). In 1986, China's utilized FDI accounted for 7.4 per cent of total FDI inflow into LDCs, while in 1994, the corresponding figure rose to 30.6 per cent. The increase of FDI inflow into China was much quicker than the rest of the world. It reflects that China is becoming an increasingly attractive investment market for MNCs.

Since 1990, the proportion of government loans and bank loans in China's total foreign capital inflow has decreased dramatically. Comparing with 1990, the proportion of government and bank loans in 1996 has dropped by 18 and 16 per cent respectively, while the proportion of FDI utilized through joint ventures and wholly-foreign-owned-enterprises rose by 20 per cent and 16 per cent (Table 10.1). Thus foreign capital inflow into China has shifted from a government- to an enterprise-based behavior, and tends to be more matured than that in the 1980s. It has alleviated the Chinese government's pressure and risk for paying back foreign debts, and has promoted Chinese enterprises to enter the global market and find their way to survive in fierce competition.

1.2.2 Size, Sectoral Distribution and Origins of FDI in China

In the 1980s, foreign investment projects were mainly in the form of government loans, and tended to be small in number and large in scale. As the Open Policy deepened, the number of foreign investment projects increased fast and the average investment amount per project dropped from around US$3.3 million in the period of the Sixth FYP to US$1.5 million in 1991. But since 1990, the scale of FDI projects has reversed (Table 10.2). Comparing with 1991, the average investment per project in 1996 increased 1.2 times and the average amount per FDI project has increased 2.2 times.

Table 10.1 Changes in Foreign Investment (FI) utilization in China, 1990-1996

Form of Foreign Investment (FI)	Utilized FI amount in 1990 (US$ billion)	Utilized FI amount in 1996 (US$ billion)	Compo-sition in 1990(%) (1)	Compo-sition in 1996(%) (2)	Difference between 1996 and 1990 (2)-(1)
Total	**10.29**	**55.80**	**100.0**	**100.0**	
1. Loans	**6.53**	**12.67**	**63.5**	**23.1**	**-40.4**
Government	2.52	3.45	24.5	6.3	-18.2
Financial institutions	1.07	3.00	10.4	5.5	-4.9
Banks	2.04	1.49	19.9	2.7	-17.1
2. FDI	**3.49**	**41.73**	**33.9**	**76.1**	**42.2**
Joint venture	1.89	20.76	18.3	37.9	19.5
Cooperative operation	0.67	8.11	6.5	14.8	8.3
Wholly-Foreign-owned	0.68	12.61	6.6	23.0	16.4
Cooperative development	0.24	0.26	2.4	0.5	-1.9
3. Others	**0.27**	**0.41**	**2.6**	**0.7**	**-1.9**

Table 10.2 Changes in average amount per project in China

Years	No. of projects signed	Total amount (US$ billion)	Average amount per project (US$ million)	# No. of FDI projects signed	# Amount of FDI projects (US$ billion)	# Average amount per FDI project (US$ million)
1981-1985	5561	18.1	3.25	5399	11.3	2.10
1986-1990	23183	63.4	2.74	22728	24.0	1.06
1991-1995	230369	409.3	1.78	229739	355.5	1.55
in which 1991	13086	19.6	1.50	12978	12.0	0.92
1992	48858	69.4	1.42	48764	58.1	1.19
1993	83595	123.3	1.47	83437	111.4	1.34
1994	47646	93.7	1.97	47549	82.7	1.74
1995	37184	103.2	2.78	37011	91.3	2.47
1996	24673	81.6	3.31	24556	73.3	2.98

FI is still mainly concentrated in manufacturing and real estates by sector. In the 1990s, foreign investment in China has diversified and began to incline towards capital-and technology-intensive industries, such as communication, energy and raw materials industries. In 1995 it was still mainly concentrated in manufacturing and real estates which accounted for 60.5 per cent and 17.3 per cent respectively. The concentration of FDI in these two sectors was even higher, accounting for 67.5 per cent and 19.5 per cent.

The country/region of origin (including the three parts of China, i.e. Hong Kong, Taiwan and Macau which are under non-socialist market economic systems) tends to be diversified. The number of countries of origin was 28 in 1985. The largest 5 accounted for 73 per cent of China's total utilized foreign investment, while the largest 10 and 20, accounted for 77 per cent and 79 per cent respectively. In 1990, the number of countries of origin rose to 36. The corresponding proportions for the largest 5, 10 and 20 countries were about 69 per cent, 80 per cent and 84 per cent respectively. In 1995, the number by countries jumped to 97, while the three corresponding figures were 70 per cent, 78 per cent and 86 per cent. Thus although foreign investment inflows into China by countries of origin tends to be diversified, it was still concentrated in the largest 5 (Table 10.3). In 1995, the largest 10 foreign investors by country/region were in rank order: Hong Kong, Japan, Taiwan, the United States, Singapore, Korea, United Kingdom, France, Canada and Italy.

Table 10.3 Rank by amount of actual foreign investment inflows to China, by country/region of origin

Absolute amount	Utilized foreign investment in 1995 US$ 48.1 billion		Utilized foreign investment in 1990 US$10.3 billion		Utilized foreign investment in 1985 US$4.5 billion	
Rank	Country/ region	Accumulated percentage	Country/ region	Accumulated percentage	Country/ region	Accumulated percentage
1	Hong Kong*	42.4	Japan	29.4	Japan	35.7
2	Japan	53.0	Hong Kong	52.5	Hong Kong	58.4
3	Taiwan	59.6	U. S.	58.3	U. S.	67.0
4	U. S.	66.1	France	64.0	Germany	70.5
5	Singapore	70.0	U. K.	69.0	U. K.	72.7
6	Rep. of Korea	72.4	Germany	72.9	France	74.5
7	U. K.	74.5	Canada	75.2	Italy	75.3
8	France	76.0	Austria	77.2	Australia	76.0
9	Canada	77.3	Italy	78.4	Canada	76.4
10	Italy	78.4	Spain	79.5	Kuwait	76.7

* Parts of China with non-socialist market economies.

The degree of concentration of FDI inflow into China by countries/regions of origin was even higher. The accumulated percentages of the largest 10 investors in 1985, 1990 and 1995 were all around 90 per cent (Table 10.4). Thus the pattern of inflow of FDI, measured by such aspect, is stable. Moreover, by country/region of origin, there exist different investment preferences. Capital from Hong Kong, Taiwan, Macau and Singapore are predominantly FDI. In 1985, their FDI accounted for 50 per cent of total FDI inflow into China, while in 1990 and in 1995, the corresponding figures rose to 56 per cent and 68 per cent respectively. As a comparison, capital from Japan, Canada, France, Italy, Sweden and Belgium are constituted by a higher proportion of loans, which is usually 40-50 per cent of their outflows into China.

Table 10.4 Rank by amount of actual FDI inflows into China, by country/region of origin

Absolute amount	Utilized FDI in 1995 US$37.5 billion		Utilized FDI in 1990 US$3.5 billion		Utilized FDI in 1985 US$2.0 billion	
Rank	Country/ region	Accumulated percentage	Country/ region	Accumulated percentage	Country/ region	Accumulated percentage
1	Hong Kong	53.5	Hong Kong	53.9	Hong Kong	48.9
2	Taiwan	61.9	Japan	68.3	U.S.	67.1
3	Japan	70.2	U.S.	81.4	Japan	83.2
4	U.S.	78.4	Germany	83.3	U.K.	86.9
5	Singapore	83.3	Singapore	84.7	France	88.5
6	Rep.of Korea	86.1	Macau	85.7	Germany	89.8
7	U.K.	88.5	Australia	86.4	Italy	90.8
8	Macau	89.7	Kuwait	87.1	Australia	91.5
9	Germany	90.7	France	87.7	Singapore	92.5
10		91.6	Holland	88.8	Canada	92.9

The regional impact of foreign investment in China will be discussed in conjunction with foreign trade later.

2. Pattern of Foreign Trade

2.1 Growth of China's Foreign Trade and its Rank in the World

Since the 1980s, especially the 1990s, and almost concurrent with the rapid increase of inflow of foreign capital, China has witnessed a significant growth in foreign trade. Her total export value in 1995 reached US$148.8 billion, ranking as the eleventh largest exporting country in the world (Fig. 10.4).

Since 1980s, the increase rate of China's import and export value has also been higher than the world as a whole. During 1980-1994, the accumulated increase rate of China's import and export value reached 3,960 per cent, while the corresponding figure for the world as a whole was only 2,024 per cent. The difference is further enlarged after 1992 (Fig. 10.5).

Figure 10.4 China's foreign export value and its rank in the world, 1953-1995

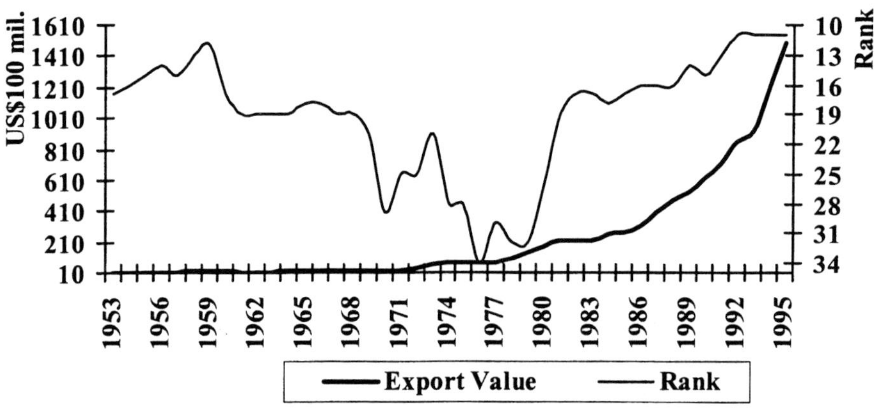

Figure 10.5 Comparison between China and the world as a whole by accumulated increase rate of the import and export value, 1980-1994

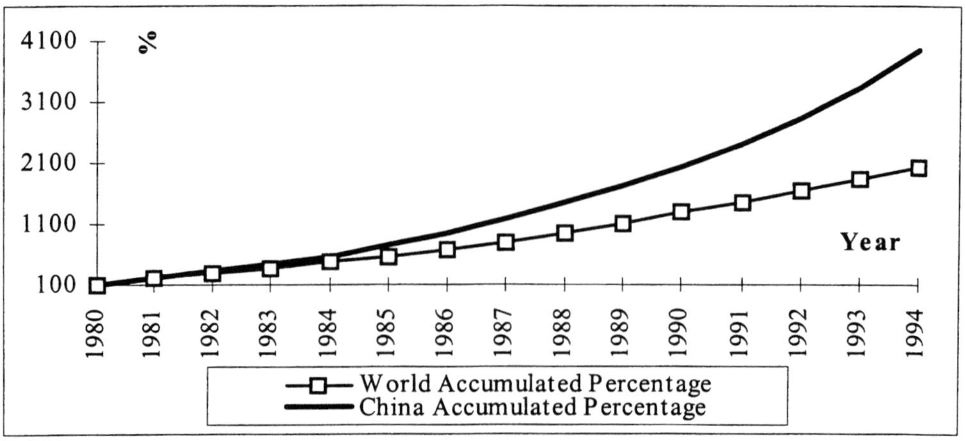

2.2 Structure of Import and Export

Since 1980, China's import and export structure has been improved. The proportion of primary products in export has dropped, while that of manufacturing goods has increased. In 1980, their ratio was 50:50, while in 1990 and 1996, the corresponding ratios became 26:74 and 14:84. In imports, the ratio of primary to manufacturing goods has remained more or less 18:82 since 1990 (Table 10.5).

Table 10.5 China's import & export value and product composition for selective years

Year	Import & Export Value US$ billion		Proportion of Primary products %		Proportion of Manufacturing products %	
	Export	Import	Export	Import	Export	Import
1980	18.1	20.0	50.3	34.8	49.7	65.2
1985	27.4	42.3	50.6	12.4	49.5	87.7
1990	62.1	53.4	25.6	18.3	74.4	81.6
1995	148.8	132.1	14.4	18.5	85.6	81.5
1996	151.1	138.8	14.5	18.3	85.5	81.7

During the Eighth FYP (1990-1995), China's export of mechanical and electrical products has registered fast growth. Their export value accounted for 18 per cent of China's total export in 1990, while in 1995, the corresponding proportion had risen to 30 per cent and had for the first time, surpassed the export value of clothing and textile products, China's traditional largest exporting sector. Thus the strategy formulated in the mid-1980s to promote the development of export-oriented mechanical and electrical products has had initial success.

2.3 Trade Partners

China's foreign trade partners numbered 157 in 1985 and it rose to 180 in 1990. After 1991, China adopted a new diversified strategy in foreign trade. Thus in 1995, the number of trade partners further rose to 228. Yet in terms of value, more than 94 per cent of China's foreign trade was still concentrated in Asia, Europe and North America. In 1995, foreign trade with these three continents accounted for 61 per cent, 18 per cent and 16 per cent respectively. In the last 10 years, China's foreign trade with Asia and North America has experienced rapid increase, while trade with Europe has slowed down relatively (Table 10.6). Trade with Africa, South America and Australia is still undeveloped.

Table 10.6 Composition of China's foreign trade value by continent

Item	Import and Export Value			Export Value		
Year	1995	1990	1985	1995	1990	1985
Value US$ billion	280.9	85.1	60.3	148.8	52.1	25.9
Composition%	100.0	100.0	100.0	100.0	100.0	100.0
in which: Asia	60.5	57.0	57.2	61.8	65.3	64.2
Africa	1.4	1.1	1.0	1.7	1.3	1.6
Europe	18.1	20.4	22.2	15.4	16.8	18.0
Latin America	2.2	2.2	3.3	2.1	1.3	2.7
North America	16.0	13.5	13.5	17.6	10.0	11.1
Oceania	1.8	2.1	2.1	1.3	0.9	0.9

Nevertheless, in terms of export value, the proportion of China's trade with Asia and Europe has actually dropped, representing a tendency of dispersion and diversification in recent years. In fact, the proportion of China's export trade with Oceania and Africa has increased though there are still room for improvement.

Similar to FDI, China's foreign trade partners show a high concentration and stability. Japan, Hong Kong and United States were all China's largest partners in 1985, 1990 and 1995. Their total trade value with China for these years accounted for 57 per cent, 58 per cent and 59 per cent respectively of China's total foreign trade value. In addition, Germany, Singapore, Russia and Britain were other important trade partners (Table 10.7).

Table 10.7 China's major trade partner by country/region, ranked by their proportion of import & export value

Rank	**Country/Region**	**1995 %**	**Country/Region**	**1990 %**	**Country/Region**	**1985 %**
1	Japan	20.5	Hong Kong	31.3	Japan	27.3
2	U.S.A.	14.5	Japan	15.2	Hong Kong	18.1
3	Hong Kong	14.1	U.S.A.	11.5	U.S.A.	11.7
4	Taiwan	6.4	Germany	4.6	Germany	5.6
5	Republic, Korea	6.0	Russia	4.6	Singapore	3.8
Above 5 accumulated		61.5		67.0		66.4
6	Germany	4.9	Singapore	2.7	Russia	3.1
7	Singapore	2.5	France	2.2	U.K.	2.4
8	Russia	1.9	Canada	2.0	Canada	1.8
9	Italy	1.8	Australia	1.9	Jordan	1.8
10	U.K.	1.7	U.K.	1.8	Australia	1.8
Above 10 accumulated		74.3		77.6		77.3

3. IMPACT OF FOREIGN INVESTMENT AND FOREIGN TRADE ON CHINA'S ECONOMIC DEVELOPMENT

Foreign trade plays an increasing role in China's economic development. In order to compare the contribution of foreign trade value to China's GDP since 1978, we firstly recalculate the trade value in terms of RMB according to the official foreign exchange rate of each year, then we compute the annual proportion of trade value to annual GDP. The result as listed in Table 10.8 shows an obvious rising trend. In 1978, the proportion was about 10 per cent, while in 1985 and 1994, the corresponding figures rose to 23 per cent and 44 per cent respectively. In 1995 and 1996, although the proportions have dropped, they still reached 40 per cent and 36 per cent respectively. Besides, since 1990, the contribution of export to GDP has been larger than that of import, except in 1993.

Changes in the proportion of foreign investment value to GDP shows that utilizing foreign capital is very important to China's economic growth. In 1985, the proportion of foreign investment value to GDP was only 1.5 per cent, while in 1990 and 1996, the corresponding figures rose to 3.4 per cent and 6.7 per cent, i.e. the contribution of foreign investment to GDP has registered a faster increase than that of foreign trade.

The analysis above shows that there is a strong relationship between foreign investment, foreign trade and economic development (GDP) in China. Here we perform an exercise to determine their degree of dependency on each other, by simple linear regression between GDP and 5 selected variables, i.e. Import and Export Value (I&E), Export Value (E), Total Investment (TI), Foreign Investment (FI) and Foreign Direct Investment (FDI). The results of the regression are shown in Table 10.9.

Table 10.8 Changes in proportion of import & export value to GDP, 1978-1996

Year	GDP RMB billion	Import & Export value(I&E) US$ billion	# Export value	Exchange rate per US$100	Proportion of I&E value to GDP %	# Proportion of Export value to GDP
1978	362	20.6	9.8	171	9.7	4.6
1979	404	29.3	13.7	171	12.4	5.8
1980	452	38.1	18.1	171	14.4	6.8
1981	486	44.0	22.0	171	15.4	7.7
1982	530	41.6	22.3	189	14.9	8.0
1983	594	43.6	22.2	198	14.5	7.4
1984	717	53.6	26.1	233	17.4	8.5
1985	896	69.6	27.4	294	22.8	9.0
1986	1020	73.9	30.9	345	25.0	10.5
1987	1196	82.7	39.4	372	25.7	12.3
1988	1493	102.8	47.5	372	25.6	11.8
1989	1691	111.7	52.5	377	24.9	11.7
1990	1855	115.4	62.1	478	29.8	16.0
1991	2162	135.6	71.8	532	33.4	17.7
1992	2664	165.5	84.9	551	34.3	17.6
1993	3463	195.7	91.7	576	32.6	15.3

1994	4662	236.6	121.0	862	43.7	22.4
1995	5826	280.9	148.8	835	40.3	21.3
1996	6780	289.9	151.1	831	35.6	18.5

Table 6.9 Coefficient of regression between selected variables

Variables	GDP	I&E	# E	TI	FI	# FDI
Gross Domestic Product (GDP)	1.0					
Import and Export Value (I&E)	18.8	1.00				
# Export Value (# E)	35.4	1.89	1.00			
Total Investment (TI)	2.6	0.11	0.06	1.0		
Foreign Investment (FI)	86.1	4.62	2.42	37.1	1.00	
# Foreign Direct Investment (# FDI)	89.3	5.13	2.73	42.8	1.19	1

There is a very strong positive relationship between the selected variables, especially between I&E, GDP, #E, FI and FDI. Among them, FDI bears the highest coefficient with other variables, implying that FDI has the closest relationship with other economic indicators and is more active than others in promoting economic development in China. Accordingly, US$1 of Export Value is related to the worth of RMB35 of GDP, while in the case of FI and FDI, the corresponding ratio are 1:86 and 1:89 respectively. Although it cannot be directly interpreted from coefficients alone that there exists causal relationships between the variables, it is reasonable to infer that Foreign Trade and Foreign Investment, especially FDI do have a significant positive impact on China's economic development. Their roles in keeping the vitality of China's economy have become more important since China has opened her door to the world and more and more integrated with the world economy.

4. Impact of Foreign Trade and Foreign Investment on China's Regional Development

4.1 Foreign Trade and Foreign Investment Utilization by Province

China has conducted her Open Policy in a step-by-step manner for almost 20 years, in which her regional differences have also been enlarged. In temporal terms, the Open Door policy has shown 3 phases and each had a different effect on different regions. The temporal-spatial distribution of the 3 phases is as followings: (1) first phase, 1979-1984, with the establishment of the four Special Economic Zones and Guangdong and Fujian as open provinces; (2) second phase, 1985-1992, with the designation of 14 Open Cities in the coastal areas; (3) third phase, after 1992, all the other provinces have been opened.

To compare the changing process of foreign investment and foreign trade by province, and to eliminate accidental error caused by data of any single year, we choose the 5-years accumulated data of the Sixth, Seventh and Eighth FYP as our database. Since these three FYPs more or less coincide with the three phases of the Open Policy, the effect of the Open Policy on provincial foreign investment and foreign trade development can be mapped fairly accurately.

Table 10.10 shows that all provinces have had increases in their absolute value of foreign investment and foreign trade for the 15 years of 1981-1995, but their shares of the total was different for different periods.

Table 10.10 Changes in proportionate share of Foreign Investment and Foreign Trade value in China by Province (%)

Provinces	Foreign Investment(FI)			#FDI			Foreign Trade			#Export		
	1981-1985	1986-1990	1991-1995	1981-1985	1986-1990	1991-1995	1981-1985	1986-1990	1991-1995	1981-1985	1986-1990	1991-1995
Beijing		5.77	3.74		10.49	3.58	3.21	2.97	2.02	3.05	2.87	1.59
Tianjin	2.61	5.63	3.96	2.86	2.52	2.96	6.51	4.28	2.30	6.70	4.45	2.33
Hebei	0.48	0.44	1.80	0.25	0.85	1.67	4.02	3.43	1.80	4.40	4.19	2.41
Shanxi	0.04	0.27	0.78	0.02	0.17	0.20	0.65	0.91	0.61	0.47	1.04	0.80
Neimenggu		0.06	0.20		0.00		0.44	0.76	0.67	0.39	0.76	0.62
Liaoning	1.77	8.72	4.79	1.04	4.79	4.19	19.41	9.80	5.88	22.34	11.68	7.24
Jilin	0.17	0.69	1.56		0.24	0.91	1.23	1.63	1.70	1.09	1.66	1.61
Heilongjiang	2.86	1.58	1.10	0.11	1.03	0.99	1.36	2.44	1.96	1.32	2.52	1.96
Shanghai	4.99	14.44	10.06	4.92	11.19	8.92	17.44	13.91	8.84	17.80	12.75	8.77
Jiangsu	4.53	5.03	9.85	0.54	3.57	11.85	6.23	6.17	6.64	6.63	6.50	3.63
Zhejiang	2.76	2.65	3.04	0.86	1.51	3.33	3.07	4.03	4.64	3.29	4.65	5.64
Anhui	0.45	1.11	1.24	0.07	0.27	1.02	1.00	1.34	1.00	0.94	1.50	1.16
Fujian	7.15	4.66	8.40	8.37	7.57	11.05	2.93	4.12	6.60	2.06	3.78	6.65
Jiangxi	0.30	0.59	1.00	0.27	0.23	0.76	0.97	1.09	0.78	1.05	1.28	1.04
Shandong	1.17	1.90	6.66	0.28	1.71	6.94	5.16	4.57	4.85	5.69	5.06	6.07
Henan	0.47	0.77	1.88	0.26	1.13	1.21	1.32	1.69	1.30	1.48	1.99	1.55
Hubei		1.88	2.15		0.76	1.75	2.01	2.37	1.72	2.07	2.73	1.88
Hunan	1.39	1.42	1.80	0.97	0.33	1.22	1.94	1.65	1.54	1.93	1.82	1.76
Guangdong	61.74	32.28	26.26	75.22	41.95	28.28	15.46	23.84	37.01	12.31	20.16	35.20
Guangxi	1.93	1.66	2.09	0.75	1.70	2.24	1.77	1.57	1.41	1.73	1.59	1.55
Hainan	1.23	1.77	3.07	1.66	3.18	3.14	0.17	1.23	1.43	0.20	0.72	0.93
Sichuan	1.25	3.45	1.46	0.38	0.76	1.97	1.03	2.24	2.01	0.85	2.33	2.06
Guizhou	0.33	0.26	0.17	0.07	0.16	0.17	0.21	0.31	0.29	0.16	0.31	0.32
Yunnan	0.04	0.13	0.61	0.07	0.19	0.48	0.76	1.25	0.90	0.67	1.18	0.94
Xizang							0.07	0.05	0.08	0.02	0.04	0.03
Shaanxi	0.77	1.83	1.04	0.63	3.17	0.76	0.41	0.90	0.91	0.33	0.92	1.06
Gansu	0.56	0.11	0.13	0.15	0.13	0.17	0.27	0.35	0.30	0.25	0.40	0.36
Qinghai			0.01			0.02	0.05	0.12	0.08	0.06	0.13	0.12
Ningxia		0.07	0.12		0.02	0.08	0.19	0.18	0.11	0.18	0.19	0.15
Xinjiang	1.02	0.84	1.03	0.25	0.37	0.16	0.68	0.78	0.62	0.57	0.80	0.58
China	100.00	100.00	100.00	100.00	100.00	100.00	100.00	100.00	100.00	100.00	100.00	100.00
East	90.34	84.95	83.72	96.75	91.03	88.14	85.40	79.93	83.42	86.18	78.41	82.00
Middle	5.70	8.36	11.70	1.71	4.16	8.05	10.93	13.90	11.28	10.74	15.28	12.39
West	3.97	6.70	4.58	1.54	4.81	3.81	3.67	6.17	5.30	3.08	6.30	5.61

In the first period, although all provinces with open port-cities in the coastal area attracted certain amount of foreign investment, Guangdong was the absolute dominant recipient due to its advantages of being the earliest beneficiary of the Open Policy and has more Special Economic Zones, as well as being adjacent to Hong Kong and Macau the main FDI sources. Its foreign investment inflow in this period accounted for 62 per cent of China's total. Fujian was the second largest recipient, accounting for 7 per cent. The other large recipients were Shanghai and Jiangsu, accounting for 5 per cent and 4.5 per cent respectively.

In the second period, China enlarged her opened areas. The relative proportion of foreign investment inflow to Guangdong and Fujian dropped to 32 per cent and 4.7 per cent, while the newly opened areas, such as Shanghai, Liaoning, Beijing, Tianjin and Jiangsu, increased their shares to between 5 and15 per cent of China'a total. Shanghai enjoyed the fastest growth, reaching 14.4 per cent. Yet, foreign investment inflow was still concentrated in a few provinces, especially in Guangdong, Shanghai and Liaoning. They together accounted for more than 55 per cent of the total.

After 1992, concentration of foreign investment by province has further dropped, when China had almost fully opened her door to the world. The proportion of the largest 4 recipients, Guangdong, Shanghai, Jiangsu and Fujian together accounted for 54 per cent of China's total FI. But since FDI has become the predominate component of foreign investment, which tends to flow into economically vigorous areas, provinces with a larger proportion of private sector ownership and higher rate of economic growth had attracted more foreign investment. For example, Jiangsu's share of FI inflow rose from 5 per cent in the second period to nearly 10 per cent in this period, while Shandong from 2 to 6.7 per cent, Fujian from 4.7 to 8.4 per cent. On the contrary, the share by provinces with higher proportion of state-owned economic sectors declined in this period. The share of Liaoning, Beijing and Tianjin dropped from 8.7, 5.8 and 5.6 to 4.8, 3.7 and 4 per cent respectively.

In terms of FDI, the concentration by province for the three periods were much higher and its distribution was more closely related to the general economic development conditions of each province. Therefore, as the proportion of FDI in the total FI rises further, the attractiveness of each province to foreign investment in future will be more dependent on their general economic environment than the preferential policy they enjoy.

Table 10.10 also shows that the regional pattern of foreign investment has a remarkable effect on regional foreign trade share. Guangdong and Fujian have the highest proportion of import and export value in China, and their proportions are also growing constantly, whereas the second tier beneficiaries, Zhejiang, Jiangsu and Shandong have also experienced rapid increase in recent years as their foreign investment inflow expands at a high speed.

In order to further compare the dynamic changes in proportion of foreign investment and foreign trade by province for the three periods, we classified the provinces into 4 types. Type 1 are provinces with proportions rising in the first period and declining in the third period; Type 2 are provinces with a reversed trend to that of type 1; Type 3 are provinces with proportions keep rising for the three periods and Type 4 are provinces with a reversed trend to that of Type 3. The classifications and the common characteristics for each type are listed in Table 10.11.

Table 10.11 Change types in proportions of foreign investment and foreign trade value to the country's total for the three periods by province and the possible reasons

Types	Provinces: according to changes in proportion of foreign investment to the country's total
Rising then declining	**(Beijing, Tianjin, Shanghai)** With economic structure being in transition. Those provinces began to selectively utilize foreign investment based on their development targets and therefore are facing unfair competition from other provinces where the preferential policies in attracting the foreign investment were less conditional; **(Liaoning, Jilin, Sichuan, Shaanxi)** Those are provinces with high proportion of state-ownership component in their economic composition, which has negative influences on their attracting foreign investment
Declining then rising	**(Zhejiang, Fujian)** With adjustment of economic structure for a certain period, those provinces draw new attention to the foreign investors; **(Guangxi, Gansu, Xinjiang)** As further exploitation of the natural resources, those provinces are becoming the new favorite of the foreign investment
Keeping rising for the three periods	**(Jiangsu, Shandong, Hainan)** With constantly vigorous economic development, those provinces always have great attractions to the foreign investment; **(Shanxi, Anhui, Jiangxi, Henan, Hubei, Hunan, Neimenggu, Yunnan, Qinghai, Ningxia)** With their individual specific advantages, those provinces are being their initial stage in utilizing foreign investment. They are the new targets of China's future market for MNCs
Keeping declining for the three periods	**(Guangdong)** With economic structure in transition and too high proportion of foreign investment utilization in the first period, the share of Guangdong is relatively declining as China gradually opened her door wider; **(Hebei, Heilongjiang, Guizhou)** With less vigorous urban economic development, those provinces relatively have poor investment environment
Types	**Provinces: according to changes in proportion of foreign trade value to the country's total**
Rising then declining	**(Shanxi, Neimenggu, Heilongjiang, Anhui, Jiangxi, Henan, Hubei, Sichuan, Guizhou, Yunnan, Gansu, Qinghai, Xinjiang)** Being in initial stage in utilizing foreign investment, the previous export products in those provinces are either unsuitable for the market or facing new competition from the coastal areas
Declining then rising	**(Jiangsu, Shandong)** With big increase of utilizing foreign investment, the products in those provinces are being replaced with new generation, which are favorable for exporting, or the import value of foreign enterprises in those provinces begin to increase **(Xizang)** A special case
Keeping rising for the three periods	**(Jilin, Zhejiang, Fujian, Guangdong, Hainan, Shaanxi)** The export-oriented economy in those provinces tends to be mature, they both import raw material and export products in the world market; Or the products in those provinces are quite special without much competition

Keeping declining for the three periods	**(Beijing, Tianjin, Hebei, Liaoning, Shanghai, Hunan, Guangxi, Ningxia)** The traditional products in those provinces are either facing fierce competition, or shrinking in market needs; Or the products of foreign enterprises in those provinces are mainly sold in the internal market with small proportion for exporting to the international market

4.2 Contribution of Foreign Investment and Foreign Trade to Regional GDP

The relative position of a province in the country may not be high in terms of its proportion of foreign investment and foreign trade value in the country's total, but the contribution of foreign investment and foreign trade to the local economy and its own economic development may be decisive. Therefore we have examined changes in the proportion of foreign investment and import and export values to the local GDP for each province (Table 10.12).

Table 10.12 shows that foreign investment and foreign trade have become more and more important to regional economic development. Their proportions to local GDP are constantly rising, showing that since the Open Policy was adopted, the export-oriented economy in all provinces has experienced a faster growth than other economic sectors. The increase in the proportion of foreign investment to GDP was remarkable, reaching 12 times in 1985-1995 on average. The proportions of foreign investment and foreign trade value to local GDP in the East were around 3 times higher than that in the Central and the West on average, but the increase in the latter were much faster. During 1985-1995, the proportion of foreign investment to local GDP in the East has increased 10 times, while in the Central region, the corresponding increase was 30 times. In the same period, the proportion of foreign trade value to local GDP in the East has increased less than 3 times, while the corresponding figure in the West was nearly 4 times.

Table 10.12 Foreign Investment and Import & Export Value as a Proportion of Local GDP by Province %

Regions	Proportion of FI value to GDP				Proportion of I&E value to GDP			
	1980	1985	1990	1995	1980	1985	1990	1995
Beijing			3.88	11.83	8.2	11.1	16.7	26.3
Tianjin	0.03	1.26	5.14	19.15	30.2	24.9	34.0	59.4
Hebei		0.11	0.24	3.18	4.9	10.5	10.4	11.6
Shanxi		0.02	0.42	2.46	0.2	4.6	5.8	10.5
Neimenggu			0.16	1.30	1.1	3.3	7.2	13.8
Liaoning	0.02	0.24	3.54	5.70	24.7	30.6	28.4	32.9
Jilin		0.07	0.68	6.67	1.9	8.1	10.7	20.1
Heilongjiang		0.14	0.79	3.11	1.0	4.2	10.0	14.2
Shanghai		0.73	4.93	17.96	24.7	26.3	47.0	64.6
Jiangsu		0.42	1.48	8.56	5.1	8.4	12.5	27.2
Zhejiang		0.44	0.86	3.65	2.5	7.7	13.6	27.3
Anhui		0.17	0.37	3.20	0.5	3.8	5.4	9.6
Fujian	0.12	2.60	3.47	16.00	9.9	16.4	29.0	56.8
Jiangxi		0.14	0.57	3.14	1.6	4.5	7.0	11.5

Shandong	0.05	0.28	0.98	5.45	11.0	13.2	13.3	21.9
Henan		0.08	0.06	2.86	1.5	2.9	5.1	7.5
Hubei			0.98	3.87	2.3	5.1	6.9	12.9
Hunan		0.34	1.54	3.34	2.9	4.4	6.0	11.1
Guangdong	1.02	4.89	6.57	18.77	17.8	28.6	53.0	145.6
Guangxi		0.73	0.64	5.01	6.6	8.5	9.6	16.7
Hainan	0.02	1.80	8.85	33.36	0.0	0.0	43.7	52.0
Sichuan		0.19	1.34	1.46	0.4	2.5	5.5	10.2
Guizhou		0.31	0.55	1.28	0.9	1.6	4.0	8.4
Yunnan		0.03	0.12	2.39	2.6	4.5	7.9	14.7
Xizang					3.4	3.0	5.2	9.7
Shaanxi		0.22	0.67	4.55	0.0	2.6	7.4	14.5
Gansu	0.02	0.11	0.09	1.11	0.9	2.4	4.0	8.1
Qinghai				0.40	0.4	1.9	4.8	8.2
Ningxia			0.99	2.72	5.0	5.2	6.3	13.7
Xinjiang	0.02	0.14	1.33	5.47	1.0	7.6	7.2	14.3
China	0.07	0.62	1.87	7.22	7.7	10.9	16.4	33.9
East	0.12	1.06	2.86	10.22	13.5	17.1	24.8	49.8
Middle		0.12	0.65	3.36	1.5	4.4	6.9	11.6
West		0.16	0.84	2.35	0.9	3.1	6.0	11.6

By province, there is a strong positive relationship between the proportion of foreign investment to local GDP and the proportion of its foreign trade value to local GDP (Fig. 10.6). By rank, provinces with foreign investment as a proportion of local GDP higher than 10 per cent in 1995 were respectively Hainan (33%), Tianjin (19%), Guangdong (19%), Shanghai (18%), Fujian (16%) and Beijing (12%). They all are earlier opened areas in the first or the second period. Guangdong has become an export-oriented area in its economic structure with import and export value as a proportion to local GDP as high as 146 per cent in 1995. Besides, a number of provinces have the proportions higher than 50 per cent, such as Shanghai (65%), Tianjin (59%), Fujian (57%) and Hainan (52%).

Figure 10.6 Foreign Investment (FI) and Import & Export (IE) Value as a Proportion of GDP by Province (1995)

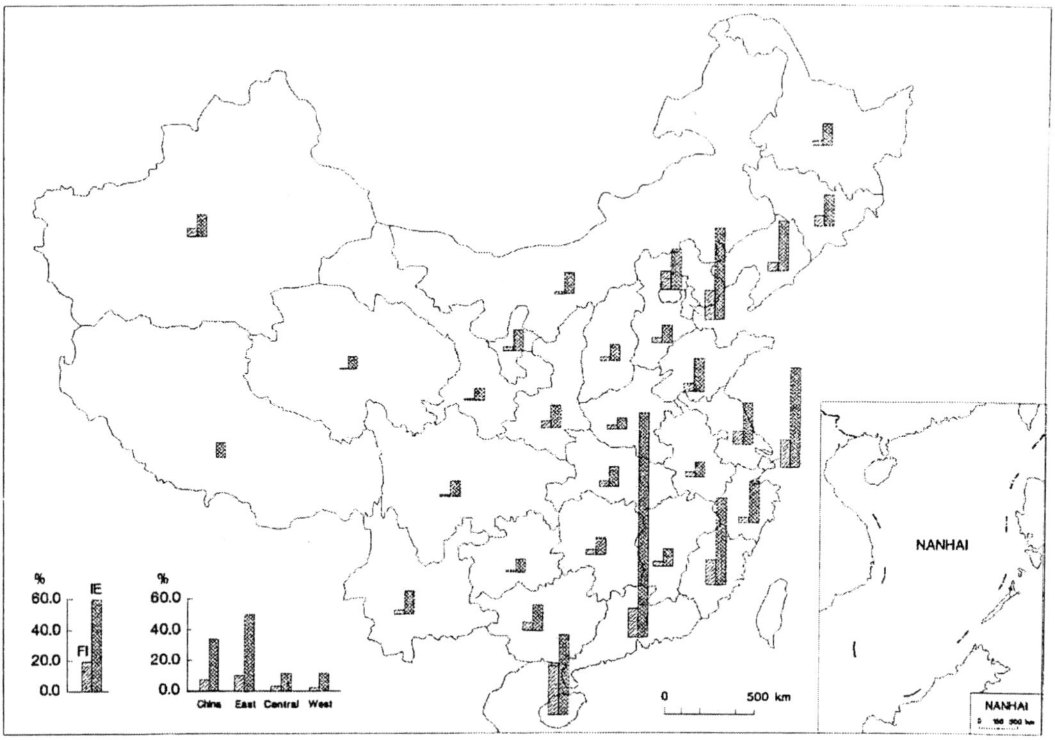

According to his studies on international investment distribution, Dunning has concluded that the volume of foreign investment attracted by a particular country is much dependent on the general economic development of the country and there exists a positive relationship between foreign investment and GDP per capita. Empirical study on China's time series data for nearly 20 years in the previous part of this chapter has also drawn a similar conclusion. For testing further this conclusion, we conduct another regression analysis on the relationship between foreign investment and GDP by province. We obtained the following three linear regression equations for the data of 1990, 1995 and 1996:

1990:	FI=-11684+0.0274GDP	(R=0.57, S=34208, F=12)
1995:	FI=-57694+0.0966GDP	(R=0.69, S=182397, F=24)
1996:	FI=-73387+0.0773GDP	(R=0.66, S=180790, F=21)

where, FI is the volume of utilized foreign investment for a particular year by a particular province.

Although the regression coefficients are not very strong, they all pass the F test, suggesting that there exists a positive relationship between the two variables. That is, the general economic development of one region does have a positive effect on its utilization of foreign investment. According to the regression result of 1996, we could roughly claim that when local GDP per capita in a province increases 10,000 units, its volume of utilized foreign investment would correspondingly increase by about 770 units.

The existence of this relationship also implies that the key to whether one region can continuously attract foreign investment is much dependent on its general economic growth, its market potential and its economic vitality. Therefore the belief that rapid development of the export-oriented economy in the coastal area of China since 1978 has been mainly benefited by the preferential policies given by the central government is not necessarily reflected all major reasons behind. Preferential policies do spur the start of the inflow of foreign investment in the East, yet to sustain attraction to foreign investment requires regional development vigor and other positive-economic factors.

4.3 Regional Bias of Major Foreign Investors

A particular home country's investment in China is likely to concentrate in certain provinces. We examine such regional patterns of the eleven major home countries/regions in 1996. They are Hong Kong, Japan, Singapore, United States, Taiwan, Britain, Republic of Korea, Germany, Australia, Canada and France which accounted for more than 80 per cent of China's total utilized foreign investment. In 1995, the corresponding proportion was 86.4 per cent. With advance economic and technological developments, most of these countries/regions are seats of the world's largest MNCs. Therefore, their investment behavior are more stable and protracted, and will have profound influence on China's future regional economic development.

Investment Location Quotient (ILQ) of the 11 major home countries are computed[2] (Table 10.13). If the ILQ is larger than 1, then there is a regional bias for a particular investor. The bigger the ILQ is, the more obvious the regional bias the investor has. By rank, the regional bias of each home country with ILQ larger than 1 in China are as follow:

- **Hong Kong:** Guangdong (1.69), Henan (1.3), Ningxia (1.26), Gansu (1.22), Fujian (1.19), Jiangxi (1.14), Guangxi (1.07)
- **Japan:** Liaoning (2.39), Ningxia (2.33), Jiangsu (1.43), Hebei (1.17), Gansu (1.16), Shanghai (1.03)
- **Singapore:** Hainan (7.86), Jiangsu (1.88), Guangxi (1.64), Hunan (1.05)
- **United States:** Beijing (1.55), Shandong (1.51), Jilin (1.31), Ningxia (1.27), Jiangsu (1.21), Zhejiang (1.04)
- **Taiwan:** Fujian (3.48), Gansu (2.58), Henan (1.57), Hebei (1.42); Jiangsu (1.23), Zhejiang (1.21); Shandong (1.17), Jiangxi (1.03)
- **Britain:** Neimenggu (5.04), Hebei (2.48), Shanghai (2.07), Guangxi (1.65), Beijing (1.43), Sichuan (1.11)
- **Republic of Korea:** Shandong (4.1), Liaoning (3.96), Jilin (2.83), Hebei (1.46), Tianjin (1.7)
- **Germany:** Beijing (1.74), Shanghai (1.48), Guangxi (1.47), Hebei (1.30), Jilin (1.08)
- **Australia:** Hunan (17.24), Shanghai (2.17), Neimenggu (1.17)
- **Canada:** Hebei (3.88), Neimenggu (3.27), Beijing (2.88), Hunan (2.35), Jilin (1.9), Guangxi (1.29), Sichuan (1.13), Tianjin (1.06), Hainan (1.05)
- **France:** Neimenggu (3.19), Beijing (1.77)

[2] ILQ=(Investment to i province from j country/total foreign investment to i province)/(total investment to China from j country/total foreign investment to China)

Table 10.13 Provincial Investment Location Quotient by China's major foreign investors in 1996

Regions	Hong Kong	Japan	Singapore	USA	Taiwan	U.K.	Korea	Germany	Australia	Canada	France
Beijing	0.85	0.62	0.58	**1.55**		**1.43**	0.80	**1.74**		**2.88**	**1.77**
Tianjin	0.27	0.43	0.16	0.76	0.41	0.39	**1.37**	0.30	0.19	**1.06**	
Hebei	0.75	**1.17**	0.57	0.90	**1.42**	**2.48**	**1.46**	**1.30**		**3.88**	
Shanxi	0.18	0.18	0.11	0.47	0.16			0.30	0.44		
Neimenggu	0.71	0.62		0.39	0.92	**5.04**			**1.17**	**3.27**	**3.19**
Liaoning	0.62	**2.39**	0.43	0.87	0.55		**3.96**				
Jilin	0.31	0.20		**1.31**	0.26	0.61	**2.83**	**1.08**		**1.90**	
Heilongjiang	N/A										
Shanghai	0.29	**1.03**	0.92	0.87	0.58	**2.07**	0.94	**1.48**	**2.17**		
Jiangsu	0.64	**1.43**	**1.88**	**1.21**	**1.23**						
Zhejiang	0.86	0.85	0.34	**1.04**	**1.21**		0.20	0.47	0.17		0.75
Anhui	N/A										
Fujian	**1.19**		0.73	0.32	**3.48**	0.48				0.18	0.33
Jiangxi	**1.14**	0.48	0.90	0.72	**1.03**						
Shandong	0.68	0.96		**1.51**	**1.17**		**4.10**				
Henan	**1.30**	0.28	0.40	0.88	**1.57**						
Hubei	N/A										
Hunan	0.98	0.38	**1.05**	0.33	0.41				**17.24**	**2.35**	
Guangdong	**1.69**	0.29	0.51	0.44	0.55	0.55					
Guangxi	**1.07**	0.17	**1.64**	0.15	0.93	**1.65**		**1.47**		**1.29**	
Hainan	0.26	0.19	**7.86**	0.48	0.20	0.27	0.44			**1.05**	
Sichuan	0.67	0.80	0.38	0.93	0.63	**1.11**	0.69		0.14	**1.13**	
Guizhou	0.72	0.11		0.41	0.19					0.98	
Yunnan	0.59		0.45	0.34	0.67						
Xizang	0.61										
Shaanxi	0.52	0.30	0.28	0.52	0.27						
Gansu	**1.22**	**1.16**			**2.58**						
Qinghai	0.83			0.93	0.32						
Ningxia	**1.26**	**2.33**	0.80	**1.27**	0.59						
Xinjiang	0.18				0.34						

The ILQ to a certain extent also reflects competition between major investors in China's regional market. If MNCs of a particular country would like to concentrate their investment in a specific area, it would mean that these MNCs may hold a view that the prospect of that area is bright, and they either want to exploit the local resources, dominate the local export-oriented economy, or try to obtain as much the market share in that area as possible. Therefore even if the investment of the home country to that area is small in terms of absolute value, yet it still shows a regional bias towards that area once her ILQ is higher than 1. To the investor, this specific area is relatively more important, because comparing to other investors, the higher ILQ means higher ratio of investment risk in that area.

Table 10.13 and Fig. 10.7 show that major foreign investors do have obvious regional biases. Hong Kong investment has a regional bias towards its adjacent areas, such as Guangdong, Guangxi and Fujian. Japanese investment prefers East Liaoning Peninsula and the Yangtze Delta. Korea investment concentrates in Shandong Peninsula and Northeast China. Taiwan prefers to invest in Fujian, Jiangsu and Zhejiang. European, United States and Australia investments have regional bias in North China and the Yangtze Delta. Singapore investment is mainly concentrated in Jiangsu and Hainan.

Figure 10.7 Regional Bias of Major Foreign Investors

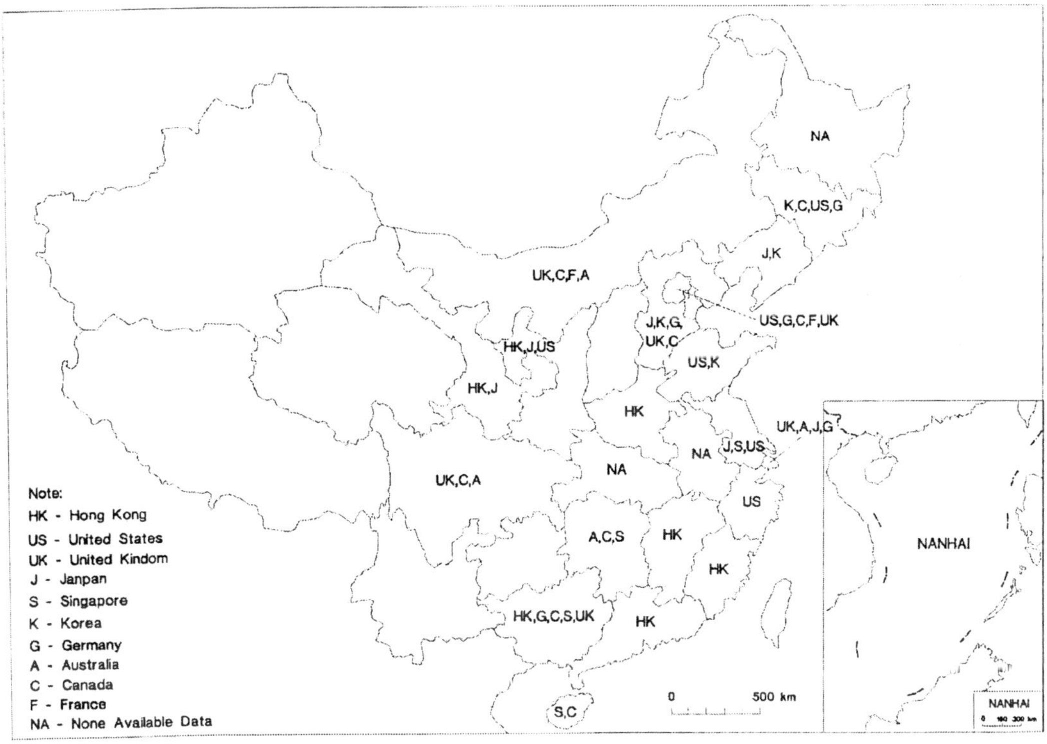

From the ILQ reading, some provinces in the Central Region, such as Jilin, Hunan, Henan, Jiangxi and Sichuan, will be new hot areas for attracting foreign investment in the foreseeable future. In addition to their relatively high absolute volume of utilized foreign investment in 1995 (1.76, 1.38, 0.92, 0.68 and 1.63 billion US dollars respectively), they also held high ILQ indices, meaning that they are becoming new centers of major investors' attention. As a consequence, economic development in these provinces will be expected to accelerate.

4.4 Sectoral Distribution of Foreign Investment by Province

The number of contracted foreign investment projects in China in 1995 was 37011, with the total contracted value of US$91.3 billion. The average investment per project was US$2.47 million. From the sectoral composition of the projects, it can be seen that foreign investment is still mainly interested the sectors of industry, real estates and commerce. In 1995, the number of contracted FDI projects in these three sectors accounted for 75 per cent, 8.9 per cent and 5 per cent of China's total respectively. But because the average investment project size in different sectors varies, their proportions in terms of the contracted values changed to 67.5 per cent, 19.5 per cent and 3.8 per cent of China's total respectively. In average investment size, the communication sector was the largest, reaching US$6.33 million, and real estates was second, reaching US$5.44 million. The trend shows that foreign investment in China has recent begun to shift towards capital-intensive infrastructure sectors, such as transportation and communication. Real estates remains the favorite sector, and its average project size is also comparatively large.

The sectoral distribution of foreign investment by province reflects, to a certain extent, the regional bias and regional investment strategy of the investors. For doing so, we calculated the average project size and the Sector Location Quotient (SLQ) by province based on data on contracted foreign investment by sector and province in 1995 (Table 10.14).

Large-scale foreign investment projects tend to locate in provinces opened early or those provinces with a relatively advanced export-oriented economy, such as Shanghai, Jiangsu, Guangdong, Fujian and Hainan. The average investment project size in these provinces were all above US$3 million, more than 20 per cent higher than the country's average.

Agriculture is an obvious resource-oriented sector. Those with agricultural SLQ greater than 1 are respectively Xinjiang, Neimenggu, Liaoning, Jilin, Shandong, Fujian, Guangxi, Yunnan and Hainan. With their specific agricultural products which can be processed at economic scale, such as livestock products, seafood or subtropical agricultural products, etc., these provinces are a predominant in utilizing foreign investment in this economic sector.

There is no big difference between provinces in industrial SLQ. Only Hebei, Shanxi, Shandong, Henan, Guangdong and Xinjiang are slightly greater than 1. On the other hand, there are 15 provinces, half of China's total provinces, with SLQ greater than 1 for construction, representing that foreign investors are generally optimistic about China's property market. Fujian, Shanghai, Jiangsu, Shandong and Hainan, are always active in capital constructions since China's opening up. Guangxi, Henan, Jiangxi, Liaoning and Jilin, tended to be more active in recent years.

Tianjin has extremely high SLQ for commerce. This may mean that being a commercial and trade center in North China from ancient times, the importance of Tianjin's role in this aspect has been recognized by foreign investors after 1978. Having been active in real estates since 1980s, Beijing, Shanghai, Fujian, Hainan and Guangxi all have high SLQ index for the real estates sector.

5. Development Strategies for Utilizing Foreign Investment

5.1 Motives and Objectives of Foreign Investment

The Motives of foreign investment into China may be generalized into 4:

1) **Resource-oriented:** The major motive of this type of foreign investment is to guarantee the supply of raw materials to the parent company through exploitation of new resources. To exploit those renewable resources, such as agricultural products, seafood and medical herbs, foreign investment generally adopts a long-term strategy.
2) **Low-cost-oriented:** This usually occurs in the export-oriented sector. For products already occupying a certain share in the international market and has matured production technology, the major motive of foreign investment is to further enhance their competitiveness and profit in the world market by cutting costs. They prefer to invest in areas which are convenient for import-export with low production cost, such as areas with cheap land and cheap labor.

Table 10.14 Average Scale per FI project and Sector Location Quotient by Province

Regions	Signed FI value billion US$	No. of signed projects	Average scale per project (10,000 US$)	Sector Location Quotient (SLQ) (ratio of FI proportion by sector in province to FI proportion by sector in China as a whole)					
				Agriculture	Industry	Construction	Communication	Commercial	Estates
Beijing	2.74	1552	176.2	0.53	**1.00**	0.25	0.27	**1.10**	**1.35**
Tianjin	3.85	1389	277.2	0.18	0.80		0.70	**6.35**	0.41
Hebei	2.25	1263	178.2	0.62	**1.16**	0.78	0.66	0.14	0.38
Shanxi	0.23	178	128.1		**1.04**	**1.10**	**5.44**		0.38
Neimenggu	0.15	162	91.4	**1.52**	0.99	**2.90**	0.85		**1.60**
Liaoning	4.00	2406	165.0	**1.81**	0.90	**2.05**			0.21
Jilin	0.86	667	128.5	**1.60**	0.90	**2.35**		0.72	0.44
Heilongjiang	0.99	844	117.1		**1.00**				
Shanghai	10.54	2845	370.5	0.19	0.84	**1.42**	0.83	0.93	**1.59**
Jiangsu	12.41	4086	303.7		0.64	**1.15**		0.57	
Zhejiang	3.25	1836	177.0		0.74				0.23
Anhui	0.48	753	64.1	N/A					
Fujian	8.91	2728	326.5	**2.49**	0.97	0.42	0.76	0.95	**1.51**
Jiangxi	0.54	522	103.4		0.59	**3.53**		**1.80**	0.63
Shandong	4.63	2709	170.7	**1.41**	**1.08**	**1.39**	0.97	0.97	0.67
Henan	0.92	805	113.8	0.46	**1.04**	**1.95**	0.34	0.45	**1.22**
Hubei	1.09	881	123.5	N/A					
Hunan	1.36	574	236.1	0.86	0.88	0.61	**1.44**	0.56	**2.99**
Guangdong	24.83	8177	303.7	0.95	**1.07**	0.56	**1.47**	**1.13**	0.75
Guangxi	1.04	571	182.5	**1.22**	0.90	**1.58**	**2.42**	0.21	**2.10**
Hainan	2.78	389	714.9	**5.16**	0.32	**4.43**		0.05	**4.15**
Sichuan	1.21	826	146.1	0.79	0.91	**1.61**	**1.17**	0.63	**1.86**
Guizhou	0.09	103	83.5	0.80	0.79	**4.19**	**2.68**	0.19	**2.52**
Yunnan	0.37	269	139.0	**1.98**	0.78			0.07	**2.81**
Shaanxi	0.42	272	152.6	0.60	0.99	**1.01**	**1.02**	0.44	0.41
Gansu	0.20	156	128.2		0.70	**6.03**		**1.92**	
Qinghai	0.02	28	85.7	N/A					
Ningxia	0.03	42	69.1	N/A					
Xinjiang	0.09	114	79.8	**1.08**	**1.04**				
China	91.28	37011	246.6	1.00	1.00	1.00	1.00	1.00	1.00

3) **Local-market-oriented:** This usually occurs in the import-substitution sector. This type of investment has a long-term target with the major goal to control as much as local market share as possible. Most MNCs prefer this strategy for products with owner-specific advantages.

4) **Long-term-profit-oriented:** This usually occurs in infrastructure. With large investment value, long periods of construction and long investment-profit cycle, this kind of investment can only be carried out through cooperation between the best and large-scale MNCs. There are 4 types of cooperation in this kind of investment at present in the world, i.e. Privatization (MNCs possess part ownership), Build-Operation-Transfer (BOT), Build-Own-Operation (BOO) and Build-Transfer-Operation (BTO). Such investment is sensitive to the political and economic stability of the host country.

There were also obvious differences between major home countries in their motives and objectives to invest in China. We shall analyze this in terms of China's major FI sources.

Hong Kong & Macau investment: It has been China's largest 'foreign' investment origin for nearly 20 years. Dominated by medium and small-scale manufacturing enterprises, most products formerly made in Hong Kong, such as toys, clothes and watches, had occupied quite a large share in the international market in the 1960s to early 1980s due to their cheap prices. But as production cost had increased rapidly in the home area, the previous advantage in price was diminishing and the profit ratio had also dropped. Thus Hong Kong had to adjust her economic structure by transferring her manufacturing to low cost areas. An investigation conducted in 1996 shows precisely that lower production cost was the first motive of Hong Kong investment that went into mainland China, while entering into China's market was only the second consideration (Table 10.15). Besides, the first 4 important location factors for a Hong Kong FDI enterprises' first operation in mainland China are: adjacent to Hong Kong, cheaper labor, better communication facilities and favorable policies, all related to the motive of lower production cost. Driven by this motive, Guangdong logically became the major destination of Hong Kong investment since 1979 and has thus become gradually dominated by such an export-oriented sector. A pattern economic cooperation/integration between Guangdong and Hong Kong, i.e. 'Front-shop-and-back-factory' has emerged in this new spatial organization, whereby production is done in Guangdong but the servicing, inventory, marketing, import and export, and other supporting activities, i.e. the 'Front-shop' roles, are performed in Hong Kong.

Taiwan investment: At present Taiwan's 'foreign' investment projects in the Mainland are still dominated by small and medium-sized enterprises. More than half of projects are below US$0.5 million in value. Those with value above US$5 million accounted for only 5.6 per cent of the total in number (Yang, 1996). Therefore the major motive of Taiwan investors in the Mainland is to obtain short-term profits. Their enterprises are by nature both export-oriented, e.g. shoe making, precision instrument, and local market-oriented, e.g. food and household appliances.

Table 10.16 shows that the average size of Taiwan investment projects in the Mainland is smaller than its FDI to other areas, i.e. US$1.82 million, while the corresponding figures for its FDI in Europe, Southeast Asia and Malaysia are US$6.31, 4.24, and 6.34 million respectively.

Table 10.15 Motives of Hong Kong investment and its location factors for first operation in mainland China

Motive description	1st choice	2nd choice	3rd choice	Weighted Score
Enjoy lower production cost	46	6	1	461
Enter into China's Market	8	25	5	292
Contribute to China's modernization	2	7	21	209
Benefit the owner's hometown	5	3	9	153
Others	1	0	0	9
Location factors for first operation in mainland China				
Close to Hong Kong	29	11	2	551
Cheaper labor	15	10	9	466
Better transport facility	9	14	13	444
Favorable policy	6	10	7	336
Suitable geographical location	7	6	3	272
Personal relation	10	2	2	262
Cultural similarity	5	4	7	238
Others	2	0	1	35

Li Xiaojian (1996), Hong Kong Direct Investment in China's Mainland: A Spatial Study with Special Reference to the Location Behavior of Investors (unpublished paper)

Table 10.16 Average size of Taiwan outward investment project by region (US$10,000)

Region	Average scale	Region	Average scale	Region	Average scale
Europe	631	China's mainland	182	Philippines	339
Southeast Asia	424	Malaysia	634	Indonesia	332
America	414	Vietnam	470	Thailand	288

Source: 1995, *Economic Forecast*, Taiwan

Japan investment: Japan's investment in China are mainly from large-scale MNCs. At present, out of the 20 largest Japanese industrial companies, 17 have invested in China (Wang, 1996). Comparing with Europe and the United States, Japan's investment bears different characteristics: (1) it is highly concentrated in electrical machinery and car parts manufacturing; (2) the investment value per project is smaller; and (3) the investment projects are quite dispersed. Take automobile industry as an example (Ren & Xing, 1996). Japan's average investment size for car assembling is US$12 million, while the corresponding values for that of Germany and France are US$70 and 58 million respectively. For car parts manufacturing, Japan's average size is the smallest among China's major foreign investors, i.e. only US$2.2 million. But at present, Japan is still China's largest automobile trade partner. It seems that direct trade rather than direct investment remains the major measure taken by Japan in exploiting China's car market. However, according to her short and medium-term planning for electrical and automobile industries during

1995-2004, Japan has placed China as her top host country for investment in the near future (Pei, 1996).

U.S. investment: Different from that of Japan, U.S. investment in China covers almost every economic sector, from marine oil exploration, airplane parts manufacturing, electronics, telecommunication, biological engineering, electrical machinery, to fast food and various services. U.S. investment to other countries are mainly carried out through merger or purchasing, while in China, the main pattern is green field investment with joint operation as the popular mode. The primary motive of U.S. investment was to enter China's vast market although it is sometimes influenced by the political factor. Just as *Business Weekly* has put it: now people no longer refer China as a potential market, but in fact the era of China's market has come. The comment from Motorola is also clear: the market of the 21st century will be in China. IMF also points out that the major attractiveness of China in utilizing FDI lies in her gigantic internal market rather than her lower production cost (Zhu, 1996). Based on this definite motive, the products made by U.S. firms in China are mainly sold in China's internal market with very low export ratios. According to a case survey in Beijing in 1992 (Chu, 1996), the export ratio of U.S. firms in China is only one third of that of other foreign investors (Table 10.17).

Table 10.17 Survey on U.S. firms in Beijing, 1992

	No. of firms in operation	Gross production Value (billion yuan)	Export value (10,000 US$)	Export value per capita (US$)	Profit per capita (yuan)	Fiscal income per capita (10,000 yuan)
Total no. of foreign enterprises	942	11.52	20797.3	1500	4200	1.32
No. of U.S. firms	129	4.47	1547.5	600	10600	3.87
Proportion of U.S. firms	13.7	38.8	7.4	33.0	252.4	297.7

In realizing their own objectives, foreign investors, have also at the same time brought along a large quantity of 'creative capitals' into China, including new technologies, research capabilities, new organization systems, advanced management skills and international market networks. They are useful to China in her attempt in setting up a modern enterprise system and market mechanisms, and are beneficial in enhancing her international competitiveness (Hu, 1996).

5.2 Barriers to the Development of China's Export-Oriented Economy

With each province doing things in its own way, there is no united internal market (Xue, 1995). As a result, duplicated foreign investment projects were often found in different areas. It not only is a waste of money but also affects the development of the chosen pillar industry. In order to attract foreign projects, some provinces also improperly provide investors too many incentives, thus impede the full use of different regional advantages. Each province tends to compete with other provinces for exporting their products at a lower price without taking national interest into consideration. Local protectionism also hinders effective merger of enterprises and slows down the upgrading process of low-quality products.

With insufficient development of new technologies, consistent export growth is undesirably affected. Although the proportion of exports in manufacturing products has been rising, the high-tech component in export products is still quite limited due to insufficient investment in new technologies and insufficient utilization of existing military technologies. Thus competition from other countries of low quality exports is still very serious.

Without compatible laws and regulations, the investment environment is undesirably affected. Since her opening up, China has formulated many laws and regulations, but these are yet adequate to meet the requirements imposed by fast changes. Some laws, especially those on loans and taxation, are still incompatible with international standards. Hence on one hand, some investors thought they are unfairly treated, and on the other hand, some dishonest investors make unlawful money by using loopholes in the law.

Insufficient attention to the shortage of qualified personnel in international trading. Up to now, China still has no special institutions to train personnel in dealing with international trade and business issues. Students graduated from foreign trade colleges may be good at handling trade issues, but not suited to solve related issues. Their knowledge is insufficient to cope with complicated modern international trade problems. China also pays too much attention to direct foreign trade while neglects the circulation market in the importing countries. Therefore she cannot get the maximum economic benefit from deepening her trade into downstream retailing activities. Du Nianshen, a general manager of a foreign trade company in the United States, has attempted to estimate such a loss to China (Du, 1995). In the recent years, China's annual export value to the U.S. was about US$20 billion, and the net profit was less than US$1 billion from such a direct trade. But if China can establish her distribution networks in the U.S. and sell directly to department stores, she can at least earn US$ 10 billion more in net profit. Moreover, China can also enlarge the market share of her products in the U.S. through the establishment of these selling networks.

5.3 Strategies for Developing the Export-Oriented Economy

According to international experience, the future success of China's export-oriented economy heavily depends on her finding the correct way to get along with foreign MNCs. Table 10.18 shows clearly that China's final competitive edge is her vast market, while for the MNCs, these are their capital and technology. A win-win situation for the two lies in the balance in the exchange of market for capital and technology. At present, some MNC products have occupied quite a large share of China's market, e.g., detergent reached 35 per cent, cosmetics 36 per cent, medicine 13 per cent, beer 20 per cent, automobiles 68 per cent, elevator 70 per cent, TV color tube 65 per cent and program controlled switchboard 90 per cent.

Table 10.18 Comparative Advantages and disadvantages of China and foreign MNCs

	Advantages	Disadvantages
China	social & political stability, high market potential, cheap labor with relative high quality, abundant natural resources, strong cultural compatibility	lack of investment and technology, low administrative efficiency, incomplete social service system and laws, competition with other Asian countries
Foreign MNCs	advance technology, sufficient capital, sophisticated management system	market is life matter, competition with other foreign MNCs

The present state of MNC involvement and the different competitive edge between MNCs and China may be starting points for strategic thinking, particular so on the following issues:

While it will be impossible for China to close her door again in future in the epoch of globalization, there is no alternative but to grow in competition with others. Her joining the WTO will not only provide a new and enlarged space for the development of her export-oriented economy, but also new challenges, especially competition in her internal market by imports and products produced locally by MNC affiliates or branches. China should make full use of the advantage as a late comer in the international market and try to localize new foreign technologies within the shortest time through copying and learning from the developed countries. As a big country in terms of population and territory, the economic development in China should be multi-dimensional.

Thus we propose the following strategic recommendations:

- **To further strengthen the power of the Central government in coordination and in adjusting regional economic development to accelerate growth of the pillar industries**

The experiences of Japan and other Newly Industrialized Countries in Asia show that the central government plays a crucial role in promoting the development of the export-oriented economy in its take-off stage (Wang, 1996). Through long-term planning and making the necessary policies, the central government can accelerate growth of pillar industries in the country and locate them properly in accordance with national interests. China is in her economic take-off stage, and faces a similar situation as the above countries did. It is imperative to strengthen the power of the central government in coordinating regional development in order to shape China's industrial competitiveness.

In the selection of pillar industries, the gradient strategy should be adopted (Lu, 1995). Pillar industries could be given different priorities and supportive policies in different periods. For pillar industries such as the automobile and high-tech industries, administrative intervention by the central government could be conducted for a certain time period so that scale of economy can be realized. The central government can also help enterprises in their utilization of foreign investment, such as negotiation with foreign investors through effective coordination, and on the selection of the right foreign partners.

- **To effectively barter for new and high technologies with a united internal market**

China's industrialization process is not yet completed. Major barriers to her economic development still lie within the production regime (Chan, 1996). Even though China proposed in the latest long-term strategic plan that the economic growth mode in future should be gradually transferred from one of an extensive to one of intensive in nature, it is still a task in the production sphere. So, China badly needs investment, advance technologies and good management. In the developed countries, the major barriers to economic growth have already been shifted to the circulation sphere. What they imperatively need is an expanding market. Therefore the exchange between technology and market can take place. China should take her market as a bargaining chip to protect China's interests and realize China's long-term development goals. For making this effective, China must re-organize herself into a united market as a pre-condition to strengthen her bargaining position.

- **To try to occupy commanding points of information and technology through construction of China's 'world cities'**

Economic globalization promotes the emergence of new economic organizations and provide new dynamics to urban development. Thus the importance of some metropolises in the world stands out, and these cities become a new type of urban center, 'the world city'. As the most important nodes in the world urban network, the world city is the place for accommodating headquarters of MNCs, the basing point of world production organizations and convergence place of information. It leads the new fashion, both in consumer type and living style, of world development (Friedmann, 1986). Therefore, it is necessary for China to concentrate the country's strength to deliberately develop Shanghai and Beijing as her future 'world cities', so that these two can quickly catch up with the world's new development frontiers and diffuse the world's information and new technologies into China's urban network. At the same time, China can also pull the country's strength to help these cities to develop their own high-level technologies for commanding information and new productive activities in the global economy.

- **To develop China's own MNCs and explore a new international market**

China is a multi-dimensional country in economic composition. Some enterprises at present have had the strength to conduct direct investment into foreign countries and operate affiliates there. In doing so, China can more effectively obtain foreign resources and exploit the global market. According to 'world Investment Report 1996', there are already Chinese MNCs among the top 50 MNCs from the developing countries. As economic growth accelerates, China should develop more of her own MNCs to optimize her industrial and production structure through the fuller utilization of internal and international resources and markets (Chu, 1996). China First Pencil Company Limited of Shanghai sets a good example. With the establishment of a new affiliate in Germany and utilizing its internalization advantage, this company promotes not only its export in equipment at the beginning of construction of the foreign subsidiary, but its also export semi-finished products at a worth of US$8 million every year, which is already equivalent to its total current export value for the whole year (Zhao, 1996). In the development of MNCs, China should also learn from the developed countries in their advanced management experience, and give MNC subsidiaries independent power in personnel management and business operation.

- **Take market as the 'dragon head' of economic development and enthusiastically to train a large number of high quality businessmen**

With the accelerated utilization of foreign investment and new technologies, industrialization in China's coastal area will be completed soon. Then its economic development will depend heavily on the development of the circulation market. Therefore more intensive international competition will soon be inevitable. So China must pay enough attention in the training of a large number of high quality businessmen and try her best to attract overseas talents who understand and appreciate both Western and Chinese culture to manage China's industrial and commercial sectors. Comparing with other types of personnel, China desperately needs bankers, entrepreneurs and businessmen.

6. Conclusions

Foreign direct investment of MNCs is the main driving force in economic globalization. China has joined the rest of the world in making use of MNCs and their FI and technologies in development. According to the official exchange rate, China's import and export value is about 36 per cent of her total GDP, showing that her economic openness has reached quite a high level. The analysis of historical data has also shown that there is a strong positive relationship between foreign investment and economic performance in China, i.e. the country has benefited from FI in both becoming more export-oriented and having had a good record of economic growth.

Rapid economic development of China's coastal area in the last 20 years has benefited even more from the Open Policies and FI inflow, but continuous development of the export-oriented economy there is due more to its general economic strength and its high adaptability to market changes. One of the major reason why there exists a big difference between the South and North within the coastal area of China is their difference in ownership of the economic sectors. Therefore a good business environment is more important than preferential policies for attracting foreign investment.

The regression analysis on the relationship between foreign investment and GDP by province also indicates that the volume of foreign investment a province had utilized depends much on its general economic development level. Thus, in 1995, the utilized foreign investment value in the East accounted for 87 per cent of China's total. It can be deducted that the East Coastal area will still be the main receiver of foreign investment in China for a long time to come.

China's Open Policies have operated in a zonal-spread manner. They have been successful and effective. Because of the tremendous regional disparity, it will be impossible for China to have a uniform spatial economy for a long time to come. Therefore, a multi-dimensional regional economic structure will last for some time. The strategies in attracting foreign investment by different provinces therefore should also be different.

Foreign investment from different origins has clear regional-biases and this will have important influence on the regional economic structure. At present, foreign investments from Europe and United States are mainly concentrated in North China and the Yangtze Delta, while Japanese and Korean investment focus in Northeast China and Shandong peninsula, Hong Kong in adjacent Guangdong, Guangxi and Fujian, and Singapore in Jiangsu and Hainan. There is also certain division in the spatial economy. South China is characterized by an export-oriented economy in utilizing foreign investment, while the Yangtze Delta by an import-substitution economy and Bohai Rim by a mix of the two.

Foreign investment from different origins bears different motives and objectives. Hong Kong and Taiwan investments are primarily for higher profits. With no big cultural difference and small scale in investment projects, their decisions for investment are much quicker, and their operations are also more flexible. Singaporean and Korean investments are usually subject to their national development strategies and therefore tend to have stronger sectoral and regional biases in order to occupy a certain market quotient in a certain sector or province. Japan has more or less regarded China as her potential future competitor and therefore likes to control or occupy China's market through direct trade rather than direct investment. Japan's FDI projects in China are usually smaller in scale and more dispersed in location and sectoral make-up. German and French investments, without much attachment to political considerations, mainly aim at China's vast market. For realizing such a long-term target, their MNCs, to a certain extent, are more willing to transfer modern and high-level technologies to China. In fact, since 1995, for the purpose of enlarging their share in China's market, some MNCs from Germany have conducted a series of large-scale systematic investment in China with an emphasis on infrastructure projects (Zhu, 1996). American and British investments mainly aim at China's market but play some heed to their government's international relations too. With multi-objectives and step-by-step operation, their decisions for investment are very cautious and usually take a long time. Without care for short term gain or loss, they like to configurate and locate their projects according to their global strategies. Take IBM as an example. As early as 1979, IBM began to sell its products to China. Then in mid-1980s, it began to set up office in Beijing and Shanghai. Only until 1992 that it opened its first joint venture in Beijing in the name of 'IBM China'. After that, IBM has established two affiliates in Shanghai and Guangzhou. Thus its operation has at last deepened in terms of activities and covers more of China's space.

The economic and spatial implication of FI in China has been great. FI and foreign trade, together with an increasingly open economy have transformed the industrialization path of China to some extent and has induced a new regional pattern of economic development inclined towards the Coastal areas. The future behavior of MNCs' and their FI's spatial and sectoral preferences in China, together with relevant Chinese policies and the prospects of her joining WTO will be significant forces in the ever evolving regional economic landscape of the country.

REFERENCES

1. Chan Huai. 1996. Ponderation from foreign investment racing for internal market. *International Trade Problems* 7. Beijing: (Chinese text).
2. Chu Xiangyin. 1996. Opportunities and barriers facing U.S. firms in widening their investment to Mainland of China. *International Economic Cooperation* 6. Beijing: (Chinese text).
3. Chu Yinxiang. 1996. The actively and effectively utilize FDI of MNCs. *International Economic Cooperation* 9. Beijing: (Chinese text).
4. De Souza, A. R. and Stutz, F. P. 1994. *The World Economy – Resources, Location, Trade and Development*, 511-622. New York: MacMillan College Publishing Company.
5. Du Nianshen. 1995. How to enter into the international market for China's foreign trade. *China and Foreign Management Introductions* 2. Beijing: (Chinese text).
6. Dunning, J. H. 1993. *The Globalization of Business*. London: Routledge.
7. Friedmann, J. 1986. The world city hypothesis. *Development and Change* 17(1): 69-84.
8. Hu Zhaoliang. 1996. *China After Open Up*, 37-47. Beijing: China Environmental Science Press (Chinese text).
9. Li Mu. 1995. On gradient strategy for the development of China's export industries. *International Commercial*. 12 August (Chinese text).
10. Pei Changhong. 1996. Outward foreign investment trend of Japan's MNCs in 1990s. *International Economic Cooperation* 2. Beijing: (Chinese text).
11. Ren Qimin, Xing Houyuan. 1996. Impacts of foreign investment on China's automobile industry. *International Economic Cooperation* 6. Beijing: (Chinese text).
12. Sit, V. F. S. and Yang Chun. 1997. Exo-urbanization: the case of the Zhujiang River Delta. *Acta Geographica Sinica* 52(3).
13. Sit, V. F. S. and Yang Chun. 1997. Foreign-investment-induced exo-urbanization in the Pearl River Delta, China. *Urban Studies* 34(4): 647-77.
14. United Nations. 1996. *World Investment Report 1996 – Investment, Trade and International Policy Arrangement*. New York and Geneva: United Nations Publication.
15. Wang Xuezhen. 1996. The experiences of some countries and regions in attracting MNCs. *International Trade Problems* 9. Beijing: (Chinese text).
16. Wang Zhile. 1996. Probe into several problems on foreign direct investment. *International Economic Cooperation* 5. Beijing: (Chinese text).
17. Xue Rongjiu. 1995. International trade development towards 21st century and foreign trade in China (part 1 and part 2). *International Trade Problems* 9 and 10. Beijing: (Chinese text).
18. Yang Zhong. 1996. Motives, characteristics and countermeasures of Taiwan's investment to Mainland of China. *International Economic Cooperation* 1. Beijing: (Chinese text).
19. Zhao Yongqing. 1996. To better utilize foreign capital in foreign country and promote the export of China's famous brand products. *International Economic Cooperation* 7 Beijing: (Chinese text).
20. Zhu Xuebao. 1996. New trends of Sino-Germany economic cooperation. *International Economic Cooperation* 4. Beijing: (Chinese text).
21. Zhu Yujie. 1996. Analysis on investment of MNCs in Beijing. *International Economic Cooperation* 11. Beijing: (Chinese text).

Chapter 11

DEVELOPMENT AND REGIONAL EFFECTS OF SPECIAL ECONOMIC ZONES AND OPEN CITIES

FAN Jie

Professor, Institute of Geography,
Chinese Academy of Science

Yok-shiu LEE

Associate Professor, Department of Geography,
University of Hong Kong

The regional economies associated with the "Special Economic Zones (SEZs)" and the "Open Cities" located in China's coastal areas have recorded, since the introduction of economic reform measures, the fastest rates of economic growth and the highest degrees of foreign-oriented economic sectors. They have become the primary bases and major conduits through which China's economy is increasingly articulated with the global markets. They have also developed into a special category of "regional growth economies" inside China and have begun to exert influence on regional economic development throughout China.

From the mid 1980s through the mid 1990s, these coastal SEZs and Open Cities have benefited from a series of preferential policies promulgated by the Central Government. These policies allow them to attract huge amount of foreign investment, import technology, engage in large-scale foreign trade activities, and upgrade their infrastructure. All of these factors have helped promote a rather rapid industrial restructuring process, leading to sustained economic growth in all of these regional economies.

Fifteen years after the establishment of the coastal SEZs and Open Cities, however, changing times have presented them with a great challenge. At year end of 1995, during a tour of the Shantou Special Economic Zone, Party Secretary General Jiang Zemin pointed out the need for the SEZs and Open Cities to further identify a new niche for their regional economies. He urged them to aspire to be the nation's models by realizing two fundamental industrial structural shifts and constructing two spiritual civilizations. Some SEZs then answered his call by developing a new development strategy - which embodied primarily modernization and internationalization. Specifically, this means the establishment of a basic socialist market economic system which is increasingly articulated with the global market system. Still other SEZs proposed to build special economic zones that are heavily science- and technology-based.

In 1996, the Central Government implemented three major reforms on the taxation system on import and export activities. All of these reforms primarily mean the reduction of the differentially favorable tax benefits accorded the SEZs and Open Cities, forcing them to search for a "second industrial discovery." Major governmental programmes have been introduced to promote a greater degree of cooperation and linkages between the eastern and the western parts of the country, and the SEZs and Open Cities have been given a critical role in such programmes. One central aspect of such programmes was the emphasis placed on relocating labor-intensive and higher value-added industries from coastal regions to the central and western parts of the country. The coastal regions were also encouraged to increase their investment in the latter (Tian 1997).

The major objectives of this chapter include (1) the delineation of the increasing scale of the SEZs and Open Cities, the growth of their economic strengths and the change in their industrial structures for the period of 1991-96; (2) the identification and analysis of the functions, structures and locational patterns of several major categories of SEZs and Open Cities; and (3) the depiction of the major policies associated with the "second industrial discovery" process proposed by the SEZs and Open Cities.

1. THE DEVELOPMENT OF THE SEZS AND OPEN CITIES UNDER THE EIGHTH FIVE YEAR PLAN PERIOD

The SEZs and Open Cities have achieved sustained record rates of economic growth and expansion during 1991-1995. With their economies significantly strengthened and their industrial structures further upgraded, their role within their own provinces' internal economic development processes is further enhanced. However, the foreign-oriented aspect of most of the economies of the SEZs and the Open Cities has declined in significance.

1.1 Economic Growth and the Role of SEZs and Open Cities in Provincial Economies

Generally speaking, the economic vitality of the SEZs and Open Cities has been strengthened throughout the years of the Eighth Five Year Plan. Their positions within their respective provinces have been further reinforced, and their per capita figures are generally higher than the provincial averages. However, since the country's entire coastal areas have by now all turned toward attracting foreign investment, the comparative advantages of the SEZs and Open Cities have clearly been curtailed.

Table 11.1 Major Indicators for Special Economic Zones and Open Cities

Unit: % share in the province

		Fujian	Guangdong	Liaoning	Hebei	Shandong	Jiangsu	Zhejiang	Guangxi
Total	1990	29.3	40.2	16.8	5.2	24.7	12.5	24.4	9
GDP	1995	34.8	46.8	23.1	6.0	31.1	12.1	29.5	5.6
Foreign Capital utilized	1990	72.5	45.4	50.1	43.6	43.9	11.5	49.0	17.4
	1995	43.4	29.2	41.7	13.1	26.6	4.5	24.9	27.2

Note: Since the data for Shantou in 1990 and 1995 are not comparable, they are included in Guangdong's figures. Moreover, data for Tianjin and Shanghai are not included in the table because they are both Special Cities directly administered by the Central Government. Using the national total as the basis, Tianjin's and Shanghai's share of total GDP would be 5.9 % and 4.1 % in 1995 respectively, and their degree of utilization of foreign capital would be 10.8 % and 9.9 % respectively.

Sources: *China's Urban Statistical Yearbook 1991*, 1996.

Since the beginning of 1988, large stretches of coastal regions were gradually opened up for foreign investment. By the time of the Eighth Five Year Plan (FYP), a belt of open areas running along the country's coastline from the south to the north was formed. That is, the center of gravity of foreign investment and foreign trade has been shifting firstly from the SEZs and Open Cities onward to other coastal areas and, secondly, from the coastal areas outward to the interior regions of the coastal provinces. As a consequence, the comparative advantages of the SEZs and Open Cities have been gradually curtailed. With the exceptions of Beihai, Zhanjiang, Wenzhou, Weihai, which have recorded some increases in the share of foreign investments in their respective provinces, Xiamen, Shenzhen, Qinhuangdao, Qingdao, Ningbo, Fuzhou have all seen their respective shares of provincial foreign investment significantly declining, some by as much as half of their earlier proportions (Urban Social Economic Investigating Team of the State Statistic Bureau 1991, 1997).

Nevertheless, SEZs and Open Cities have generally maintained respectable economic growth rates. Within the period of the Eighth FYP, the per capita GNP figures in most of the SEZs and Open Cities have been maintained at the top of their respective provincial records.

On the basis of their share increases of respective provincial total GNP and their role in the respective provincial economies, the SEZs and Open Cities have been grouped into the following categories:

Shenzhen and Dalian: Where the position of SEZs and Open Cities has been significantly elevated within their provinces

With their shares of their respective provincial GNP recording a five percent increase over the Eighth FYP period, Shenzhen and Dalian are the two coastal SEZs and Open Cities that have registered the largest jump in their economic status. While Shenzhen's standing stood out at a consequence of rapid economic expansion of the entire Guangdong Province, Dalian's rise was attributed to its occupation of a key position amidst a relatively slow-growth Liaoning Province.

Fujian, Zhejiang, Shandong: Provinces where major progress has been made in expanding the scale of SEZs and Open Cities

Considerable progress has been made in the development of SEZs and Open Cities in Fujian, Zhejiang and Shandong provinces. For instance, the percentage share of provincial GDP of the following cities has improved by more than two per cent: Xiamen, Fuzhou, Wiehai, Yantai, Ningbo, and Wenzhou. These cities have strengthened their role as regional growth centres and have facilitated regional economic growth.

Provinces where the development of SEZs and Open Cities has been relatively slow and their respective role in the provincial economies has been unclear or has declined

The position of coastal Open Cities in Hebei and Jiangsu provinces within their provincial economies has recorded very limited gains and has even suffered from a mild decline. There are two major reasons: First, Qinhuangdao, Lainyungang, and Nantong were not regarded as key cities in their respective provincial economies; and secondly, they were not regarded as key growth nodes in the Eighth FYP.

In the case of Beihai in Guangxi Province, its role as a regional growth centre has also been limited because it was at the "take-off" stage. The expansion of its city scale and the construction of its infrastructure were only being started during the Eighth FYP. The rate of increase in the utilization of foreign capital has increased by only 10 per cent over a five year period, signifying that Beihai is still at an early stage of development as a regional growth centre.

In Guangdong Province, given that the rate of development in areas outside of the SEZs and Open Cities has been quite considerable, the role of the latter has been circumscribed. Although Guangzhou and Zhuhai have registered above provincial-average growth figures, their relative gains have been quite limited (Guangdong Province Planning Committee, 1996).

1.2 Changes in the Scale of the Cities

During the period of the Eighth FYP, the scale of the SEZs and Open Cities has expanded rapidly. For instance, Shenzhen is the only city that has been able to translate urban development into sectoral growth. This is evidenced by the fact that the rate of urban population growth in Shenzhen has been higher than the rate of increase of the size of the city's built up areas. This is due to the fact that Shenzhen has achieved a higher level of development with regard to its urban scale and industrial structure during the 1980s. Major challenges for Shenzhen in the 1990s include the modification and upgrading of its industrial structure.

Two major types of coastal SEZs and Open Cities have gained from expanding their scale. The first type include cities such as Qingdao and Tianjin, which contain a sizeable urban population to begin with. They have been able to control the population size of the city's built-up areas, promote re-structuring of the city's inner areas, and put emphasis on developing their satellite towns (Qingdao City Planning Committee 1996). The second type includes cities such as Xiamen and Dalian, which have imposed very strict controls over their urban population growth. In the process of conducting urban renewal projects, they have also been very careful in attending to the rationalization of their internal urban functions and the development of new neighbourhood areas. All of these actions have contributed to a co-ordinated development of tourism, commercial, trade, industrial and transportation activities. Fuzhou and Guangzhou also belong to this category.

There is another type of cities that have only started to develop rapidly right after the start of the Eighth FYP and they include Wenzhou and Beihai. They have recorded the most rapid rate of development among all coastal SEZs and Open Cities. Moreover, their urban population and built-up areas have grown at about the same rate, indicating their success in creating new industrial employment opportunities the same time they developed new urban areas within the cities. In Wenzhou, the main development strategy has shifted from "making a living from outside the city" to "building the productive base within the city. " Instead of "remaining in the countryside and engaging in family workshop production, " many in Wenzhou have "entered the city to engage in industrial and commercial activities, " leading to a rapid expansion of the urban economy. In the case of Beihai, the development opportunities were produced by the development of the ocean channel created in the Southwestern part of the province.

Finally, the following cities have also recorded high rate of urban growth: Weihai, Yantai, Shantou, Nantong, and Shanghai. This group of cities are distinguished from the other types by one aspect: the growth rates of their built-up areas are much higher than that of their urban population. The development of new urban areas in these cities is tightly linked with their respective ongoing industrial restructuring processes. For instance, in Shanghai, the industrial restructuring process is manifested in the decline of the textile and heavy industries, on the one hand, and in the emergence of tertiary and technology-intensive industries. Therefore, changes in the structural composition of the city's population have been influenced by economic changes associated with the industrial restructuring process. The Pudong area, site of numerous large-scale development projects, has also become the key location for Shanghai's newly emerging industries (Hua 1997). There is, nevertheless, the possibility that the de facto population size of these cities is much greater than the de jure population because of immense number of floating population who do not hold regular household registration in the cities.

Table 11.2 Rate of Growth of Special Economic Zones and Coastal Open Cities

		Shenzhen	Shantou	Zhuhai	Xiamen	Dalian	Qinhuangdao	Tianjin	Yatai	Qingdao	Lianyungang	Nantong	Shanghai	Ningbo	Wenzhou	Fuzhou	Guangzhou	Zhanjiang	Beihai
Annual rate of increase during the 8th Five Year Plan (%)	Urban population (10,000)	58	10	2	5	3	12	1	36	3	5	15	11	3	42	7	4	7	55
	Built-up area (sq.km.)	13	68	20	16	29	5	4	66	5	7	47	25	2	66	17	19	27	48
The actual situation in 1995	Urban population (10,000)	99	104	35	67	255	63	594	149	218	57	61	957	114	113	148	385	121	48
	Built-up area (sq.km.)	88	87	56	58	218	65	359	107	104	48	58	390	62	74	68	259	58	28

Sources: China's Urban Statistical Yearbook 1991, 1996.

1.3 Changes in Industrial Structures

The industrial structures of coastal SEZs and Open Cities have undergone significant changes at the same time as the size of their economies expanded. There are basically two major types of industrial structures: those that emphasize the secondary sector and those that pursue the tertiary sector.

First, there are twelve cities that have accelerated the development of tertiary industries. One of the major characteristics found in these urban economics is that the share of manufacturing activities in total GDP has declined whereas the share of tertiary activities has increased. In Xiamen, Shantou, Dalian, Qingdao, Shanghai, Ningbo, and Fuzhou, the rate of development of tertiary industries is much faster than that of the secondary industries and the share of tertiary industries in total GDP has exceeded 40 percent. The role of these cities as regional growth centers in their respective regional economies is clearly strengthened.

Secondly, there are six cities where manufacturing activities have remained the dominant sector and they include Zhuhai, Qinhuangdao, Lianyungang, Wenzhou, Guangzhou, and Zhanjiang. They share several common characteristics: rising share of secondary industries in total GDP, low level of urban development, and limited amount of foreign market-oriented activities. In the Eighth Five Year Plan, the basic strategy for these cities remains one of raising the extent and quality of industrial activities.

Shenzhen is the only city among all the coastal SEZs and Open Cities with a rather steady urban industrial structure and a comprehensive development of the urban economy.

2. Growth of Major Types of Development Areas

In the process of accelerating the development of coastal SEZs and Open Cities, several major categories of development areas have been developed to help promote the growth of the former's urban economies and upgrade their industrial structures.

2.1 Economic Technology Development Areas

In May 1984, at Deng Xiaping's suggestion, the Central Government, as a follow-up to its decision to establishing the SEZs, decided to open up fourteen coastal industrial cities as well. This has led to an opening up of a large stretch of China's coastal areas, which have become the country's frontier land for economic reforms.

Although the fourteen coastal industrial cities have maintained good conditions conducive to opening themselves up for foreign investments, they are still deficient in certain aspects of development in attracting and fully utilizing foreign investments. Therefore, the Central Government allowed them to utilize clearly delineated pieces of land within their city boundaries to develop new "economic technology development areas (ETDAs), " which were given preferential terms to attract foreign investments, to construct advanced industrial projects, to increase their earnings of foreign exchange, and to become the "windows" through which these cities could be fully developed into foreign trade-oriented economies.

In September 1984, the first ETDA project, located in Dalian, was approved by the State Council. During the Eighth FYP, with the exception of Wenzhou and Beihai, all the coastal Open Cities have designated ETDAs within their jurisdictions. As a countermeasure to compensate for the earlier favorable bias given to the SEZs, the ETDAs have been a critical programme in promoting the opening-up process of the fourteen coastal Open Cities. By the end of the Eighth FYP, there were a total of 32 ETDAs in China and 17 of these were found in the coastal Open Cities.

2.1.1 ETDAs Gaining Ground During the Eighth FYP period

By the 1990s, the fourteen ETDAs have developed into sites for constructing modern industrial development, attracting foreign investment and developing export trade activities. The speech made by Deng Xiaping in 1992 during his tour of the southern provinces has helped accelerate the country's opening up process and economic construction. The ETDAs have also taken advantage of the new opportunities in making further gains in economic reforms. At the same year, the number of investment projects by overseas investors, the amount of agreed upon foreign investment, and the actual amount of committed foreign investment have all increased by twofold. Moreover, there is a trend among foreign investment projects towards larger-scale and high-technology oriented ones. Industrial production, export-trade activities and economic efficiency have all gained at the same rate.

2.1.2 ETDAs Playing a Critical Role in Economic Reforms

The construction of ETDAs is a major form of opening up to the outside world. The ETDAs maintain very attractive conditions that are conducive to foreign investors coming to China. The ETDAs' share of foreign investments in Dalian, Tianjian, Yantai, Qingdao, Ningbo, Fuzhou, and Zhanjiang, as a percentage of total foreign investments in each of these cities, has been estimated at between 20 to 30 percent. The ETDAs have also become an important basis for expanding foreign trade. In 1993, a total of ten ETDAs in the coastal provinces have recorded export trade figures exceeding US$100 million. In a descending order of the amount of such figures, the top cities include Dalian, Tianjin, Guangzhou, and Shanghai. Moreover, the export products manufactured in the ETDAs are mostly produced by foreign firms and the extent of value added is substantial. As such, the ETDAs have played a very important role in helping their host cities to move forward along the path of opening up and attract foreign investments. Furthermore, in the development process of the ETDAs, the following aspects have been subject to market-oriented reform measures: investment system, enterprise system, labor utilization system, land-use system, internal management system, and the social welfare system.

Table 11.3 Basic Situation of Economic Technology Development Areas in Coastal Open Cities

	Date of approval	Planned area (sq. km)	Accumulative number of foreign enterprises	Foreign capital utilized (in US$ '000 million)	
				Agreed upon	Actual
Dalian development area	1984.9	20	1118	29	15
Tianjin development area	1984.6	33	2135	28	-
Qinhuadao development area	1984.10	10	184	5	1.8
Qingdao development area	1984.10	15	498	8.1	2.5
Yatai development area	1984.10	10	592	6.1	2.4
Waihai development area	1992.10	5.72	133	1.4	0.1
Lianyungang development area	1984.12	3	181	2.2	0.5
Nantong development area	1984.12	4.62	188	4.7	1.4
Shanghai Hongqiao development area	1986.9	-	-	11.24	
Shanghai Minhang development area	1986.8	2.13	122	-	5.3
Shanghai caohejing development area	1988.6	5	145	5	-
Ningbo development area	1984.10	3.9	522	20.14	4
Wenzhou development area	1992.3	5.11	96	0.66	-
Fuzhou development area	1985	4.4	392	10.3	3
Guangzhou development area	1984.12	9.6	340	10.8	4
Guangzhou Nansha development area	1993.3	10	90	-	4
Zhanjiang development area	1984.11	9.2	258	9.6	2

The rapid development of the ETDAs has helped promote growth of the host urban economies, upgrading of product structure, and improvement of production technology and management system. During the Eighth FYP, the ETDAs were gaining importance in the host urban economies, accounting for an average of about 20 percent of the city's total industrial output value and making up 40 percent of the year-to-year increase of industrial output value. Generally speaking, industrial projects in the ETDAs are technology-intensive and their products are highly competitive in the market place. Hence, the ETDAs have been instrumental in helping their host urban economies to restructure their economies.

2.1.3 Differentials Between the Development of ETDAs in Different Regions

There several common principles followed in the development of ETDAs: utilizing the existing infrastructure provided by the host urban economy, compensating for the deficiencies of the host urban economy, and raising the amount of value-added and product technology level. All of these tactics are combined to achieve the goal of promoting an export-oriented economy. It has been ten years since the ETDAs were first established and we can now notice major differentials arising between and among the various ETDAs with regard to the level of development, internal structure, and the relationship with the host economy.

First of all, the rate of development of the ETDAs has been spectacular. Since their introduction, the overall average rate of annual growth, in terms of the total amount of fixed asset investment, the amount of GDP generated, and the export trade figures, has been recorded at about 70 per cent. Nevertheless, there are major variations in the rate of development among the ETDAs. For instance, there are some major differences in terms of the rate of growth between Dalian and Qingdao, both are important coastal Open Cities located on the northern arc of the Bohai rim. In the case of Dalian, the size of the ETDA has expanded from 20 square kilometers to 37 square kilometers. Its industrial output is about one-fifth of Dalian's total (Chen, 1996). In the case of Qingdao, however, as a result of a less-than-ideal location, the development of the ETDA there has been lagging behind.

Although foreign investment makes up the bulk of total investment in the ETDAs, investment coming from other Chinese provinces and cities has also been prominent. The sources and scale of foreign investment in the ETDAs have been varied. For the ETDAs located in the southeastern coastal areas, most of the investment originates in Hong Kong and Taiwan. In the Changjiang river delta region, European and North American firms provide the bulk of the investment in the ETDAs. In the Shandong peninsular and the Liandong peninsular, Korean investment and Japanese investment are predominating the local ETDAs respectively.

Although manufacturing industry has been the most predominant type of investment in the ETDAs, the investment envelope has been expanded to include other types of activities such as real estate, information, transportation and communications, energy, and commercial outlets. There are major similarities with regard to the industrial structure of the manufacturing sector in different ETDAs, i.e., all are predominated by electronics, machinery, and textile. Nevertheless, there are also industries that have been developed to take advantage of local resources, such as aquaculture in Lianyungang and Qingdao, raw materials industry in Ningbo, and pharmaceutical industry in Tianjian.

Table 11.4 Pillar Industries in Selected Economic Technology Development Areas

	Types of Pillar Industries
Dalian development area	Electronics, mechanical engineering, textile, light industry, chemical engineering, pharmaceuticals, construction materials and garments
Tianjin development area	Electronics, mechanical engineering, biological pharmaceuticals, food industry, construction materials, and light industry
Qingdao development area	Electronics, mechanical engineering, textile, chemical engineering, construction materials, aquaculture and food industry
Nianyungang development area	Mechanical engineering, electronics, textile, food industry, aquaculture, micro-chemical engineering and packaging industry
Ningbo development area	Petroleum chemical engineering, textile, food industry and steel industry
Guangzhou development area	Construction materials, electronic components and instruments, chemical engineering, packaging, food industry and textile

2.2 High-Technology Industry Development Areas

In an effort to develop a cutting edge industrial group, the State Council, from 1988 to the end of the Eighth FYP, approved the establishment of 52 state-sanctioned high-technology industry development areas (HTIDAs). Of these 52 HTIDAs, eleven were located in the following coastal SEZs and Open Cities: Dalian, Tianjian, Weihai, Qingdao, Shanghai, Fuzhou, Xiamen, Guangzhou, Shenzhen, Zhuhai, and Nanhai. The HTIDAs have played a very important role in helping to upgrade the production technology as well as the industrial structure of the coastal SEZs and Open Cities (Gu, 1996).

Unlike the high technology parks in foreign countries where the emphasis has been placed on developing only one particular kind of high technology industry, the HTIDAs in China have always focused on two or more types of high technology industries at the same time. The most common types of high technology industries developed in China's HTIDAs include electronic communications, the integration of photo-electronics and photo-machinery, and new materials technology. As a consequence, there is a serious problem of duplication of efforts among some HTIDAs.

Nevertheless, from the perspective of the major approaches adopted for their detailed development, distinctive differences can be discerned between and among the HTIDAs. AS such, there are three major types of HTIDAs:

Innovative type

This type of HTIDAs are located near universities and research centres and rely upon the supply of trained technology expertise from the latter to develop innovative products and to build up the basis for new industries in the host urban economies. The Tianjian HTIDA is one such example.

Development type

Located in traditionally important industrial sites, this type of HTIDAs draw upon the strength of large- and medium-scale industrial enterprises to develop new products and new engineering processes, using new types of high technology to reform the traditional industries and to promote the upgrading of industrial structure. The Caohejing HTIDA in Shanghai belongs to this group.

Export processing type

Taking advantage of a coastal location and the opening policy and making use of foreign capital and technology, this third type of HTIDAs emphasize the development of foreign trade-oriented high technology enterprises. One such prominent example can be found in the Hainai HTIDA.

Generally speaking, each of the HTIDAs located in the coastal SEZs and Open Cities is organized around just one focal point of research centres in the host urban economy which is usually dominated by a concentration of existing industries. Examples of such an arrangement include Qingdao, Weihai, Xiamen, Shenzhen, and Hainai. Nevertheless, there are also some HTIDAs that are located in more than one site in the host city. For instance, the Dalian HTIDA is actually made up of three different sites located in different parts of the city, and so is Tianjian's. In the case of Shanghai, the HTIDA is made up of two different sites.

2.3 Bonded Zones

In May 1990, the State Council approved the country's first Bonded Zone (BZ) to be established in Waigaoqiao, Shanghai. After that, a total of fourteen other BZs were established in the following cities: Tianjin, Dalian, Shatoujiao and Futian in Shenzhen, Ningbo, Guangzhou, Zhangjiagang, Haikou, Xiangyu in Xiamen, Fuzhou, Qingdao, Shantou, Zhuhai, and Yangpu in Hainan (Liu, 1996). Many of these have already been approved by the Customs Office and went into operation.

With the exception of the Zhangjiagang BZ, all the BZs are located within the coastal SEZs and Open Cities. Since the BZs were allowed to offer even more beneficial terms than the SEZs to foreign investors, they were therefore regarded as the second round of expanding the open door policy in the 1990s. In a nutshell, the BZs are playing a significant role in the internationalization and modernization of the coastal SEZs and Open Cities.

2.3.1 BZs are Located Near Water Transport Facilities and Take Up Limited Amount of Land

The BZs generally benefit from good port facilities and good transportation networks. Moreover, they only take up a limited amount of land, which makes it easy for the purposes of controlling the investment environment. The smallest BZ is found in Shatoujiao, which occupies only 0.2 square kilometer, whereas the largest BZ is located in Waigaoqiao with a size of 10 square kilometer.

2.3.2 BZs Aim at Achieving Higher Level of Development Goals

While one of the principles in developing the BZs is to emphasize the uniqueness of each of them, they share one basic common strategy in stipulating the need to accentuate the role of the tertiary industry and new industries. Moreover, an emphasis has been given to the development of export trade activities in the overall industrial structure, such as manufacturing industry, commerce, finance, and information. In recent years, the following actions have been taken: a substantial amount of investment were committed into building the infrastructure as well establishing the offices for customs, storage and transportation, and immigration services; importation of selected high technology; establishment of both domestic and international financial organizations to create an open financial market system; deepening the degree of openness of the mechanism of tax concessions, promotion sales of finished products, as well as information services; and the development of the information industry.

2.3.3 Initial Impacts of the BZs

Construction of the BZs has followed the highest international standards in developing modern tax concessions areas in the shortest possible framework. For instance, the Guangzhou BZ was approved by the State Council on May 13, 1992. The founding stone was laid on July 8 the same year and the BZ went into full operation on May 10 the following year. The Waigaoqiao BZ in Shanghai followed a strategy of simultaneously constructing the site and recruiting potential investors at the same time.

Since their establishment, the BZs have been able to attract a considerable amount of foreign investment. By the end of 1996, the BZs have altogether attracted investors coming from over 90 countries and territories. About 6,000 projects, with a total investment of US$10 billion, were approved.

The BZs have also enjoyed a rather efficient system of management. An overall management committee for each BZ has been established, with its subordinate services firms providing assistance to foreign investors, who can process all the necessary requirements with a one-stop office visit to the committee. Labor supply, water, electricity, and communication facilities are all provided systematically by the labor services company and the public utilities company respectively.

Not surprisingly, the BZs still face a host of problems, the most significant of which pertains to the contradiction between the BZs' need for flexibility and the rigidity imposed by the macro-institutional and bureaucratic constraints. For instance, foreign trade activities are still to a great extent monopolized and administratively controlled. The approach and methods used by the Customs Office are out-of-date and they create major bottlenecks in the efficient movements of goods and the development of export-processing industries. Moreover, the degree of openness in the financial, foreign exchange, and insurance services sector is still very limited, with some detrimental effects on the internationalization process of the Bzs.

2.4 Special Open Areas

There are three Special Open Areas in China, two of which are located in the coastal areas and they include Pudong Development Zone in Shanghai and Yangpu Development Zone in Hainan province.

3. Development Strategy and Regional Effects

Since the 1990s, the Chinese government has carried out a dualistic development strategy. Firstly, it has pushed the development of the coastal SEZs and Open Cities to a higher level of development through the establishment of economic technology development areas, high technology development areas and the bonded zones. At the same time, it has decided to curtail the extent of beneficial treatment received by the coastal areas and apply similar kinds of beneficial policies to the interior parts of the country. As a result, the advantageous position of the coastal areas was adversely affected and, starting at the end of the Eighth FYP, they have re-adjusted their overall development strategy that aims at creating a "second round of growth"

3.1 The Changes in Major Macro-conditions for Development

The government's Open and Reform policy has by now been extended to the entire territory of the country, moving beyond the coastal SEZs and Open Cities and reaching cities that are located along the coastline, major rivers, border areas, and those that serve as the seats of interior provincial authorities. All of these other cities could also benefit from the same set of preferential policies and programmes formerly enjoyed exclusively by the SEZs and Open Cities.

For instance, during the period of the Eighth FYP, in addition to the already existing SEZs and Open Cities, the total number of counties and cities opened up reached 245. This means that a belt of opened-up cities and localities can be found stretching from the south to the north and they have begun to penetrate into the interior parts of the coastal provinces. At the same time, there were thirteen opened-up cities established along China's border areas. They included cities found in interior provinces such as Heilongjiang, Nei Mongol, Xinjiang, and Yunnan. Moreover, a total of nine cities and seventeen counties located next to major rivers have also been designated as opened-up cities, making up a belt of opened-up localities along an east-west axis.

The preferential policies and programmes enjoyed exclusively by the coastal SEZs and Open Cities have been gradually minimized or removed altogether. For instance, the following measures were introduced to reduce the amount of benefits received by the SEZs:

- Instead of keeping half of the industry and commercial tax generated by the trading activities of the SEZs and Open Cities, the SEZs have to forward all the revenue to the Central Government.
- In 1992, the SEZs' right to reserve a proportion of foreign exchange for their own use was taken away.
- In 1994, the SEZs' right to financial self-determination was replaced by a nation-wide standard tax system.
- In 1995, the SEZs' right to pay only half of the import tax of materials made of foreign-origin was discontinued at the end of the year.

3.2 Major Adjustments in the Overall Development Strategy

3.2.1 Major Actions During the Eighth FYP

The development of the coastal SEZs and Open Cities entered a new stage with impetus during the Eighth FYP, especially after Deng Xiaping's visit to the South. Foreign capital has become a very important source of fixed assets capital investment in the coastal SEZs and Open Cities. The structure of export products has been continuously adjusted, and the degree of reliance on foreign trade for economic development has been raised further.

Moreover, while the rate of marketization of the economy has been accelerated, the economic institutions and operational system have undergone fundamental changes. Furthermore, new investment in basic infrastructure projects has helped improve the functional operation in the cities as well as raise the efficiency of the SEZs and Open Cities.

On the basis of the continuously readjusting state-owned enterprises, tertiary industries, rural township enterprises, development zones, foreign invested enterprises have all helped contribute towards a rapid and sustainable economic growth trend. Attention has also been given to the construction of spiritual civilization and the beautification of the eco-environment.

Nevertheless, the changes taking place in the economic system and the production method have been rather slow in coming. The readjustments of the industrial structure, product structure, and the structure of enterprises have been very slow and limited. The economic strength of most cities is still very limited and they have not been able to exert economic influence beyond much of their jurisdictions. Many cities still suffer from inadequate infrastructure and this has severely limited their potential for further growth.

3.2.2 Major strategic thinking in the Ninth FYP

The coastal SEZs and Open Cities have to face the challenge brought on as a result of the Central Government's decision to curtail preferential policies accorded to them in the 1980s. Moreover, they will have to compete in the unified national market place with producers from other localities (State Information Center, 1996). Furthermore, given that China is contemplating to join the World Trade Organization, there is urgent need for the SEZs and Open Cities to readjust their strategic thinking and to realize a "second round of industrial innovation."

The major actions that can help further develop the economies of the SEZs and Open Cities include (a) a shift from the policy of the past—a reduction of taxes and increased financial support - to a new policy of renovating the institutional system, (b) further reinforce the SEZs' role as "experimental zones" to build up a socialist market economic system. All of these require a fundamental change in the enterprise system, the factors of production supply system, and the role of the government.

There is a need to shift from simply improving the hardware and providing convenient conditions to emphasizing the need to develop an investment environment that is conducive to developing local functional capabilities. There is also a need to shift from a focus on increasing the amount of supply of factors of production to an emphasis placed on technology-based growth. The quality of managers and laborers has to be upgraded. All of these will help change the nature of economic growth and development and will help enhance the role of the SEZs and Open Cities as regional growth centres.

There is a need for the SEZs and Open Cities to shift from a primary dependence on foreign investment to stress the need to turn to the hinterland for sources of further growth, to participate in international cooperation and competition at a much higher level, to establish strong linkages with the global market, and to reinforce their role as the "window" to the outside world.

There is, finally, a need to realize the coordinated development of the "two civilizations", to implement a sustainable development strategy which will take into consideration the resource base and the integrity of the eco-environment.

3.3 Regional Effects

The most critical challenge faced by the coastal SEZs and Open Cities in undertaking a "second round of industrial innovation" lies in the efforts to strengthen their linkages with the interior parts of the country, to explore inter-regional cooperation, to play their role as regional growth centres, and to enhance their overall comparative advantages to increase their competitiveness. The regional effects of the SEZs and Open Cities vary greatly between and among themselves.

3.3.1 Playing the role as "dragon head" in large economic regions: Dalian, Tianjin, Shanghai, Guangzhou

These cities have been China's most important industrial bases and port cities. Except for Dalian which has become a prominent regional economic center during the opening up period, the other three cities have been traditionally important regional economic centers.

Dalian. There is a consensus among all the provinces in northeastern China to turn Dalian into a regional economic center for the northeastern regional economy. With its function as a foreign trade port being continuously improved, Dalian is serving to lead the export-oriented development of the northeastern region.

Tianjin. As the largest coastal city in the northern part of China, Tianjin has decided to make itself the financial and trading center for the northern economic region. In particular, Tianjin serves a crucial role in the export-oriented economy in the Circum-Bohai region.

Shanghai. Although serving as China's most important international port city after Hong Kong, whether or not it should also become the country's financial center is however still open to debate. In any event, Shanghai's role as the regional economic center for the Zhangjiang delta has been reconfirmed by the Central Government in the Ninth Five Year Plan.

Guangzhou. Serving as China's southern regional economic center since the establishment of the People's Republic, Guangzhou's economic strengthen has been greatly enhanced in the past ten years.

3.3.2 Playing the role as the window to the outside global market for inter-provincial economic regions: Xiamen, Beihai, and Lianyungang

Before economic reforms, the functional role of these three cities was very limited. They are all searching for a way to establish inter-provincial cooperation to help open up opportunities for them to serve as the windows for regional foreign trade activities.

Xiamen. In 1996, an inter-provincial economic cooperation region was set up among the south-western part of Fujian province, the eastern part of Guangdong province, and the south-eastern part of Jiangxi province, with Xiamen serving the regional economic center for this huge hinterland. It is also envisioned that Xiamen will serve as a central city for increased trading activities with Taiwan. At the moment, the Fujian provincial authority has already decided to replace Fuzhou with Xiamen as the provincial economic center (Fujian Province Planning Committee, 1992).

Beihai. Although its overall economic strength was somewhat limited to begin with, Beihai has played an increasingly important role in export-oriented linkages for its hinterland which covers provinces in the south and southwestern parts of China. As such, its role as a regional economic center has provided Beihai the impetus for urban construction and urban economic growth, particularly during the Eighth Five Year Plan.

Lianyungang. In discussions relating to the construction of a land bridge between Europe and Asia, Lianyungang has always been referred to as the eastern end of such an inter-continental link. Taking advantage of such a potential role, it has readjusted its functional role and has become a port city to serve an inter-provincial economic region. At the moment, it has already played a certain role in providing external linkages for the region along the Penghai-Lanxin railway.

3.3.3 Playing the role of a provincial level regional center: Qingdao

As a provincial level regional center for Shandong province, Qingdao has focused on developing its export-oriented economy as well as moving up the ladder of industrial structure. All in all, its role as a regional center has been further reinforced.

3.3.4 Playing the role of sub-provincial level regional centers: Yantai, Wenzhou, Ningbo, Qinhuangdao, Nantong, Shantou, and Zhanjiang

4. CONCLUSION

The SEZs and Open Cities have played an important role in promoting regional economic development within their respective provinces where they are located. In the early 1990s, then, the Chinese decided to curtail the extent of beneficial treatment received by the coastal areas by applying similar kinds of policies to the interior parts of the country. Thus, after a period of sustained record rates of economic growth in the first half of the 1990s decade, most of the economies of the SEZs and Open Cities has declined in significance. Nevertheless, the further development of the coastal SEZs and Open Cities has now been maintained by a newly designed strategy of promoting a "second round of industrial innovation" through the establishment of economic technology development areas, high technology development areas and the bonded zones. The future growth prospect of the coastal SEZs and Open Cities will greatly depend on their ability to successfully implement this new development strategy.

REFERENCES

1. Fujian Province Planning Committee. (ed.) 1992. *Comprehensive Planning and Development of Coastal Areas of Fujian Province*. Beijing: China Planning Publishing Society (Chinese text).
2. Gu Chaolin. 1996. The types and directions of development of high technology parks in China. *Economic Geography* 16(1): 9-16 (Chinese text).
3. Guangdong Province Planning Committee. (ed.) 1996. *Make New Achievement, Toward a New Century*. Guangzhou: Guangdong Economic Publishing Society (Chinese text).
4. Hua Jianming. (ed.) 1997. *The Present and Future of Shanghai's Economic Development*. Shanghai: Economic Management Publishing Society (Chinese text).
5. Liu Qun. 1996. The operation and development scenarios of China's tax concession areas. *China's State of Affairs and State Capacity* 6 (Chinese text).
6. Qingdao City Planning Committee. (ed.) 1996. *Qingdao Toward the Twentieth-first Century*. Qingdao: Ocean Publishing Society (Chinese text).
7. State Information Center. (ed.) 1996. *China's Economic Outlook*. Beijing: China Planning Publishing Society (Chinese text).
8. "The Ninth Five-Year Plan and the Outline Long-term Objectives in the Year 2010". In Chen Yaobang (ed.) 1996. *Accelerate the Two Transformations, Revitalize the Regional Economy in North-East*. Beijing: Reform Publishing Society (Chinese text).
9. Tian Huiming. 1997. *An analysis of the Economic Hotspots of China's Economy in the 1990s*. Beijing: China City Publishing Society (Chinese text).
10. Urban Social Economic Investigating Team of the State Statistic Bureau. 1991. *China's Urban Statistical Yearbook 1991*. Beijing: China Statistical Publishing Society (Chinese text).
11. Urban Social Economic Investigating Team of the State Statistic Bureau. 1997. *China's Urban Statistical Yearbook 1997*. Beijing: China Statistical Publishing Society (Chinese text).

INDEX